Marine and Brackish Bryozoans from the Southern Bight of the North Sea

Hans De Blauwe

Marine and Brackish Bryozoans from the Southern Bight of the North Sea

Identification Guide for Belgium, the Netherlands and the North Sea Coast of Germany

Springer

Hans De Blauwe
Royal Belgian Intitute of Natural Sciences
Dudzele, Belgium

ISBN 978-3-031-92512-2 ISBN 978-3-031-92513-9 (eBook)
https://doi.org/10.1007/978-3-031-92513-9

With Dutch-language titles: "Mosdiertjes van de Zuidelijke bocht van de Noordzee"

Translation from the Dutch language edition: "MOSDIERTJES VAN DE ZUIDELIJKE BOCHT VAN DE NOORDZEE" by Hans De Blauwe, © The Author 2009. Published by Vlaams Instituut voor de Zee (VLIZ). All Rights Reserved.

This Springer imprint is published by the registered company Springer Nature Switzerland AG
The registered company address is: Gewerbestrasse 11, 6330 Cham, Switzerland

Preface

A Dutch version of this book was published in 2009. It was used extensively by Western European researchers working in marine biomonitoring programs and often consulted and cited in articles on bryozoans worldwide. The sand that is predominant on the sea bottom of the studied area is usually poor substrate for moss animals. The fact that such a large number of species are described in this work is the result of an in-depth study of more than 20 years of observation (tidal mark material, specimens collected on the hard substrate of breakwaters, port structures, windmill piles, gas platforms, shells, and stones) and of the review of the historical collections in Brussels (Royal Belgian Institute of Natural Science (RBINS)).

Over time, there was a rising demand for an English-language version. We are now about 15 years further. Due to increasing interest in this animal group, many observations were made at that time, which led to new distribution data. Several new species have also been introduced via shellfish culture, on ship hulls or other vectors. The use of the scanning electron microscope made it possible to distinguish and describe new species for science. Nine species have been described as new in this English update.

These species descriptions and distribution data make the book indispensable for anyone working with bryozoans anywhere in the world. All species are included in identification keys, and illustrated with drawings or photographs made or chosen in such a way that identification can be carried out with a binocular loupe. For a number of species it is almost necessary to use a scanning electron microscope though.

For 25 years, I have been able to enjoy the understanding of my family for my long stay at the binocular and the computer and the encouragement of many members of the Strandwerkgroep Belgium and Strandwerkgemeenschap in The Netherlands. Thanks to all who shared information, material, and photos and for giving permission to use it to contribute to a better understanding of this undervalued group of animals. Thanks to A. Naber, Waterdienst van Rijkswaterstaat, for giving permission for the use of the data related to *Triticella flava*.

It is almost impossible to list all the people that contributed to the end result. Over the last two years, the identification keys have been tested during workshops in France, Belgium, the Netherlands, and Germany. Questions were asked and lots of observations were passed on from these countries. Many sent photos that I was allowed to use in this publication. I could always contact members of the International Bryozoology Association with my questions. Thanks to all who supported and encouraged me. The Royal Belgian Institute of Natural Science (RBINS) in Brussels let me revise their collection and gave me access to the scanning electron microscope. The Flanders Marine Institute (VLIZ) supported me with the literature from their library, published my first book (De Blauwe 2009), and gave me access to their scanning electron microscope and light microscope with stacking facilities.

I am very grateful to René Vanoutryve(†) for his work at the light microscope to make color photos of my collection and to Julien Cillis of the RBINS for most of the black-and-white photos with the scanning electron microscope.

Dudzele, Belgium Hans De Blauwe

Competing Interests The author has no competing interests to declare that are relevant to the content of this manuscript.

Acknowledgments

I am very grateful for the scientific editing performed by Dr. Dennis P. Gordon, National Institute of Water & Atmospheric Research, Kilbirnie, Wellington (order Cheilostomatida), Dr. Paul D. Taylor, Department of Earth Sciences, Natural History Museum, London, Scientific Associate (order Cyclostomatida), Dr. Thomas Schwaha, University of Vienna, Department of Evolutionary Biology, Vienna (order Ctenostomatida), Andrew (Andrey N.) Ostrovsky, Professor, Saint Petersburg State University, and Research Associate, University of Vienna (reproduction); and Dr. Jean-Georges Harmelin, Station Marine d'Endoume, GIS Posidonie, OSU Pytheas, Aix-Marseille University, F-13007 Marseille (family Cribrilinidae).

Contents

1 General Introduction ... 1
Sources ... 1
Taxonomy ... 1
Resources ... 1
Illustrations ... 2
Bryozoa or Hydrozoa? ... 4
Identification Key to the Hydrozoa and Bryozoa ... 4
Bryozoan Colony Appearance ... 4
Anatomical Orientation of Bryozoan Zooids ... 5
General Morphology ... 7
Reproduction, Establishment, and Dissemination ... 7
Collection ... 10
Examination and Identification ... 10
Conservation ... 11
Classification and Key to the Orders ... 11
References ... 13

2 Cyclostomatida ... 15
Classification ... 15
Specific Morphology ... 16
Key to the Genera ... 18
Systematic Part ... 21
Filicrisia geniculata (Milne Edwards, 1838) ... 22
Crisidia cornuta (Linnaeus, 1758) ... 22
Crisia aculeata Hassall, 1841 ... 25
Crisia denticulata (Lamarck, 1816) ... 26
Crisia eburnea (Linnaeus, 1758) ... 26
Crisia sp. ... 29
Stomatoporina incurvata (Hincks, 1859) ... 29
Tubulipora aperta Harmer, 1898 ... 33
Tubulipora liliacea (Pallas, 1766) ... 33

Tubulipora lobifera Hastings, 1963 36
Tubulipora phalangea Couch, 1844 37
Tubulipora plumosa Thompson in Harmer, 1898 37
Oncousoecia dilatans (Johnston, 1847) 38
Plagioecia patina (Lamarck, 1816) 40
Plagioecia sarniensis (Norman, 1864) 42
Diplosolen obelium (Johnston, 1838) 42
Desmeplagioecia amphorae Harmelin, 1974 44
Annectocyma major (Johnston, 1847) 46
Entalophoroecia deflexa (Couch, 1842) 46
Disporella hispida (Fleming, 1828) 49
Patinella verrucaria (Linnaeus, 1758) 49
References 51

3 Ctenostomatida 55
Classification 55
Specific Morphology 57
Key to the Genera 58
Systematic Part 62
Alcyonidium albidum Alder, 1857 64
Alcyonidium condylocinereum Porter, 2004 64
Alcyonidium diaphanum (Hudson, 1778) 67
Alcyonidium gelatinosum (Linnaeus, 1761) 69
Alcyonidium hirsutum (Fleming, 1828) 70
Alcyonidium hydrocoalitum Porter, 2004 72
Alcyonidium mamillatum Alder, 1857 73
Alcyonidium mytili Dalyell, 1848 (=*Alcyonidioides mytili* Dalyell, 1848) 74
Alcyonidium parasiticum (Fleming, 1828) 76
Alcyonidium polyoum (Hassall, 1841) 77
Alcyonidium ambulans (Remane, 1936) (=*Monobryozoon ambulans* Remane, 1936) 78
Flustrellidra hispida (O. Fabricius, 1780) 80
Arachnidium lacourti d'Hondt and Faasse, 2006 81
Arachnidium fibrosum Hincks, 1880 83
Panolicella nutans Jebram, 1985 84
Anguinella palmata van Beneden, 1845 86
Tanganella mülleri (Kraepelin, 1887) 87
Victorella pavida Saville-Kent, 1870 89
Walkeria uva (Linnaeus, 1758) 89
Triticella flava Dalyell, 1848 92
Triticella pedicellata (Alder, 1857) 92
Farrella repens (Farre, 1837) 93
Hypophorella expansa Ehlers, 1876 96
Penetrantia concharum Silén, 1946 98

Immergentia suecica Silén, 1947 98
Spathipora sertum Fischer, 1866 101
Spathipora sp. 102
Amathia spinosa (Linnaeus, 1758) 102
Amathia citrina (Hincks, 1877) 104
Amathia gracilis (Leidy, 1855) 107
Amathia cf. *gracilis* sp. 1: Pink Embryos 107
Amathia cf. *gracilis* sp. 2: Yellow Embryos 107
Amathia imbricata (Adams, 1800) 108
Amathia lendigera (Linnaeus, 1758) 110
Amathia tertia (Winston and Hayward, 2012) 112
Buskia nitens Alder, 1857 113
References 115

4 Cheilostomatida 119
Classification 119
Specific Morphology 125
Key to the Genera of Cheilostomatida 131
Uniserial Chains of Autozooids 131
Bi- or Multiserial Colonies 133
Erect Colonies 133
Encrusting Colonies 136
Encrusting Colonies with an Opesia in the Frontal Shield 136
Encrusting Colonies with Calcified Shield, Except the Orifice 142
Encrusting Colonies with a Shield of Fused Flattened Spines 150
Systematic Part 152
Aetea anguina Linnaeus, 1758 152
Aetea sica (Couch, 1844) 153
Aetea truncata (Landsborough, 1852) 154
Scruparia ambigua (D'Orbigny, 1841) 156
Scruparia chelata (Linnaeus, 1758) 158
Eucratea loricata (Linnaeus, 1758) 159
Membranipora membranacea (Linnaeus, 1767) 161
Biflustra tenuis Desor, 1848 163
Biflustra grandicella (Canu and Bassler, 1929) 164
Jellyella tuberculata (Bosc, 1802) 167
Jellyella eburnea (Hincks, 1891) 169
Conopeum chesapeakensis (Banta, Perez, and Santagata, 1995) 171
Conopeum reticulum (Linnaeus, 1767) 171
Conopeum seurati (Canu, 1928) 174
Electra cf. *asiatica* Grischenko, Dick and Mawatri, 2007? 175
Electra monostachys (Busk, 1854) 176
Electra pilosa (Linnaeus, 1761) 179

Mass Strandings of *Electra pilosa* in 1965/66 and 2020/21 179
Einhornia crustulenta (Pallas, 1766) 181
Pyripora catenularia (Fleming, 1828) 184
Aspidelectra melolontha (Landsborough, 1852). 184
Flustra foliacea (Linnaeus, 1758). 187
Chartella papyracea (Ellis and Solander, 1786). 188
Securiflustra securifrons (Pallas, 1766) 189
Hincksina flustroides (Hincks, 1877) 191
Callopora craticula (Alder, 1856) 193
Callopora discreta (Hincks, 1862) 194
Callopora dumerilii (Audouin, 1826). 195
Callopora lineata (Linnaeus, 1767) 197
Callopora rylandi Bobin and Prenant, 1965. 197
Crassimarginatella solidula (Hincks, 1860). 200
Cauloramphus spinifer (Johnston, 1832) 200
Parellisina curvirostris (Hincks, 1862). 201
Tegella unicornis (Fleming, 1828) 203
Amphiblestrum auritum (Hincks, 1877) 205
Amphiblestrum flemingii (Busk, 1854) 205
Ammatophora nodulosa (Hincks, 1877) 208
Bugula neritina (Linnaeus, 1758). 213
Bugulina avicularia (Linnaeus, 1758) 214
Bugulina flabellata (Thompson in Gray, 1848) 215
Bugulina fulva (Ryland, 1960) 217
Bugulina simplex (Hincks, 1886) 217
Bugulina stolonifera (Ryland, 1960) 218
Bugulina turbinata (Alder, 1857) 221
Crisularia plumosa (Pallas, 1766) 221
Bicellariella ciliata (Linnaeus, 1758). 224
Beania mirabilis Johnston, 1840. 224
Caberea boryi (Audouin, 1826) 225
Cradoscrupocellaria ellisi (Vieira and Spencer Jones, 2012) 229
Cradoscrupocellaria reptans (Linnaeus, 1767) 231
Scrupocellaria scrupea Busk, 1852 232
Scrupocellaria scruposa (Linnaeus, 1758). 234
Tricellaria inopinata d'Hondt and Occhipinti Ambrogi, 1985. 235
Tricellaria ternata (Ellis and Solander, 1786) 237
Setosella vulnerata (Busk, 1860) 238
Steraechmella buski Lagaaij, 1952 238
Cellaria fistulosa (Linnaeus, 1758). 241
Cellaria salicornioides Lamouroux, 1816 242
Cellaria sinuosa (Hassall, 1840). 243
Collarina gautieri Harmelin, 2019 244
Cribrilina cryptooecium Norman, 1903 246
Cribrilina punctata (Hassall, 1841) 246

Cribrilina belgica nov. sp. 249
Juxtacribrilina mutabilis (Ito, Onishi and Dick, 2015) 251
Cribrilaria arrecta Bishop and Househam, 1987. 254
Cribrilaria bifida (d'Hondt, 1970) 256
Cribrilaria directa (Bishop and Househam, 1987). 256
Cribrilaria innominata (Couch, 1844) 258
Cribrilaria venusta (Canu and Bassler, 1925) 259
Genus *Glabrilaria* Bishop and Househam, 1987 260
Glabrilaria corbula Bishop and Househam, 1987 261
Puellina modica Bishop and Househam, 1987. 262
"Puellina" monostachys nov. sp. 264
"Puellina" nana Reverter-Gil and Fernandez Pulpeiro, 1996 264
Puellina praecox Bishop and Househam, 1987 267
Distansescharella seguenzai Cipolla, 1921 269
Distansescharella spirorbicola nov. sp. 270
Figularia figularis (Johnston, 1847). 272
Membraniporella nitida (Johnston, 1838) 273
Hippothoa divaricata Lamouroux, 1821 274
Hippothoa flagellum Manzoni, 1870 276
Celleporella hyalina (Linnaeus, 1767) 277
Plesiothoa gigerium (Ryland and Gordon, 1977). 278
Trypostega venusta (Norman, 1864). 280
Chorizopora brongniartii s.l. 281
Haplopoma graniferum (Johnston, 1847). 283
Haplopoma impressum (Audouin, 1826) 283
Reptadeonella insidiosa (Jullien, 1903) 285
Reptadeonella violacea (Johnston, 1847). 287
Palmiskenea skenei (Ellis and Solander, 1786). 287
Porella concinna (Busk, 1854) 290
Porella chadleyi nov. sp. 291
Escharella gilsoni De Blauwe, 2006. 293
Escharella immersa (Fleming, 1828) 294
Escharella labiosa (Busk, 1856). 296
Escharella variolosa (Johnston, 1838) 296
Escharella ventricosa (Hassall, 1842) 299
Escharoides bishopi De Blauwe, 2006 300
Escharoides coccinea (Abildgaard, 1806) 302
Escharoides mamillata (Wood, 1844). 303
Neolagenipora collaris (Norman, 1867). 304
Oshurkovia littoralis (Hastings, 1944) 307
Umbonula ovicellata Hastings, 1944 307
Prenantia cheilostoma (Manzoni, 1869). 310
Smittina affinis (Hincks, 1862) 310
Smittina crystallina (Norman, 1867) 311

Smittina landsborovii (Johnston, 1847) 313
Smittoidea prolifica Osburn, 1952 315
Smittoidea reticulata (MacGillivray, 1842) 315
Parasmittina trispinosa (Johnston, 1838) 318
Phylactella labrosa (Busk, 1854) 318
Hippoporina pertusa (Esper, 1796) 320
Metroperiella cf. *biformis* (Zhang and Liu, 1995) 322
Pentapora foliacea (Ellis and Solander, 1786) 324
Schizomavella (Calvetomavella) biancae Pica and Berning, 2022 . . . 327
Schizomavella (Schizomavella) auriculata (Hassall, 1842) 328
Schizomavella cristata (Hincks, 1879) 330
Schizomavella (Schizomavella) cornuta (Heller, 1867) 330
Schizomavella (Schizomavella) hastata (Hincks, 1862) 333
Schizomavella (Schizomavella) linearis (Hassall, 1841) 333
Schizomavella (Schizomavella) teresae Reverter-Gil and Fernandez Pulpeiro, 1996 336
Watersipora subatra (Ortmann, 1890) 338
Terwasipora complanata (Norman, 1864) 338
Schizoporella hesperia Hayward and Ryland, 1995 *or S. magnifica* Hincks, 1886 341
Schizoporella japonica Ortmann, 1890 342
Schizoporella unicornis (Johnston in Wood, 1844) 345
Schizoporella cf. *variabilis* (Leidy, 1855) 345
Schizoporella sp. 347
Schizobrachiella verrilli (Maturo and Schopf, 1968) 350
Fenestrulina sp. 352
Fenestrulina delicia Winston, Hayward and Craig, 2000 352
Escharina haywardi Zabala, Maluquer and Harmelin, 1993 356
Escharina johnstoni (Quelch, 1884) 356
Escharina vulgaris (Moll, 1803) 358
Herentia hyndmanni (Johnston, 1847) 360
Phaeostachys spinifera (Johnston, 1847) 361
Pacificincola perforata (Okada and Mawatari, 1937) 363
Pacificincola hozawai (Okada, 1929) 366
Pacificincola cf. *hozawai* with up to Five Long Spines on the Ovicell 368
Primavelans glabricollaris Min, Chae, Yang, Noh, Lee and Seo, 2021 369
Cryptosula pallasiana (Moll, 1803) 370
Hagiosynodos latus (Busk, 1856) 372
Microporella vanoutryvei nov. sp. 376
Microporella dirupoi nov. sp. 378
Microporella kindi nov. sp. 379
Microporella annae nov. sp. 380
Microporella helgolandensis nov. sp. 382

Cellepora pumicosa (Pallas, 1766) 384
Buffonellaria muriella Berning and Kuklinski, 2008 387
Celleporaria brunnea (Hincks, 1884) 387
Celleporina decipiens Hayward, 1976 390
Celleporina hassallii (Johnston, 1847) 392
Celleporina cf. *pygmaea* (Norman, 1868) 392
Celleporina sp. 393
Lagenipora lepralioides (Norman, 1868) 394
Omalosecosa ramulosa (Linnaeus, 1767) 395
Turbicellepora avicularis (Hincks, 1860) 397
Turbicellepora magnicostata (Barroso, 1919) 399
Hippoporidra lusitania Taylor and Cook, 1981 399
Rhynchozoon bispinosum (Johnston, 1847) 402
Schizotheca divisa (Norman, 1864) 402
Schizotheca fissa (Busk, 1856) 405
References 405

Glossary 415

Index 419

Abbreviations

BCS	Belgian continental shelf
DDNZS	Duik De Noordzee Schoon (Expedition North Sea)
IfAÖ	Institut für Angewandte Ökosystemforschung GmbH
NHMUK	National History Museum, London
NIOZ	Koninklijk Nederlands Instituut voor Zeeonderzoek (Royal Netherlands Institute for Sea Research)
RBINS	Royal Belgian Institute of Natural Sciences
SEM	Scanning electron microscope
VLIZ	Vlaams Instituut voor de Zee (Flanders Marine Institute)

Chapter 1
General Introduction

Abstract This chapter cites the taxonomy followed, the sources of information, and the names of photographers who kindly allowed their photographs to be used. Furthermore, the difference between Hydrozoa and Bryozoa is briefly explained, which is not so obvious for beginners. The general morphology of bryozoans is explained, and the classification of the three extant orders is given. A brief explanation follows about reproduction, collection, research, and preservation. Finally, there is an identification key to the orders.

Sources

Taxonomy

Classification of the species is based almost completely on that provided by https://www.marinespecies.org/ and http://bryozoa.net/

Resources

This book is based on De Blauwe (2009). Almost all marine and brackish-water species living in the southern bight of the North Sea, between the Pas de Calais and the German/Danish border, are included, as well as the species that wash up on the beaches, which come largely from the English or French Channel coast or even further afield, attached to drifting seaweed or plastic.

The following sources were used to determine the occurrence of the species:

- The author carried out a revision of the RBINS collection, of which the immense "Gilson Collection," collected between 1899 and 1914, dominates.
- All published and unpublished observations of the author from the area are included here.

H. De Blauwe, *Marine and Brackish Bryozoans from the Southern Bight of the North Sea*, https://doi.org/10.1007/978-3-031-92513-9_1

– Interpretation of literature data provided a wealth of information from which a fairly complete picture was distilled of the historical situation. Changes in the distributions are noticeable for a number of species. Observations from the biological journals “De Strandvlo” and “Het Zeepaard” or from members of the Strandwerkgroep (Belgium) and the Strandwerkgemeenschap (the Netherlands) were checked where possible to complete the picture. For the Netherlands, the publications of Faasse and De Blauwe (2004) and Faasse et al. (2013) were used. Ancient literature often does not state whether the colonies were collected alive and in situ or were washed ashore.
– Reports of Alcyonidiidae were treated very critically, and many have been omitted. Only those observations that seem reliable and relevant are mentioned from the nineteenth-century literature.
– Previous records are from De Blauwe (2009). Since 2009, many new observations have been made and much literature has been published on the taxonomy and occurrence of the species in the region. These are included here.
– A bryozoan workshop in Heligoland in November 2023 and April 2024 yielded many distributional records of German species.

Illustrations

Many photographs and drawings made especially for De Blauwe (2009) have been retained. The drawings were converted from bmp to svg to improve their quality. They have been supplemented with new photographs. Much of the photographed material was collected in Normandy and Brittany. These were chosen for inclusion when more beautiful material was collected there, or when indigenous material was not available.

Almost all photographs were made by the author from material in his own collection; these specimens will eventually be deposited in the RBINS. I am very grateful to René Vanoutryve† for allowing me to use his equipment for many color photographs. Since 2024, I have had the opportunity to use a Leica M205 stereomicroscope with a maximum of 160× magnification, at the LifeWatch Marine Observatory (VLIZ). Most scanning electron micrographs were made in the RBINS, under the guidance for many years of Julien Cillis and latterly Laetitia Despontin. Since 2024, I had the opportunity to use the SEM at the VLIZ. I am grateful to Jonas Mortelmans and Mattias Bossaer for their guidance in using the apparatus in VLIZ.

I am indebted to many people who lent me the following photos:

2.1: *Tubulipora aperta* (Marc Cochu)
2.13a: *Tubulipora aperta*, ooeciostome, Brittany, France (Marc Cochu)
3.1b: *Amathia citrina*, zooids on kenozooidal stolon (Britta Kind)
3.1c: *Flustrellidra hispida* with spinozooids (Britta Kind)
3.6a: *Alcyonidium albidum*, off the Walcheren coast (D. Tempelman in Faasse et al. 2013)

3.7a: *Alcyonidium condylocinereum*: beached at Nieuwpoort, 2005 (Filip Nuyten)
3.7b: *Alcyonidium condylocinereum*: colony on *Ensis* sp. (Francis Kerckhof)
3.7e: *Alcyonidium condylocinereum*: colony on *Hinia reticulata* (Arnold Wijker)
3.11: *Alcyonidium hydrocoalitum*, Vlissingen (Marco Faasse)
3.16: *Alcyonidium ambulans*: Heligoland, 2022 (Thomas Schwaha)
3.18b: *Arachnidium lacourti*: colony from Nieuwe Waterweg 2006 (Marco Faasse)
3.19a: *Arachnidium fibrosum:* (Britta Kind)
3.22b: *Tanganella mülleri*: showing brooded embryos in the vestibular wall (Thomas Schwaha)
3.24b: *Walkeria uva*: colony from Germany (Britta Kind)
3.25a: *Triticella flava*: on claw of *Upogebia deltaura* from Oyster grounds (David Tempelman)
3.26a: *Triticella pedicellata*: from Tiefe Rinne, 2022 (Britta Kind).
3.38a: *Amathia lendigera*: colony beached at Petten, the Netherlands, 2016 (Rien de Ruijter)
4.2: Schematic image (Dr. Claus Nielsen)
4.17b: *Aetea anguina*: fig. From Cook (1977)
4.28a: *Jellyella eburnea* (Arie Twigt)
4.30a: *Conopeum reticulum* on *Mytilus edulis* (marina Zeebrugge 2008) (Michel Decleer)
4.32a: *Electra* cf. *asiatica* zooids with few pseudopores (Jan Leitinger)
4.35a: *Electra pilosa:* spherical colony at Katwijk, 3 December 2020 (Jacoline van Duijn)
4.35b: *Electra pilosa:* mass stranding at Texel on 7 December 2020 (Sytske Dijksen)
4.39d: *Flustra foliacea*: encrusting colony with avicularia and few kenozooids (Rien de Ruijter)
4.41a: *Securiflustra securifrons*: colony beached at Texel in 2021 (Rob Dekker)
4.43: *Callopora craticula*: Heligoland (Britta Kind)
4.51a, b: *Tegella unicornis*, Heligoland (Britta Kind)
4.57a: *Bugula neritina*: colony in the Grevelingen (Ad Aleman)
4.66a: *Beania mirabilis*: colonies on beached plastic, Texel 2017 (Sytske Dijksen)
4.102a: *Celleporella hyalina*: Prinses Amalia windpark (Marco Faasse)
4.120a: *Escharoides mamillata*: Ploubazlanec, France, Februari 2023 (Florence Gully)
4.123b: *Umbonula ovicellata*: colony from Verdelet, France (Florence Gully)
4.133a: *Metroperiella* cf. *biformis*: colony at Stavanger, Norway (Marco Faasse)
4.142a: *Watersipora subatra*: colonies at Brittany, France (Marco Faasse)
4.142b: *Watersipora subatra*: Brittany, France (Marco Faasse)

4.149c: *Schizobrachiella verrilli*: zooid with two avicularia, Tiefe Rinne (Britta Kind)
4.160a: *Pacificincola hozawai* from Oostende spuikom, Belgium (Francis Kerckhof)
4.161a: *Primavelans glabricollaris*: map showing the two locations where this species has been collected (Schüler et al. 2022)
4.172e: *Celleporaria brunnea*: from Banyuls-sur-mer, France, Mediterranean, May 2024, SEM (Gabin Droual, Nicolas Gayet)
4.182a: *Rhynchozoon bispinosum*: Brittany, France, 2023 (Florence Gully).

Bryozoa or Hydrozoa?

For beginners, the distinction between Hydrozoa (phylum Cnidaria) and Bryozoa is not easy. The tentacles of hydroids capture and anesthetize the prey with stinging cells and bring them to the mouth opening. Bryozoans have a ring of tentacles that carry cilia, not stinging cells. These cilia initiate a flow of water from which suspended particles can be filtered. The tentacles can move individually. When examining fresh substrates (mussels, pebbles, algae, etc.) in seawater, sometimes bryozoans are discovered unexpectedly when the characteristic tentacle crowns are protruded.

Identification Key to the Hydrozoa and Bryozoa

1. (a) Tentacles not retractable into an enclosure (Fig. 1.1a). .... Athecate hydroid
   (b) Tentacles in a whorl and retractable into an enclosure (Fig. 1.1b, d). 2
2. (a) Tentacles with stinging cells not creating a water flow (Fig. 1.1b) .... Thecate hydroid
   (b) Tentacles with cilia creating a flow of water (Fig. 1.1c, d). .... Bryozoan

Bryozoan Colony Appearance

The appearance of bryozoan colonies is extremely varied. They can be calcified or not, the colony encrusting or erect, firm or flexible, a few millimeters to 30 cm in size. The color varies from species to species, and brightly colored embryos can give the colony a characteristic color during the reproductive period.

Fig. 1.1 (**a**) Athecate hydroid with club-shaped tentacles. (**b**) Thecate hydroid. (**c**, **d**) Bryozoan tentacle crowns

Anatomical Orientation of Bryozoan Zooids

Understanding the orientation of individuals (zooids) in bryozoan colonies is very important (Fig. 1.2a):

- Basal: The surface of the zooid applied to the substratum
- Lateral: The sides
- Frontal: The top surface containing the opening
- Proximal: Closest to the origin of the colony
- Distal: Toward the growing edge of the colony

Generally, with only a few exceptions elsewhere in the world, the orifice is situated near the distal end of the feeding zooid (autozooid).

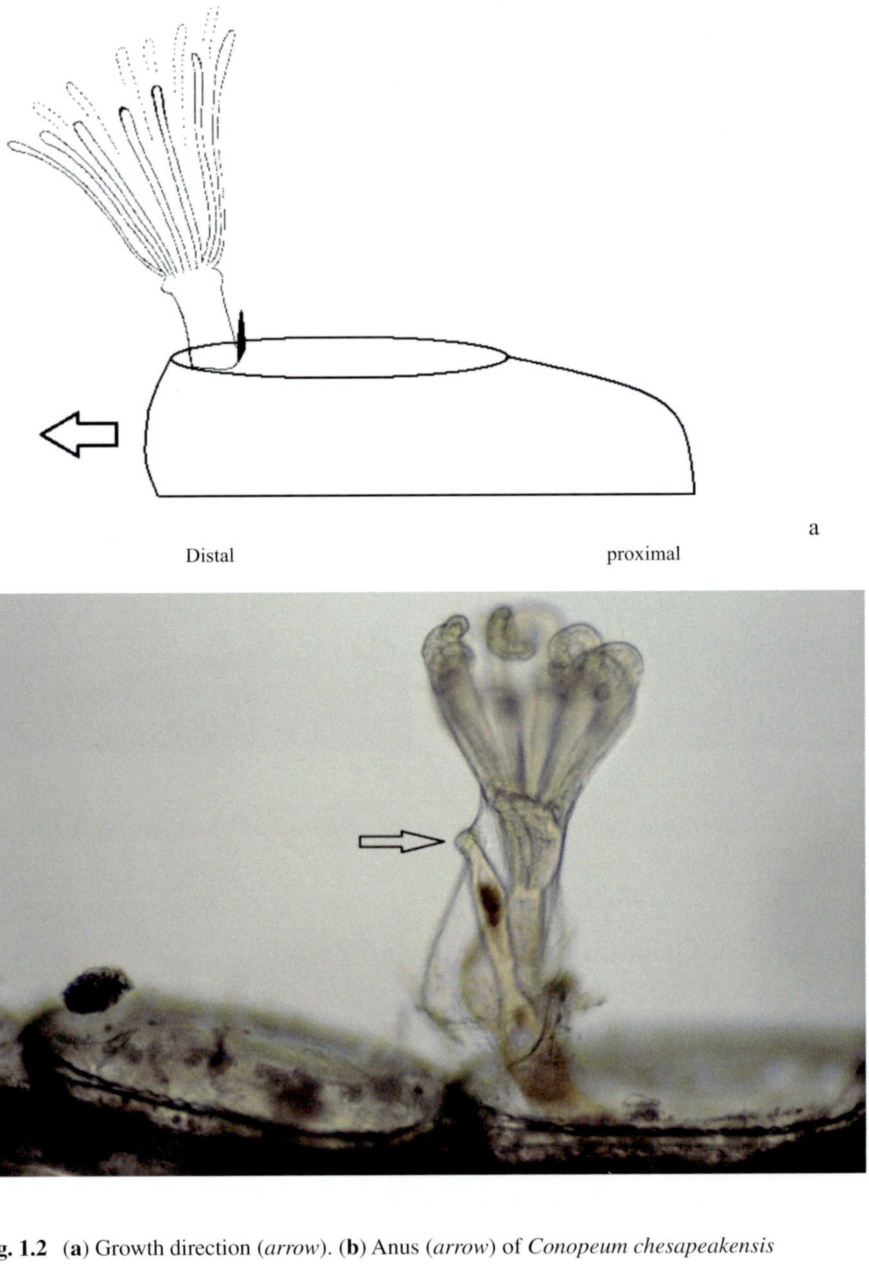

Fig. 1.2 (**a**) Growth direction (*arrow*). (**b**) Anus (*arrow*) of *Conopeum chesapeakensis*

General Morphology

A colony consists of a group of individuals (modules) that are interconnected with each other. Each feeding module (autozooid) (Fig. 1.2a) consists of a protective body wall (cystid) containing a retractile polypide. The cystid contains an opening through which the polypide can be partially protruded for feeding. A polypide consists of a tentaculate lophophore, a U-shaped gut, a muscular system, reproductive organs, and a nervous system. In the center of the tentacle crown is the mouth; the anus (Fig. 1.2b) is situated outside the tentacle crown. Polypides can degenerate into "brown bodies." The cystid may persist after the death of the polypide, especially if it is calcified, often developing a new polypide inside.

Specific morphological characters are explained in the chapter for each order.

We distinguish three groups of bryozoans in terms of structure:

Ctenostomatida have no calcified exoskeleton and no operculum. The autozooids arise from a stolon (Fig. 1.3a) or directly from a parent zooid (Fig. 1.3b).

Cyclostomatida (Fig. 1.3c) have a cylindrical calcified cystid and are not closed by an operculum.

Cheilostomatida (Fig. 1.3d–i) have a box-like or sac-like calcified cystid. Cheilostomatida can be divided into five main morphological groups (these groups are morphological grades and are not evolutionary monophyletic divisions):

Autozooids with an opesia (large opening frontally) covered by a membranous frontal wall. The orifice and its closing operculum are set in this membrane (Fig. 1.3d, e).

Autozooids with a membranous frontal wall underlain by the internal calcified wall (cryptocyst of Setosellidae, Microporidae, and Cellariidae). The orifice and its closing operculum are set in the frontal membrane (Fig. 1.3f).

Autozooids with an opesia overlain by a basket-shaped shield of fused flattened spines (Fig. 1.3g).

Autozooids with a frontal surface completely calcified, except for the opening closed by the operculum. The calcified wall is not covered by a membrane (gymnocystal shield of Hippothooidea, Fig. 1.3h).

Autozooids with frontal shields completely calcified, except for the opening closed by the operculum. The calcified wall is covered by a frontal membrane, with a slit-like space (hypostegal coelom) between it and the frontal shield (Fig. 1.3i, umbonuloid shield; Fig. 1.3j, lepralioid shield).

Reproduction, Establishment, and Dissemination

Bryozoan colonies are hermaphrodites. They make male and female sex cells, simultaneously or at different times. Embryos are incubated in specialized brood chambers (Fig. 1.4a), within the zooids themselves (Fig. 1.4b), or in Cyclostomatida in swollen gonozooids (Fig. 1.4c).

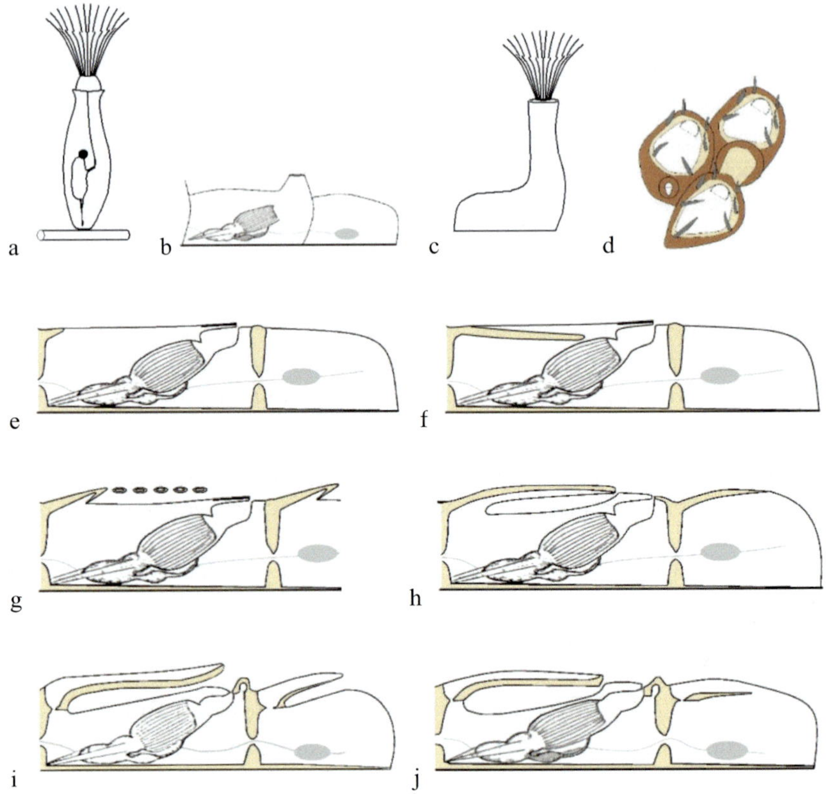

Fig. 1.3 (**a**) Ctenostomatida, autozooids arise from a stolon. (**b**) Ctenostomatida, autozooids arise directly from the parent zooid. (**c**) Cyclostomatida have a cylindrical calcified exoskeleton. (Modified and redrawn from Hayward and Ryland 1998). Cheilostomatida. (**d**) Autozooids with an opesia covered by a membrane, frontal view. (**e**) Idem, longitudinal section of autozooid and bud. (**f**) A membranous frontal wall underlain by the internal calcified wall (cryptocyst). (**g**) Autozooid with an opesia overlain by a basket-shaped shield of fused flattened spines. (**h**) Autozooid with frontal surface completely calcified. (**i, j**) The calcified wall itself covered by a membrane, with a space between the two layers (**i**) umbonuloid and (**j**) lepralioid shield

Most cheilostome species have ciliated (coronate) larvae that are brooded until their release, do not feed, and live only a short time (hours or days) in the plankton before settling. Consequently, the distance of dispersal possible by individual larvae is limited.

Exceptions to this include the Membraniporoidea and some species from the genus *Alcyonidium*. These species release cyphonautes larvae that can survive for a few weeks or months and feed while in the plankton. As a result, larvae can disperse over great distances and therefore species such as *Electra pilosa* and *Conopeum reticulum* with cyphonautes larvae are the first to colonize new substrates such as recent shipwrecks.

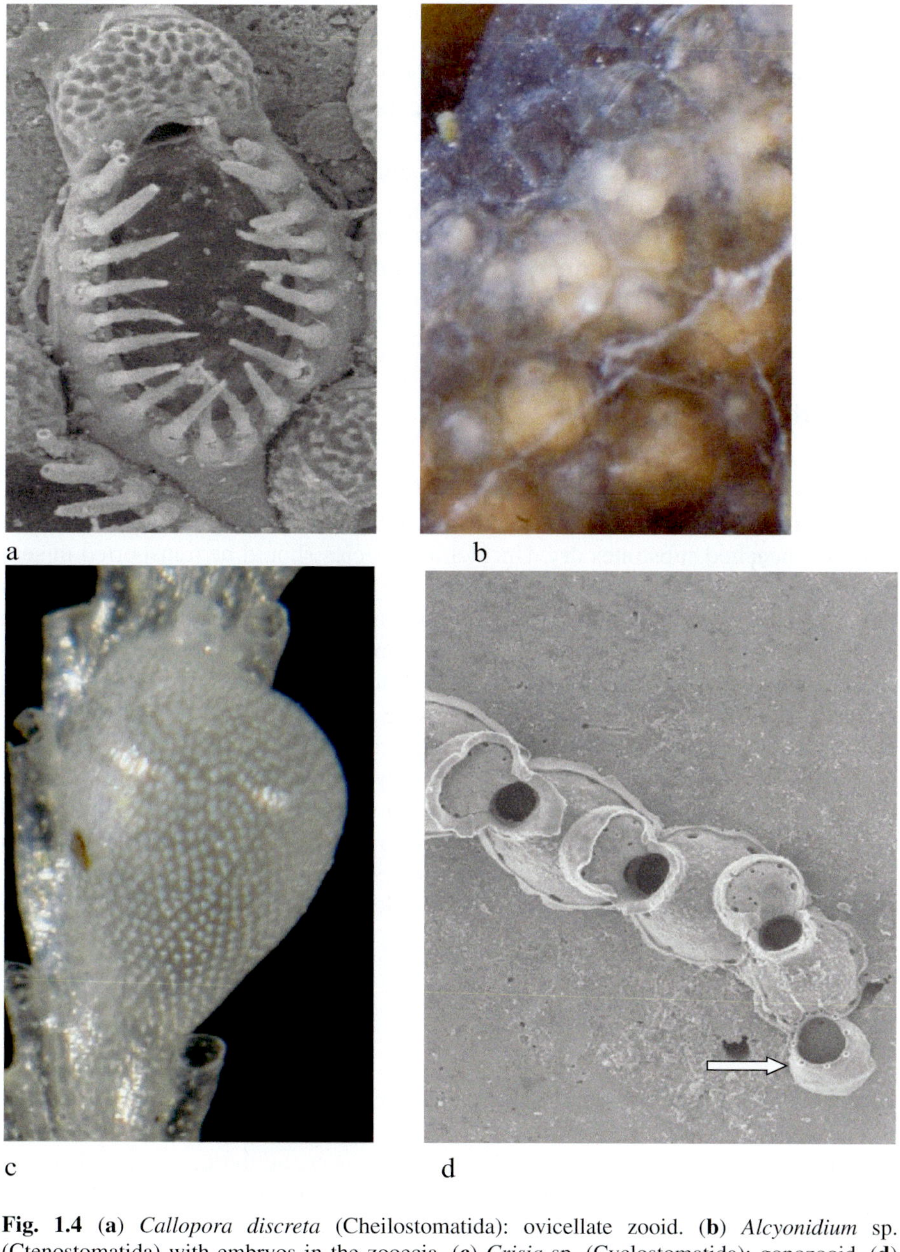

Fig. 1.4 (**a**) *Callopora discreta* (Cheilostomatida): ovicellate zooid. (**b**) *Alcyonidium* sp. (Ctenostomatida) with embryos in the zooecia. (**c**) *Crisia* sp. (Cyclostomatida): gonozooid. (**d**) *Neolagenipora* (Cheilostomatida): ancestrula (*arrow*)

An attached colony that moves with its substrate can of course cover great distances. Fertile colonies, for example, can be carried about on floating algae, plastic objects, or ship hulls. The large amount of plastic in the sea is undoubtedly contributing to the spread of bryozoans today. Recreational boating transfers species from one marina to another.

The larva settles on a suitable substrate and forms a first zooid or ancestrula (Fig. 1.4d). This ancestrula is usually different in shape from the later zooids and can be a useful aid for identification. The ancestrula typically buds 1, 2, or 3 first daughter zooids, which in turn expand the colony by further budding. Colonies are built of tens to hundreds of genetically identical zooids.

Collection

Scientific samples are often transported in 70% alcohol, seawater-formalin, or special liquids for DNA research. A beachcomber with an interest in bryozoans can transport beached substrates dry. Uncalcified species should be transported in seawater or alcohol. Pruning shears are useful for cutting the most interesting pieces out of plastic, so you do not have to carry large amounts. Bush-shaped colonies can be removed from the substrate with a knife. Handle them gently to preserve important features, e.g., spines. The use of brushes to clean the colony is therefore not recommended for species with spines. Rinsing under running water is sometimes sufficient. In dry, calcified species, annoying grains of sand can be removed under magnification with a moist needle to which they adhere. Dirt can be removed from calcified colonies by soaking them in a bleach solution for a while (about half an hour, depending on the firmness of the colony). Rinse well afterward. Bleach removes membranes and tissues but unfortunately also uncalcified joints, spines, or even autozooids in some weakly calcified species.

Examination and Identification

A magnification of 30× is very helpful. In some species, photography with a scanning electron microscope is necessary.

Uncalcified colonies are examined in seawater, as are all living colonies. In fresh, living colonies, one can easily observe the number of tentacles and the color of the tentacles, eggs, embryos, tissues, etc.

Calcified colonies are viewed as dry, or completely submerged. Damp colonies that have just been removed from the water are difficult to identify due to the reflective glittering of the remaining water.

Keep in mind that many features may have disappeared from washed-up or long-deceased colonies. Spines are usually found in young zooids at the colony edge; in older colony parts, they are often broken and lost. The frontal surface may be worn off from sandblasting. Species that have a membrane over the frontal surface covering the opesia or frontal calcification may have lost that membrane.

Lighting is very important during the investigation. Lighting from above does not always provide good results. Side lighting casts shadows to enhance features, useful when counting the spines. When photographing specimens, the distal end should be at the top and the predominant direction of lighting from the top left.

Casts can be made of the few species that drill into shells. Brush the shell with liquid polyester on the inside with a brush. Leave to dry, preferably on a vibrating surface such as the air pump of an aquarium. Then, place the shell in hydrochloric acid (30% solution) in a Petri dish. When the shell has dissolved, you can carefully rinse off the remaining polyester fleece. The cast of the zooids and kenozooids can then be studied under binocular magnification. Store with care as these casts are very delicate.

Conservation

Calcified species can be kept dry after rinsing with fresh water. With colonies on algae, it must be taken into account that the algae shrink when drying, while the attached bryozoans do not, so that they will fracture. It is better to keep such colonies in 70% alcohol. Colonies without calcification are best stored in 4% neutralized formalin made with seawater. Important finds belong in a natural history museum. The amateur can keep the colonies, properly labeled, temporarily in rubbing alcohol from the supermarket or pharmacist and donate them later to a museum.

Classification and Key to the Orders

Class Stenolaemata: Zooidal wall calcified and tubular

Order Cyclostomatida: Orifice circular or polygonal, at the end of a small tube

Class Gymnolaemata: Zooidal wall calcified or not. Zooids tubular, bottle-shaped, or box-shaped

Order Ctenostomatida: Zooidal walls not calcified, membranous. Zooids cylindrical, bottle-shaped, or flattened. Orifice terminal or frontal in encrusting species, not closed by an operculum but closed by pleated collar in most. Heterozooids absent, or only kenozooids

Order Cheilostomatida: Zooidal walls calcified, box-shaped. Orifice frontal and closed by a proximally hinged operculum. Heterozooids usually present, commonly diverse; suborders based upon frontal calcification and mechanism of lophophore protrusion

Fig. 1.5 (**a**) Encrusting colony of Cyclostomatida. (**b**) Part of erect colony of Cyclostomatida. (**c**) Encrusting colony of Ctenostomatida. (**d**) Erect colony of Ctenostomatida. (**e**) Encrusting colony of Cheilostomatida, with calcified frontal shield. (**f**) Encrusting colony of Cheilostomatida, showing opesiae; frontal membranes are visible in some zooids

1. (a) Zooidal wall is not calcified (Fig. 1.5c, d). .... Chapter 3 Ctenostomatida (p. 55)
   (b) Calcification is always present, at least in the side walls. 2
2. (a) Orifice at the distal end and round or polygonal, wall completely calcified, tubular (Fig. 1.5a, b). ....Chapter 2 Cyclostomatida (p. 15)

(b) Frontal or subterminal orifice, usually closed by an operculum; the autozooid is box-shaped, frontally calcified (Fig. 1.5e), or membranous (Fig. 1.5f). .... Chapter 4 Cheilostomatida (p. 119)

More about bryozoan morphology, reproduction, classification, and ecology relevant to neontologists can be found in Cook et al. (2018a, b), Hayward (1985), Hayward and Ryland (1985, 1998, 1999), Ostrovsky (2013), Schwaha (2020), and (Taylor 2020).

References

Cook PL, Bock PE, Gordon DP, Weaver HJ (eds) (2018a) Australian Bryozoa volume 1: biology, ecology and natural history. CSIRO Publishing, Melbourne

Cook PL, Bock PE, Gordon DP, Weaver HJ (eds) (2018b) Australian Bryozoa volume 2: taxonomy of Australian families. CSIRO Publishing, Melbourne

De Blauwe H (2009) Mosdiertjes van de Zuidelijke bocht van de Noordzee: Determinatiewerk voor België en Nederland. Vlaams Instituut voor de Zee (VLIZ): Oostende

Faasse MA, De Blauwe H (2004) Faunistisch overzicht van de mariene mosdiertjes van Nederland (Bryozoa: Stenolaemata, Gymnolaemata). Nederlandse Faunistische Mededelingen 21:17–54

Faasse M, Van Moorsel GWNM, Tempelman D (2013) Moss animals of the Dutch part of the North Sea and coastal waters of The Netherlands (Bryozoa). Nederlandse Faunistische Mededelingen 41:1–14

Hayward PJ (1985) Ctenostome bryozoans: keys and notes for the identification of the species. Synopses of the British fauna (new series), E.J. Brill, W. Backhuys, London, p 33

Hayward PJ and Ryland JS (1985) Cyclostome bryozoans: keys and notes for the identification of the species. Synopses of the British fauna (new series), E.J. Brill, W. Backhuys, London, p 34

Hayward PJ and Ryland JS (1998) Cheilostomatous Bryozoa: 1. Aeteoidea – Cribrilinoidea: notes for the identification of British species, 2nd edn. Synopses of the British Fauna (new series), Field Studies Council: Shrewsbury, p 10

Hayward PJ and Ryland JS (1999) Cheilostomatous Bryozoa: 2. Hippothooidea – Celleporoidea: notes for the identification of British species, 2nd edn. Synopses of the British Fauna (new series), Field Studies Council, Shrewsbury, p 14

Ostrovsky AN (2013) Evolution of the sexual reproduction in marine invertebrates: example of gymnolaemate Bryozoa. Springer Verlag, Dordrecht, p 356

Schüler L, Leitinger J, Faasse M, Kind B, De Blauwe H (2022) First record of the recently described Korean bryozoan Primavelans glabricollaris Min, Chae, Yang, Noh, Lee and Seo, 2021 (Cheilostomatida: Pacificincolidae) from the North-east Atlantic. Bioinvasions Records 11(3):659–666. https://doi.org/10.3391/bir.2022.11.3.08

Schwaha T (2020) Handbook of zoology, phylum Bryozoa. De Gruyter. https://doi.org/10.1515/9783110586312

Chapter 2
Cyclostomatida

Abstract This chapter discusses the order Cyclostomatida found in the southern bend of the North Sea, between Cap Gris-Nez in France and the German-Danish border. Twenty species have been identified. In addition to an identification key, the species are briefly described and illustrated, and the known distribution data is listed in detail.

Classification

Family Crisiidae

Filicrisia geniculata (Milne Edwards, 1838)
Crisidia cornuta (Linnaeus, 1758)
Crisia aculeata Hassall, 1841
Crisia denticulata (Lamarck, 1816)
Crisia eburnea (Linnaeus, 1758)
Crisia sp.

Family Stomatoporidae

Stomatoporina incurvata (Hincks, 1860)

Family Oncousoeciidae

Oncousoecia dilatans (Johnston, 1847)

Family Tubuliporidae

Tubulipora aperta Harmer, 1898
Tubulipora liliacea (Pallas, 1766)
Tubulipora lobifera Hastings, 1963
Tubulipora phalangea Couch, 1844
Tubulipora plumosa Thompson in Harmer, 1898

H. De Blauwe, *Marine and Brackish Bryozoans from the Southern Bight of the North Sea*, https://doi.org/10.1007/978-3-031-92513-9_2

Family Plagioeciidae

Plagioecia patina (Lamarck, 1816)
Plagioecia sarniensis (Norman, 1864)
Diplosolen obelium (Johnston, 1838)
Desmeplagioecia amphorae Harmelin, 1974

Family Annectocymidae

Annectocyma major (Johnston, 1847)
Entalophoroecia deflexa (Couch, 1842)

Family Lichenoporidae

Disporella hispida (Fleming, 1828)
Patinella verrucaria (Linnaeus, 1758)

Specific Morphology

There is a big difference in the appearance of cyclostome colonies depending on the species. The colony can be encrusting (Fig. 2.1a) or erect (Fig. 2.1b), rigid or flexible, ranging from a few millimeters to 30 cm in size. Cyclostomatida have a cylindrical calcified skeleton. The zooidal orifice is not closed by an operculum.

Colony members are divided into two categories or polymorphs: autozooids and heterozooids. Autozooids include the ancestrula and subsequently budded zooids, having a tentacle crown and being able to feed. Heterozooids have no feeding capability and include gonozooids, kenozooids, and nanozooids.

Embryos are incubated in the swollen gonozooids, and larvae are released via the ooeciopore, the opening of the ooeciostome (Fig. 2.2). Both ooeciostome and ooeciopore have species-specific shapes: oval, flattened, or flared. The ooeciopore is often slit-like (Ostrovsky 1998). Gonozooids have mostly distinct non-calcified areas (pseudopores) in their zooidal wall and seem perforated, but they are not true perforations. Autozooidal walls have much fewer pseudopores.

Kenozooids form root-like structures (rhizoids, Fig. 2.1c) for the attachment of the colony or spines (Fig. 2.3b). Nanozooids are small polymorphs with a single tentacle, probably used to sweep detritus and small organisms off the colony surface.

Jointed branches in the colonies in the family Crisiidae consist of calcified parts (internodes) and cuticular joints making colonies very flexible.

a

b c

Fig. 2.1 Cyclostomatida. (**a**) *Tubulipora aperta*, slit-like ooeciopore (*arrow*). (Photo by M. Cochu). (**b**) *Crisia eburnea*: joints (*arrows*) between internodes. (**c**) *Crisia denticulata*: jointed rhizoids. Brittany (France)

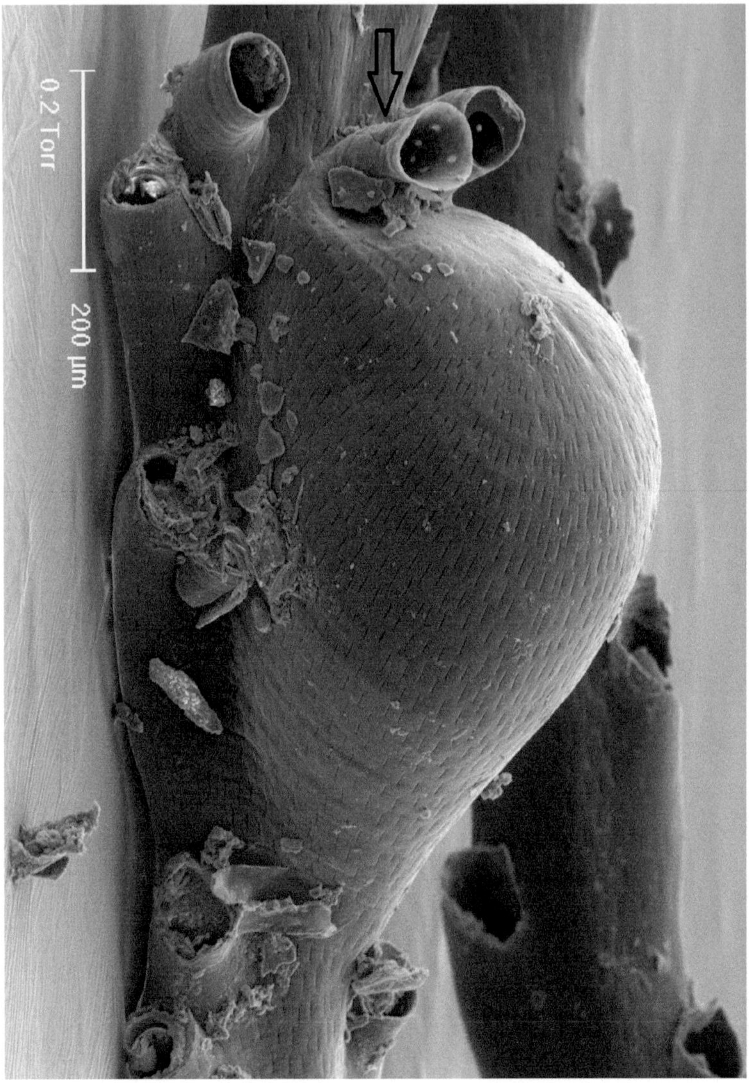

Fig. 2.2 *Crisia* sp. showing gonozooid with ooeciostome (*arrow*) and slit-like pseudopores. Le Conquet (France)

Key to the Genera

1. (a) Colony is erect and branching. Branches are slender; zooids are arranged in one or two rows, with or without spines. Internodes consist of one or more zooids, and internodes are separated by uncalcified joints (Figs. 2.1b and 2.3a). 2
   (b) Colony is encrusting or erect. If erect and branching, colonies are not jointed. 4

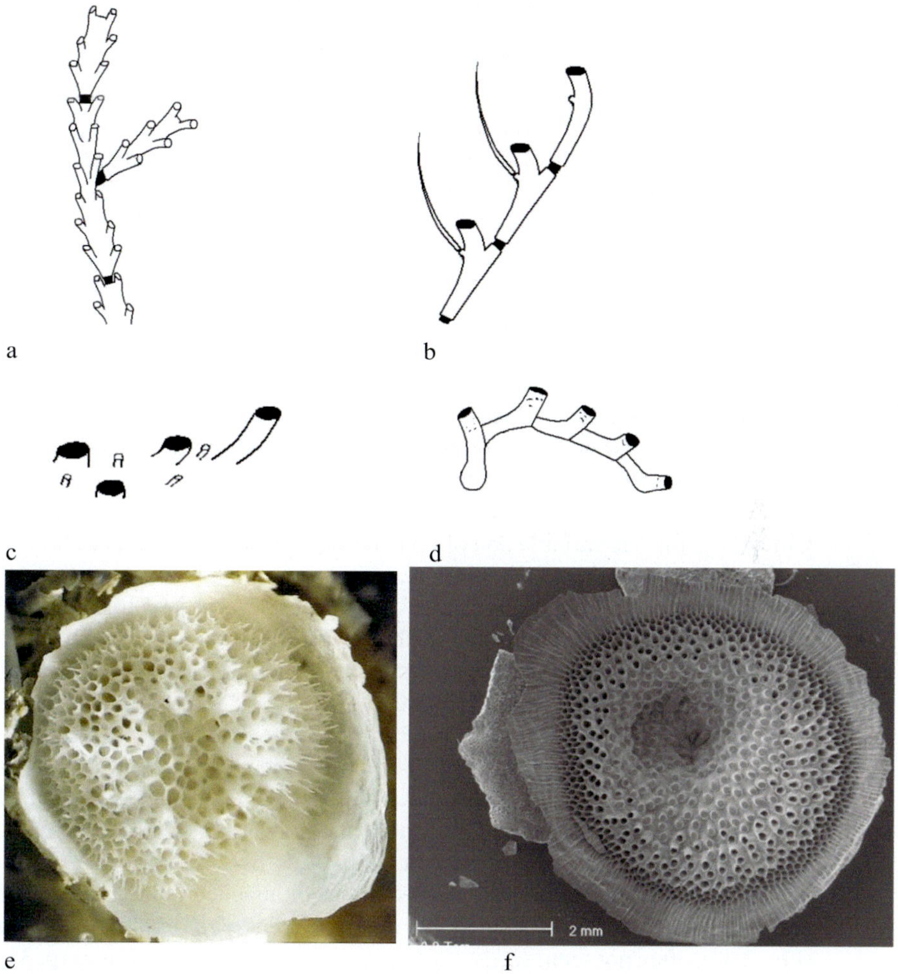

Fig. 2.3 (**a**) Internodes separated by uncalcified joints. (**b**) Each zooid with a long spine. (**c**) Zooids and nanozooids. (**d**) *Stomatoporina incurvata*. (**e**) *Disporella hispida* on beached plastic, Belgium. (**f**) *Plagioecia patina* on beached on plastic in Wenduine (Belgium)

2. (a) An internode is composed of a single zooid, and an uncalcified joint is present before and after the zooid (Fig. 2.3b). 3
   (b) Internodes are composed of more than one zooid (Fig. 2.3a). .... *Crisia* (p. 24)
3. (a) Each zooid bears (or had) a long spine (Fig. 2.3b). .... *Crisidia cornuta* (p. 22)
   (b) No spines. .... *Filicrisia geniculata* (p. 22)
4. (a) Zooid interstices are polygonal spaces that are partially or completely closed. Peristomes often end in two or three points (Fig. 2.3e). Gonozooid aperture is an ordinary tube or flaring funnel. .... *Lichenoporidae* (p. 48)

(b) There are no polygonal spaces between the zooids. Peristomes have a round aperture. 5

5. (a) Each autozooid is accompanied by a nanozooid (Fig. 2.3c). .... *Diplosolen obelium* (p. 42)
   (b) There are no nanozooids between the autozooids. 6
6. (a) Colony is fully attached. Usually, the colony consists of a curved row of zooids, the orifices all facing the outside of the curvature (Fig. 2.3d). Sometimes several rows of zooids lie against each other, in which case the orifices of a row of zooids remain directed to the same side. .... *Stomatoporina incurvata* (p. 29)
   (b) Orifices are not always oriented sideways. 7
7. (a) Colony flat or cup-shaped. Old zooids are sealed by a terminal calcified diaphragm (Fig. 2.3f). .... *Plagioecia* (p. 40)
   (b) Zooids lack a terminal calcified diaphragm. 8
8. (a) Colony is flat and thin. Gonozooid is teardrop-shaped and not penetrated by autozooidal peristomes (Fig. 2.4a). .... *Oncousoecia dilatans* (p. 38)
   (b) Gonozooid form is variable and often penetrated by autozooidal peristomes. 9
9. (a) Colony is encrusting and fan-shaped. Sometimes several fans form a more rounded colony. At the colony edge, the zooids form groups of two to four. Ooeciopore is round, the same size as the autozooid aperture, and directed upward on a short ooeciostome (Fig. 2.4b). .... *Desmeplagioecia amphorae* (p. 44)
   (b) Zooids are not in groups of three or four, and ooeciopore is not directed upward. 10
10. (a) Colony is often purple, fully attached and lobed, or with erect lobes. Peristomes are usually congregated in alternating rows. Gonozooid is swollen and conspicuous. Ooeciostome is close to or attached to an autozooidal peristome, broad funnel-shaped or a short tube with an inverted cap (Fig. 2.4c), often not seen. .... *Tubulipora* (p. 32)
    (b) Colony is adnate. Zooids are arranged in narrow branching rows. 11
11. (a) A distal branch and a lateral branch arise from the ancestrula. Zooids are arranged in transverse rows of two to four, partially fused at their bases. Ooeciostome is a curved tube, and ooeciopore is horizontal slit-shaped (Fig. 2.4d). .... *Annectocyma major* (p. 46)
    (b) From the ancestrula arises a first zooid, and from it arise two zooids that grow in opposite directions and thus form the first branch. Possibly single row at first, later typically two- or three-rowed, developing into multi-row lobes at the ends of the branches. The colony sometimes forms erect, cylindrical branches (often broken off). Peristomes are not fused at their bases. Ooeciostome is a long slender tube (Fig. 2.4e).. .... *Entalophoroecia deflexa* (p. 46)

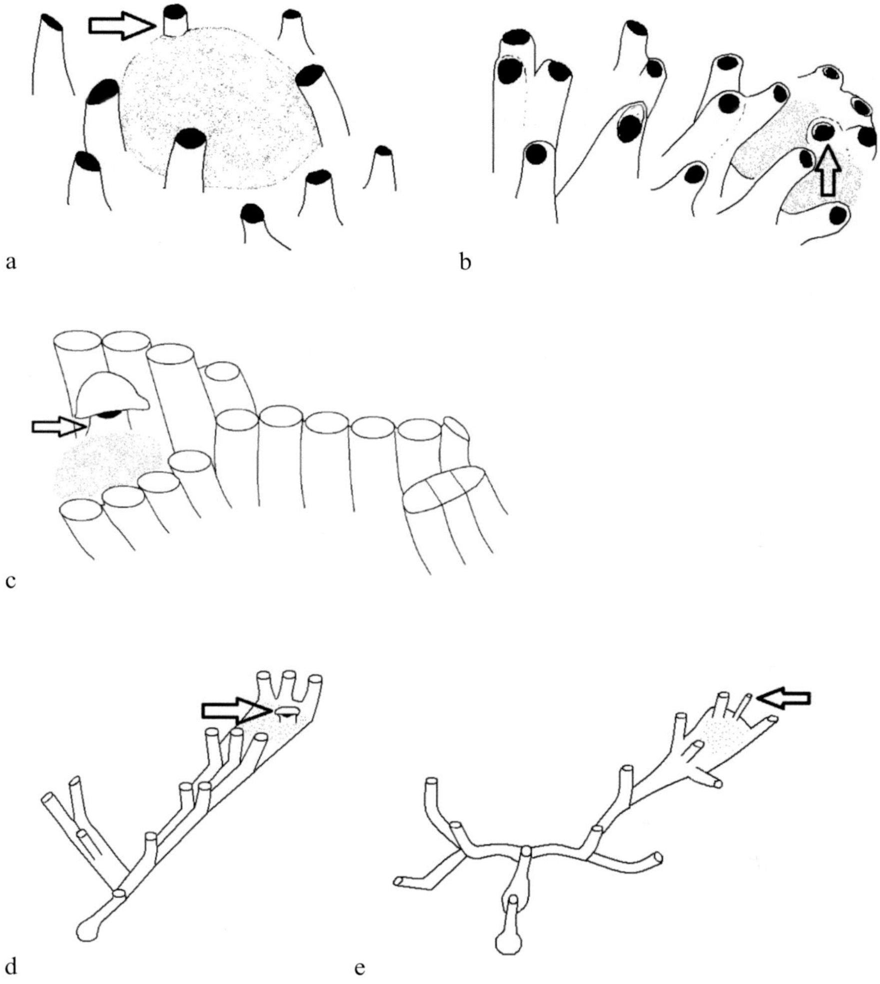

Fig. 2.4 (**a**) *Oncousoecia dilatans* with gonozooid. (**b**) *Desmeplagioecia amphorae*. (**c**) *Tubulipora*. (**d**) *Annectocyma major*. (**e**) *Entalophoroecia deflexa*

Systematic Part

Family Crisiidae

Colony is erect, branching, and articulated. Branches consist of internodes, made up of one or two rows of zooids. The internodes are connected to each other by uncalcified connection joints (nodes).

The number of zooids per internode varies from 1 to more than 20. The gonozooids are club-shaped, drop-shaped, or almost spherical and are located between the zooids.

Gonozooids can be recognized by their high density of pseudopores and their bulbous brood chambers.

The rhizoids consist of kenozooids which, like the internodes, are separated by joints. They serve to adhere colonies to the substrate and resemble rootlets. Some species have long spines, which consist of one or more kenozooids.

Genus *Filicrisia*

Filicrisia geniculata (Milne Edwards, 1838)

Description Colony is erect and branching. Internodes consist of one zooid, rarely more, apart from internodes with gonozooids, which consist of three to five zooids. Zooids are very long and slender, straight, curved, or angled. Zooid orifices are round and distally directed. Nodes between the internodes are pale in color. The polypide has eight tentacles. Gonozooid is in the second, third, or fourth position along the internode. Gonozooid is club-shaped and fully attached to the next zooid. Ooeciostome is almost terminal, forming a short tube. Ooeciopore is transversely oval, directed forward and upward. No spines (Fig. 2.5).

Occurrence There are no reports of native material from Belgium or the Netherlands. However, the species occurs rarely on beached buttons of the alga *Himanthalia* in Belgium (De Blauwe 2000, 2005; Vanhaelen et al. 2006) and the Netherlands (Bloklander and Leenhouts 1948; Verkuil 1998; de Ruijter 2012a, 2018b).

Genus *Crisidia*

Crisidia cornuta (Linnaeus, 1758)

Description Colony is often a dense tuft, up to 15 mm high. Internodes consist of one zooid bearing a pair of subopposite branches, or a branch and a long spine. The spine is made of kenozooids separated by joints. Fertile internodes consist of one gonozooid and are pear-shaped and free-growing with two branches at their base. Autozooids are horn-shaped with round distally directed orifices. Polypide has nine tentacles. Ooeciostome is tubular, facing forward and upward. Terminal ooeciopore is round (Fig. 2.6).

Occurrence

Belgium In the collection of the RBINS, there is a colony on *Flustra foliacea* dredged from Westhinder Bank (51°24′ N, 2°28′ E) on July 21, 1904. Loppens (1906) describes this species as rare on algae and shells. It rarely washes ashore on *Himanthalia* buttons in Belgium (De Blauwe 2005), including Zeebrugge in

Fig. 2.5 *Filicrisia geniculata*. (**a–c**) On beached *Himanthalia* button, Oostduinkerke (Belgium)

November 2023 (own obs. unpublished). Several colonies were collected during the DDNZS campaign in June 2021 on the wreck of the S.S. Ptarmigan at 51°41.332′ N, 02°25.087′ E (van Leeuwen 2023).

The Netherlands Maitland (1851) reports this species from Walcheren on washed-up algae and bryozoans, Lacourt (1949) found colonies on beached cork, in the National Museum of Natural History in Leiden, collected in Noordwijk, Scheveningen, and Terschelling. It is also recorded on washed-up plastic (de Ruijter 2014c, 2015b, 2018a, 2020, 2021, 2023).

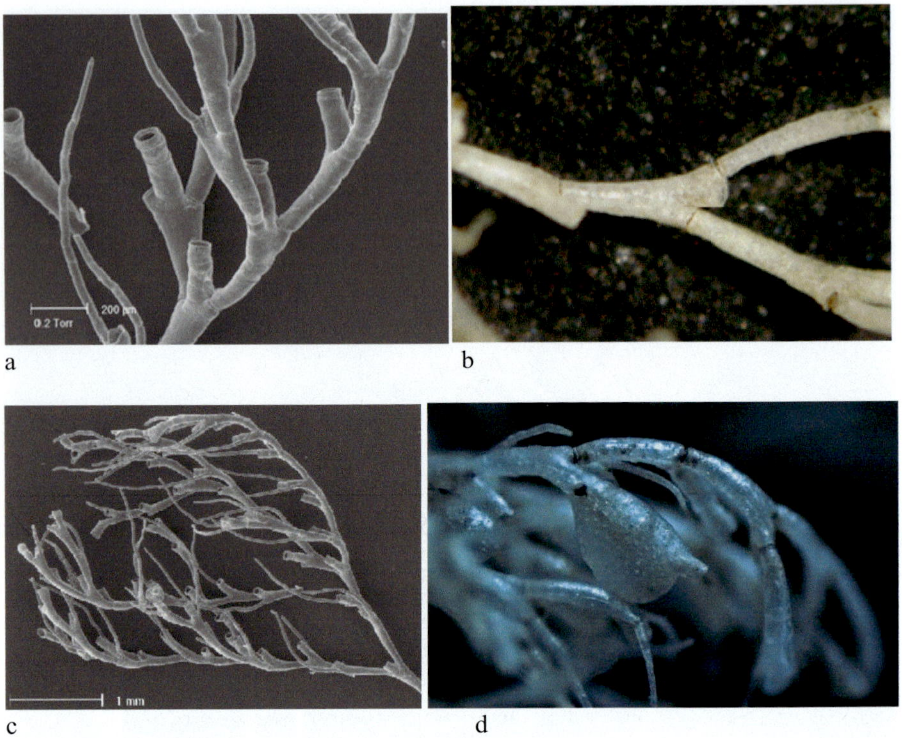

Fig. 2.6 *Crisidia cornuta.* (**a**) Internodes. (**b**) Spines are lost. (**c**) Washed ashore at De Haan (Belgium). (**d**) Gonozooid with terminal ooeciostome. Wreck of "Ptarmigan"

Germany Reported from the German North Sea (Harms 1993; Zettler et al. 2018, pers. comm. B. Kind).

Genus *Crisia*

1. (a) Internodes usually consist of a maximum of nine zooids, apart from fertile gonozooids, which can consist of more than nine zooids.......................... 2
   (b) Internodes generally consist of 11 or more zooids. 3
2. (a) Internodes usually consist of five to seven zooids and are strongly bent. Zooids never bear spines. .... *Crisia eburnea* (p. 26)
   (b) Internodes often consist of more than seven zooids in the distal parts of the colony and can be very weakly bent. Many zooids bear a long spine, but this is often broken off. .... *Crisia aculeata* (p. 25)
3. (a) Joints are black, except on young branches. Rhizoids (root-like attachments) also show black connections between the kenozooids. .... *Crisia denticulata* (p. 26)
   (b) Joints are yellow to brown, never black. .... *Crisia* sp. (p. 29)

Crisia aculeata Hassall, 1841

Description Forms delicate tufts up to 2 cm high. Internodes are very weakly bent, often consisting of five or seven zooids, more distally. Branching occurs at the first or second position on the internode, and sometimes at a higher position at the extremities of the colony.

On some internodes, one or more zooids may bear long slender spines composed of kenozooids separated by joints. The spines are bent forward. Such spines are usually present on the first zooid of the internode.

Joints are yellow or colorless in young colony parts. Peristomes are curved toward the frontal surface. Orifice is round, without a pointed outgrowth next to it. Gonozooids are in the third to fifth position of the internode and are teardrop-shaped. The ooeciostome was inconspicuous in our material (Fig. 2.7).

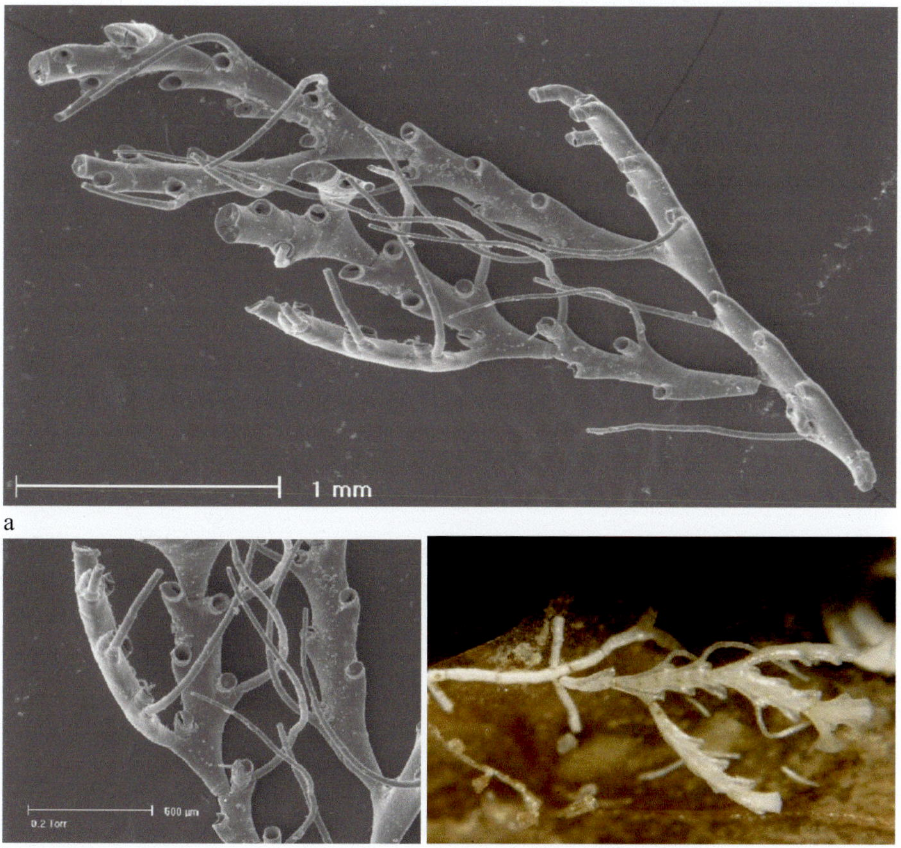

Fig. 2.7 *Crisia aculeata.* (**a**–**c**) Colony washed up in Southern England

Occurrence Washes ashore on *Abietinaria abietina* and on plastic.

Belgium A colony was found on the wreck of the Kilmore in 2003 and 2004 (Zintzen 2007) and on the wreck of S.S. Ptarmigan in 2021 (van Leeuwen 2023).

The Netherlands Lives in the most seaward parts of the Oosterschelde and the Westerschelde. Sampled in 2015 on shipwreck 2801 (55°12′46″ N, 03°32′46″ E) on the Dogger Bank (Faasse et al. 2016). Recorded on DDNZS nets in 2016 at the Niponia wreck (no. 2305, 52°44.520′ N, 3°58.702′ E) (van Leeuwen et al. 2017).

Germany Reported from the German North Sea (Harms 1993; Zettler et al. 2018; pers. comm. B. Kind), and as an epizoic on *Flustra foliacea* from the Heligoland Deep Trench or Tiefe Rinne (Bitschofsky 2013).

Crisia denticulata (Lamarck, 1816)

Description Colony forms an erect tuft up to 2.5 cm high. Internodes consist of 15 or more zooids except near the base of the colony. Joints are inky black, except when newly formed. Internodes with an odd number of zooids usually carry a side branch, and those with an even number of zooids usually do not. The side branches are typically in the fifth to ninth position on the internode. Many rhizoids with black joints are present. Zooids have a flat surface, except for a forward-facing peristome. Orifice is round, sometimes with a pointed protrusion next to it. The gonozooid is located high on the internode and is teardrop-shaped. Ooeciostome is an inconspicuous, very short tube (Fig. 2.8).

Occurrence

Belgium In 1904, a colony was collected at the West-Hinder lightship, and in 1905, in the Strait of Dover (RBINS collection). Loppens (1906) rarely found this occasional species on *Flustra foliacea* and on seaweed washed ashore. Rarely, it occurs on the Belgian coast on beached *Cystoseira baccata* and plastic (Zeebrugge, December 2000).

The Netherlands Maitland (1851) reports a colony on *Flustra foliacea* washed ashore in Westkapelle. It has been found on washed-up plastic at Westkapelle in 2004 and also at Ouddorp (de Ruijter 2014c).

Germany Reported from the German North Sea (Harms 1993; Zettler et al. 2018).

Crisia eburnea (Linnaeus, 1758)

Description Colony forming a dense tuft, up to 2 cm high. The branches are characteristically curved inwardly. Internodes usually consist of five to seven zooids. Side branches generally arise on the first zooid in an internode, alternately on the

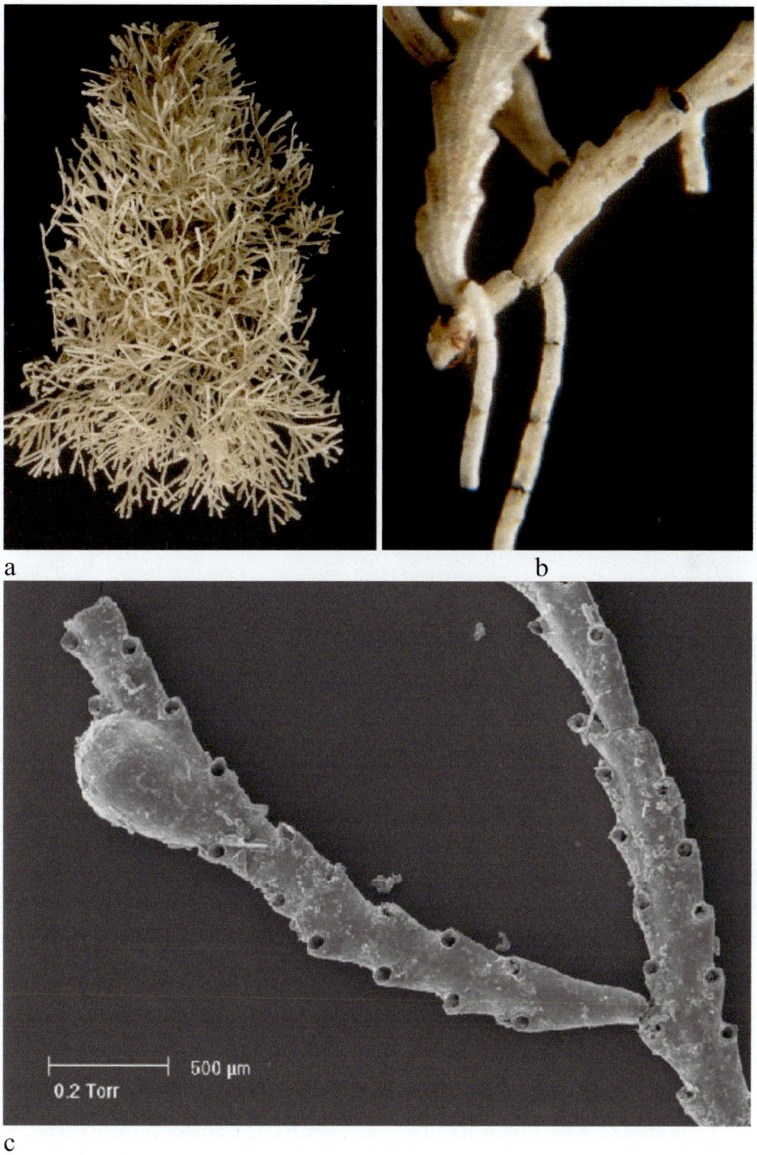

Fig. 2.8 *Crisia denticulata.* (**a**) Colony, St-Jacut (France). (**b**) Internodes and rhizoids with black joints. (**c**) Internode with gonozooid

left and right of the main branch. Joints are yellow or colorless in young internodes, and brown in older colony parts. Peristome is short and curved, with the orifice facing forward. Sometimes, a small pointed outgrowth is next to the opening.

The gonozooid is teardrop-shaped and replaces the second or sometimes the third zooid in the internode. The ooeciostome is a gradually tapering tube, curved and bent forward, and the ooeciopore is round or transversely oval (Fig. 2.9).

Fig. 2.9 *Crisia eburnea*. (**a–c**) Colony beached at Zeebrugge (Belgium)

Occurrence

Belgium Loppens (1906) considers this species to be common on *Flustra foliacea* washed ashore. It often washes up on *Abietinaria abietina*, *Flustra foliacea*, plastic, and *Himanthalia* buttons (De Blauwe 2005; Vanhaelen et al. 2006).

The Netherlands Lacourt (1949) reports colonies washed up on *Flustra foliacea*, hydroids, and cork along the entire coast. Maitland (1851) mentions sites, such as Zandvoort, Katwijk, and Scheveningen, probably on washed-up algae and *Flustra foliacea*. It was collected during the DDNZS campaign in June 2018 on the wreck of the Somali. It is often found washed up on plastic and *Himanthalia* buttons (de Ruijter 2013a, b, 2014a, b, 2015a).

Germany Reported from the German North Sea (Armonies et al. 2018; Harms 1993; Zettler et al. 2018, pers. comm. B. Kind) and as epizoic on *Flustra foliacea* from the Heligoland Deep Trench or Tiefe Rinne (Bitschofsky 2013).

Remark Rob Dekker found a very small colony of *Crisia* sp. between *Tricellaria inopinata* at the north side of the Zeelandbrug in the Oosterschelde, approximately 8 m deep, in October 2024. It has a maximum of seven zooids per internode. Two incomplete gonozooids are present, and side branches arise on the third zooid (not from the first zooid) in an internode.

Crisia sp.

Description Internodes usually consist of more than 11 zooids, and the joints are never black. Hayward and Ryland (1985) lists three species from around the British Isles. *Crisia ramosa* Harmer, 1891 is a species of shallow coastal waters common in the southwest of England; *Crisia calyptostoma* Hayward and Ryland, 1985 is a deepwater species known from the western approaches of the English Channel; and *Crisia klugei* Ryland, 1967 is a northern species reaching the Skagerrak. The shape of the ooeciostome, the peristome diameter, the shape of the peristome, and the density of the pseudopores should lead to a reliable determination of the material recorded here, which may belong to one of these three species (Fig. 2.10).

Occurrence

Belgium Colonies with more than 11 zooids per internode and yellowish joints have been found washed up at Zeebrugge (December 2000) on a net and at Wenduine (September 2001) on plastic. This washed-up material had no gonozooids, which compromises identification. The finding of *C. ramosa* reported in De Blauwe (2003) should be considered as *Crisia* sp.

Family Stomatoporidae

Genus *Stomatoporina*

Stomatoporina incurvata (Hincks, 1859)

Description The colony is small and completely attached to the substrate. The zooids are arranged in one or two rows. Young colonies make a 90° turn and then grow further either to form a circle with a diameter of 2 mm or continue in a straight line with two rows of zooids, the branch reaching a length of up to 8 mm as found on Westhinder Bank. Sometimes a colony bifurcates, with the two branches curving away from each other in opposite directions, with the apertures in one branch opening to the left and in the other to the right. Typically, the zooids have their apertures

Fig. 2.10 *Crisia* sp. beached at Wenduine (Belgium)

in curved rows on the outside of the circle. The distal third of the zooids is turned abruptly sideways, the openings facing sideways. On top of the zooid, there is a knobby ridge, which means that the peristome is not round in cross-section as in the other Cyclostomatida in the area (Fig. 2.11).

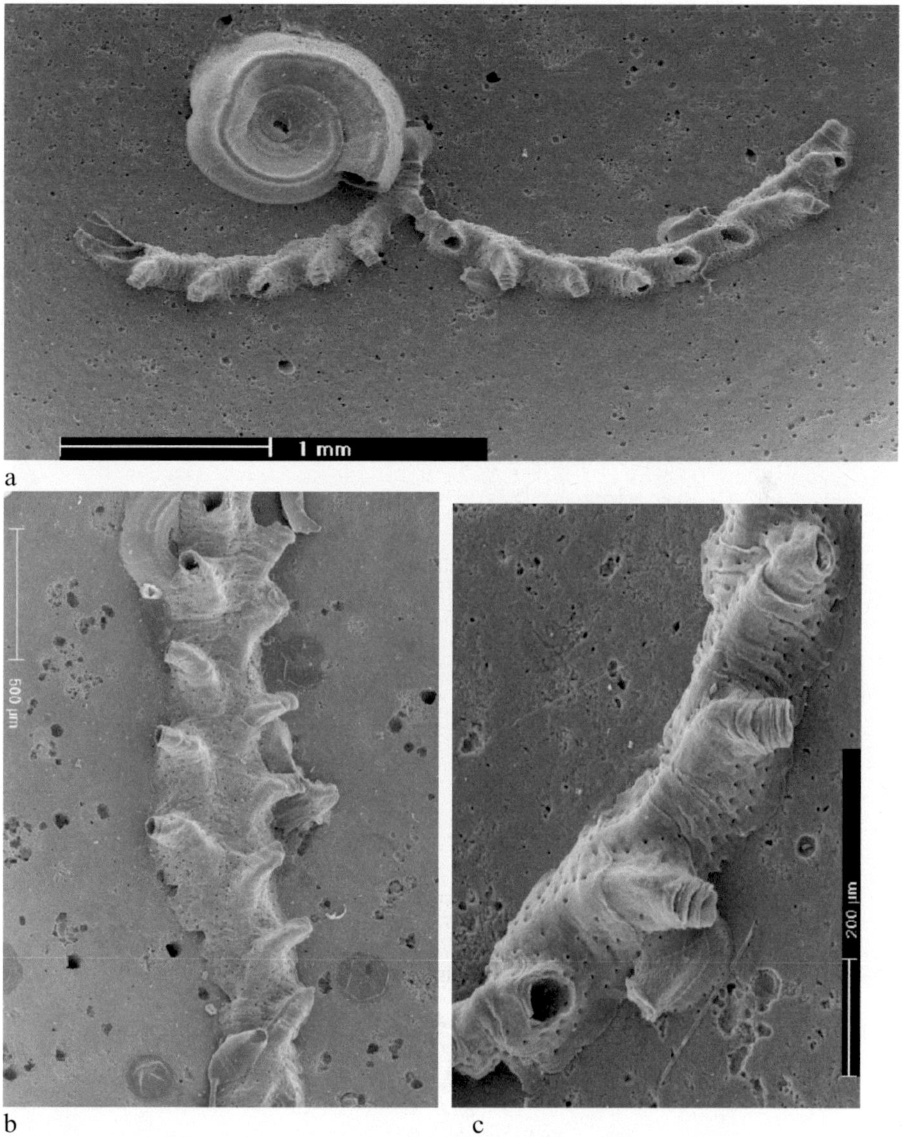

Fig. 2.11 *Stomatoporina incurvata*. (**a**) In *Cerastoderma edule*, Kwinte Bank. (**b**) Biserial colony, Westhinder Bank. (**c**) Detail, Kwinte Bank (Belgium)

Occurrence

Belgium Found on the inner side of bivalve shells. There are colonies in the RBINS, fished from the Hinder Banks in the period 1904–1906. In May 2004, a colony was found on a shell from Kwinte Bank, and in June 2005, several colonies were found on shell valves from Hinder Banks (De Blauwe 2009).

Family Tubuliporidae

Genus *Tubulipora*

The shape of the gonozooid opening (ooeciopore) at the end of the ooeciostome is very useful for identification. Unfortunately, this feature is often not visible, which prohibits identification at the species level. Hayward and Ryland (1985) list seven species, of which five are reported in our area. Other species can wash up on algae or plastic.

1. (a) Autozooids are arranged in uniserial connate rows (Fig. 2.12a) 2
   (b) Autozooids are not in connate series. Ooeciostome is a tube opening upward (Fig. 2.12b, c). 4

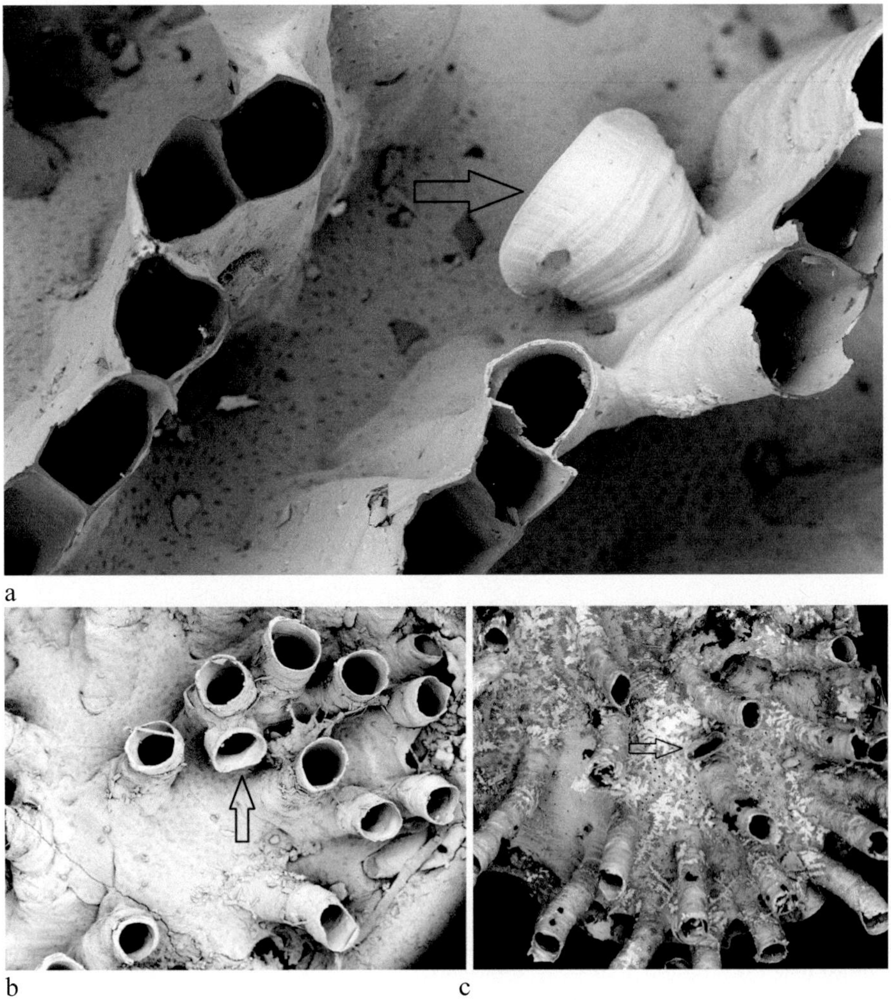

Fig. 2.12 (**a**) *Tubulipora liliacea* Le Conquet (France). (**b**) *Tubulipora lobifera* on beached plastic, Zeebrugge (Belgium). (**c**) *Tubulipora aperta*, Germany

2. (a) Ooeciostome is usually on the distal side of an autozooid, with the wide distal cap facing down or sideways. 3
   (b) Ooeciopore opens directly upward. .... *Tubulipora plumosa* (p. 37)
3. (a) Ooeciopore is as wide as a zooidal aperture. ....
   .... *Tubulipora liliacea* (p. 33)
4. (b) Ooeciopore is half of a zooidal aperture. ....
   .... *Tubulipora phalangea* (p. 37)
5. (a) Colony is thick with robust lobes. Peristomes are partially connate and short. Ooeciostome is a short tube opening upward; ooeciopore is as wide as an autozooidal aperture. .... *Tubulipora lobifera* (p. 36)
   (b) Colony is delicate, with long peristomes in alternating longitudinal rows. Ooeciostome is broadly funnel-shaped and folded, with the ooeciopore opening upward. .... *Tubulipora aperta* (p. 33)

Tubulipora aperta Harmer, 1898

Description Colony is encrusting, lobate, and oval. Autozooids are arranged quincuncial; peristomes are not connate, up to 1.5 mm long, with circular apertures. The gonozooid is expanded; the ooeciostome is independent of autozooidal peristomes but sometimes squeezed between them. It is flared, with the ooeciopore opening directly upward (Fig. 2.13a *arrow*), often with a folded edge (Fig. 2.13).

Occurrence

The Netherlands One record on beached plastic (De Blauwe and de Ruijter 2015), and several colonies on the plastic entrance of a lobsterpot beached at Petten in November 2023 (coll. de Ruijter; id. De Blauwe, unpublished).

Germany Recorded from the German North Sea (Zettler et al. 2018; pers. comm. B. Kind).

Tubulipora liliacea (Pallas, 1766)

Description Colony is lobate; calcification is fragile, translucent, and light purple in color. It is encrusting or partly erect, or almost completely erect on limited substrata such as hydroids. Autozooidal peristomes are arranged in connate rows of 4–12, resembling a pan flute, with distal free tops usually broken off. Peristome length varies, up to 2 mm high.

Fig. 2.13 *Tubulipora aperta.* (**a**) Ooeciostome (*arrow*), Brittany. (France, Photo by Marc Cochu). (**b**) Part of colony with ooeciostome, Germany

Gonozooid is distal in the colony and densely pseudoporous. Ooeciostome is on the distal face of the second or third peristome from the branch axis, is short, as wide as a peristome, with a wide distal cap, pointing downward or laterally; ooeciopore is oval. *Tubulipora phalangea* (p. 37) has a similar ooeciopore, but it is half as wide as a peristome (Fig. 2.14).

Occurrence Often washes up on plastic, nylon nets, and *Himanthalia* buttons in Belgium and the Netherlands.

Belgium Loppens (1906) found that this species was common on Belgian beaches on *Hydrallmania falcata*, *Bugulina*, and old shells.

Fig. 2.14 *Tubulipora liliacea.* (**a**) From a fishing net, Le Conquet (France). (**b**) Peristomes and ooeciostome

The Netherlands Lacourt (1949) listed finds from Westkapelle to Den Helder. Heerebout (1970) lists two sites (washed-up?): Katse Hoek and Breskens. Reports from the Netherlands are considered uncertain in Faasse and De Blauwe (2004) due to a lack of gonozooids.

Germany Recorded from the German North Sea (Harms 1993; Zettler et al. 2018; pers. comm. B. Kind; pers. obs. 2023, unpublished).

Tubulipora lobifera Hastings, 1963

Description Colony is completely adnate, pale purple in color, and regularly flower-shaped in outline because of its symmetrical lobes. Lobes are 2–4 mm wide and up to 4 mm apart. Colony is with a high profile, although the peristomes are rather short and often broken. At the colony edge, the colony quickly tapers to the substrate.

Adnate parts of the zooids are short, broad, and convex, separated by grooves. Peristomes grow together for half their length, a narrowing annulated distal part (often broken) growing freely. Zooids are usually in rows of three or five.

Gonozooid is not conspicuously swollen, occupying most of the lobe width. Ooeciostome is almost in the midline of gonozooid, short, partially adnate to a peristome or between two peristomes, funnel-shaped with a folded edge. Ooeciopore is oval, directed upward, and wider than a peristome (Fig. 2.15).

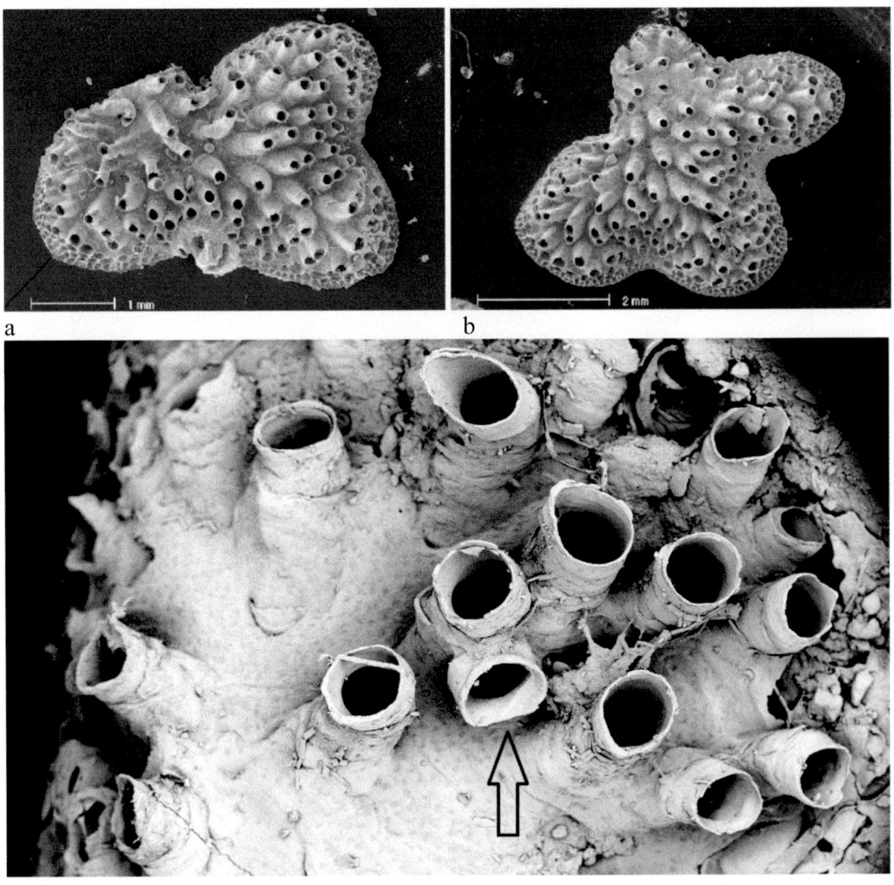

Fig. 2.15 *Tubulipora lobifera*. (**a**, **b**) Colonies on beached plastic, Zeebrugge (Belgium). (**c**) Peristomes and ooeciostome (*arrow*), Tiefe Rinne (Germany)

Occurrence

Belgium Colonies found washed ashore on plastic (Zeebrugge, 2000) and in a *Glycymeris glycymeris* valve found in a fishing net (Koksijde, 2023).

The Netherlands Collected three times on beached plastic since 1999 (De Blauwe 2009; De Ruijter 2015a and 2018a).

Germany Tiefe Rinne, 2023 (pers. obs. unpublished).

Tubulipora phalangea Couch, 1844

Description Colony is adnate, occasionally with semi-erect lobes, and pale purple in color. Autozooids are arranged in radiating uniserial connate rows of up to 14 peristomes. Ooeciostome is closely applied to an autozooidal peristome. Ooeciopore is less than half the width of an autozooidal aperture and is obscured by a basally or laterally directed narrow hood (0.1 mm wide) (Fig. 2.16).

Occurrence

The Netherlands Lacourt (1949) mentions a colony on beached *Saccorhiza bulbosa* in Scheveningen and possibly three colonies on cork in Noordwijk (National Museum of Natural History in Leiden).

Germany Collected on shells in the Tiefe Rinne in 2023 (obs. Kind and De Blauwe, unpublished).

Tubulipora plumosa Thompson in Harmer, 1898

Description Colony is encrusting, occasionally with semi-erect lobes. Autozooidal peristomes are often arranged in connate rows of two to four, resembling a pan flute, or in fascicles. Their distal ends are usually broken off, and their length is variable, up to 3 mm. The ooeciostome is a short, wide flattened tube near the zooidal peristome. Ooeciopore opens directly upward, with the lip broadly flared and folded.

Occurrence On algae and shells.

Belgium On a beached *Himanthalia* button in November 2023 (pers. obs. unpublished).

Germany Recorded from the German North Sea (Zettler et al. 2018, pers. comm. B. Kind).

Fig. 2.16 *Tubulipora phalangea*. (**a**) Colonies on dead shell. (**b**) Ooeciostome (*arrow*) half as wide as peristomes, with narrow hood. Tiefe Rinne (Germany)

Family Oncousoeciidae.

Genus *Oncousoecia*

Colony is encrusting, with a densely pseudoporous and bulbous gonozooid roof that is not penetrated by autozooidal peristomes. The ooeciostome is short and straight, not adnate to an autozooidal peristome. The ooeciopore is terminal or subterminal, circular or transversely elliptical, and a little smaller than an autozooidal aperture (Fig. 2.17).

Oncousoecia dilatans (Johnston, 1847)

Formerly reported as *Eurystrotos compacta* (Norman, 1866) in Hayward and Ryland (1985) and De Blauwe (2009).

Fig. 2.17 *Tubulipora plumosa.* Colony area showing flared ooeciostome (*arrows*). Germany

Description (after Taylor and Zaton 2008): The colony forms a thin, flat, translucent white crust, shimmering to chalky. The growth margin of the colony is minimal. Autozooids are elongate, with frontal walls that are flat proximally and gently convex distally. Pseudopores are regularly spaced, small, teardrop-shaped, and pointed distally. Apertures are subcircular, measuring 0.06–0.11 mm in diameter, and distributed more or less quincuncially, not clustered or connate. Preserved peristomes are short to moderate in length. Terminal diaphragms have scattered pseudopores that occlude some apertures and buds at inactive growing edges.

Ancestrula has a large protoecium, about 0.17 mm in diameter, with a single row of tiny pseudopores around the perimeter. The distal tube is short, 0.09–0.15 mm long, straight or curved. One distal autozooid is budded from the ancestrula.

Gonozooids are bulbous; their margins are indented by neighboring autozooids, but the roof is not pierced by peristomes; the frontal wall is densely pseudoporous, with pseudopores identical in shape to those of autozooids. The ooeciostome is a short tube, and the ooeciopore is terminal or subterminal, slightly transversely elliptical, and narrower than an autozooidal aperture (Fig. 2.18).

Occurrence In Belgium and the Netherlands, on washed-up plastic.

Belgium A colony was collected off the Belgian west coast in 1899 and two specimens on *Chlamys* in 1905 (RBINS collection).

France On boulders near Cap Gris-Nez in June 1992 (Davoult et al. 1999).

a

b

Fig. 2.18 *Oncousoecia dilatans*. (**a**) Colony on beached plastic. (**b**) Gonozooid with ooeciostome (*arrow*), on beached plastic, oblique view

Family Plagioeciidae

Genus *Plagioecia*

1. (a) Older zooids are closed by a diaphragm. .... *Plagioecia patina* (p. 40)
   (b) Older zooids are transformed into a secondary nanozooid. .... *Plagioecia sarniensis* (p. 42)

Plagioecia patina (Lamarck, 1816)

Description Colony forms a flat disc that is fully adherent to the substrate or cup-shaped structure raised from the substrate near its edge. Often, a number of daughter subcolonies form a composite colony. Subcolonies are usually 4–7 mm in diameter, though sometimes larger. Calcification is white and shimmers in living colonies. Zooids are radiating from the colony center to the periphery in alternating rows. Zooids are largely hidden, sometimes with an erect peristome, especially in deep cup-shaped colonies. Orifice is oval or round and closely spaced.

Gonozooids are often at the colony margin and are always wider than long. Swollen, densely covered with very small pseudopores, and completely enclose peristomes up to their orifice. The ooeciostome (not found in our material) is slender, directed upward or slightly curved proximally, and the ooeciopore is round (Fig. 2.19).

Occurrence

Belgium About ten specimens were collected between 1898 and 1910 on the Hinder Banks and one specimen in 1902 at "Spleet" (51°25′ N, 3°15′ E). Four colonies were found on boulders and shells of the Hinder Banks. Living colonies occur on shipwrecks off the Belgian coast (Zintzen 2007; van Leeuwen 2023). It often occurs on washed-up objects.

Fig. 2.19 *Plagioecia patina.* (**a**) White colonies on the underside of a rock, Brittany (France). (**b**) Composite colony, Pointe du Rozel (France). (**c**) Peripheral colony area, some peristomes closed by calcified diaphragm

The Netherlands Lacourt (1949) found colonies on cork in the Dutch natural history museums in Leiden and Amsterdam, collected from Scheveningen, Wassenaar, Katwijk, Noordwijk, and Den Helder. Five colonies were found on Texel on rope washed ashore in February 2006 (de Ruijter 2006), on nylon rope (de Ruijter 2011), and on plastic (de Ruijter 2012b, 2014b, 2015a, 2020). Samples were collected in situ on the Cleaver Bank in 2002 (van Moorsel 2003), near Westkapelle (Faasse et al. 2013), and on the shipwreck of the "Tubantia SS" (UTM 487399 5,742,147) (Lengkeek et al. 2013). Samples were taken in 2015 from shipwreck 2801 (55°12′46″ N, 03°32′46″ E) on the Doggers Bank (Faasse et al. 2016).

Germany On shells from the Tiefe Rinne in 2023 (pers. obs. unpublished).

Plagioecia sarniensis (Norman, 1864)

Description Colony is fully attached, thin and flat, circular or irregularly lobed, with a flat growth margin. It is translucent white and often wider than 2 cm. Daughter colonies are often present at the growth margin. Zooids are arranged in alternating rows radiating from the colony origin. Horizontal zooid parts are flat, with an inconspicuous border. Peristome length is variable, often longer than 1 mm. Older zooids are sealed with a calcified diaphragm with a short tube resembling a miniature chimney. In this case, the autozooid is transformed into a secondary nanozooid, with a miniature orifice through which a shortened unciliated tentacle is protruded (Schack et al. 2018).

Gonozooids near the colony boundary are always wider than long and completely enclose a number of peristomes. Ooeciostome distal, short, cylindrical. Ooeciopore round (Fig. 2.20).

Occurrence

Belgium Colonies were washed ashore attached to plastic, in 2000 and 2001 at Zeebrugge, in 2011 at Bredene, and again in 2024 at De Panne.

The Netherlands De Ruijter (2014b) reports a colony on beached plastic near Egmond aan Zee in 2013, and again in 2017 (de Ruijter 2018a).

Germany Reported by Harms (1993) from the vicinity of Heligoland.

Genus *Diplosolen*

Diplosolen obelium (Johnston, 1838)

Description Colony forms a thin, flat crust, translucent to chalky white, often lobed and wider than 1 cm.

Fig. 2.20 *Plagioecia sarniensis.* (**a**) On beached plastic at Zeebrugge (Belgium). (**b**) Gonozooid and ooeciostome (*arrow*). (**c**) Autozooid transformed to a secondary kenozooid with its peristome closed by diaphragm, with secondary miniature peristome

Zooids are arranged in regular alternating rows. Horizontal zooid parts are rather flat. Peristome is slender, often longer than 1 mm. Each autozooid is accompanied by a nanozooid (dwarf zooidal polymorph). The nanozooid sits on the proximal part of the autozooid and has a very narrow peristome. The nanozooid has one tentacle that wipes the surface. Gonozooids are often large, and their ooeciostome is short and cylindrical, with a partially recurved distal end. Ooeciopore is round (Fig. 2.21).

Occurrence Found infrequently washed up on plastic in Belgium and the Netherlands (de Ruijter 2015a).

Fig. 2.21 *Diplosolen obelium.* (**a**) Colony from fishing nets in Le Conquet (France). (**b**) Zooids and nanozooids

Genus *Desmeplagioecia*

Desmeplagioecia amphorae Harmelin, 1974

Description Colony forms a lobate crust; multiple lobes can give the colony a more rounded shape. The flat colony is quite thick. There is no noteworthy plate-shaped growth margin, but a border of zooidal buds. At the edge, groups of two or four zooids grow together at their base. The horizontal zooid parts are not clearly distinguishable. The peristomes grow together a little but separate at their ends.

Gonozooids are distal in the colony, between three zooid groups, and are not perforated by autozooidal peristomes. The distal ooeciostome is a short tube that faces upward, with a circular ooeciopore. Round ooeciopore has approximately the same diameter as a zooidal orifice.

Ancestrula with a basal disc and a peristome that is curved to the left or right (Fig. 2.22).

Occurrence

Belgium A few colonies were found in shell valves on the Hinder Banks in 2005.

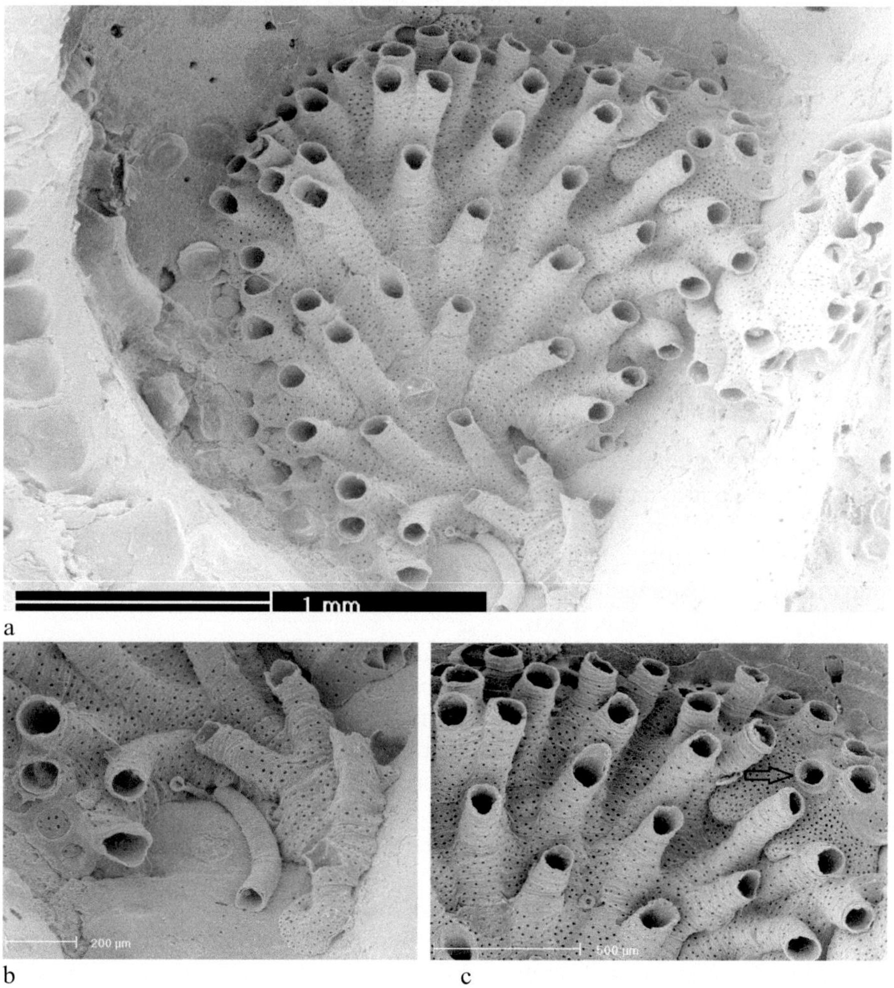

Fig. 2.22 *Desmeplagioecia amphorae* growing in a dead shell from the Westhinder Bank (Belgium). (**a**) Colony. (**b**) Ancestrular region. (**c**) Colony area showing peristomes and ooeciopore (*arrow*)

Family Annectocymidae.

Genus *Annectocyma*

Annectocyma major (Johnston, 1847)

Description Colony forms a fully attached branching network of slender lobes. Erect cylindrical colony parts, with zooids facing all around, are rare. Zooids are arranged in two to five rows, widening to eight or more rows near a branch. Peristomes are almost perpendicular to the frontal plane.

Gonozooid is variable in shape and inconspicuous. The ooeciostome is a slender compressed tube that bends horizontally; the ooeciopore is slit-like.

Two branches usually arise from the ancestrula: a distal branch and a lateral branch that grows in the opposite direction. Both branches start with one zooid and then widen into three rows of zooids (Fig. 2.23).

Occurrence

Belgium A colony was found on a clam shell fished from the Hinder Banks in 2005

Genus *Entalophoroecia*

Entalophoroecia deflexa (Couch, 1842)

Description It is very similar to *Annectocyma major*, but it often forms erect colony parts. The colony forms an attached network of very slender lobes that branch dichotomously at regular intervals. The colony is initially uniserial but later becomes biserial before widening to six to eight zooids, usually with gonozooids at each branch end. Upright branches are club-shaped, up to 2 mm in diameter and 1 cm in height. Peristomes alternate in spiral whorls, under 60° with the branch axis, opening all around the branch axis.

The ancestrula consists of a disk surrounded by a ring of pseudopores and a straight tube at an angle of 60–90° to the substrate. From the first zooid formed after the ancestrula, two zooids arise, which form the first branch. These two branches grow apart at an angle of 180°. Subsequent branches diverge at 90°. The branches end broadly with often a prominent gonozooid after four or five branches after the ancestrula.

Peristomes are perpendicular to the frontal surface, alternating, and not converging at their bases.

Gonozooids can occur anywhere in the colony. They are very convex and teardrop-shaped, densely covered with pseudopores, and a few autozooidal peristomes grow through the gonozooid. In erect colony parts, gonozooids occupy the entire distal widening, and many peristomes grow through them.

Fig. 2.23 *Annectocyma major* collected on plastic in Brittany (France). (**a**) Part of the colony with on the left a young colony *Disporella hispida*. (**b**) Ancestrular region of the same colony. (**c**) Part of the colony with ooeciostome (*arrow*)

The ooeciostome is in the center of the gonozooid or more distally, close to a peristome base, but not attached to it. It is slender cylindrical, and straight. The ooeciopore is round (Fig. 2.24).

Occurrence

Belgium Infrequently found washed up on plastic (De Blauwe 2003). Two colonies were found in June 2005 on shells from the Hinder Banks.

The Netherlands Found on a fossil bivalve (de Ruijter 2013b). De Ruijter (2014b, 2015b, 2019) reports the washing-up of a colony on plastic.

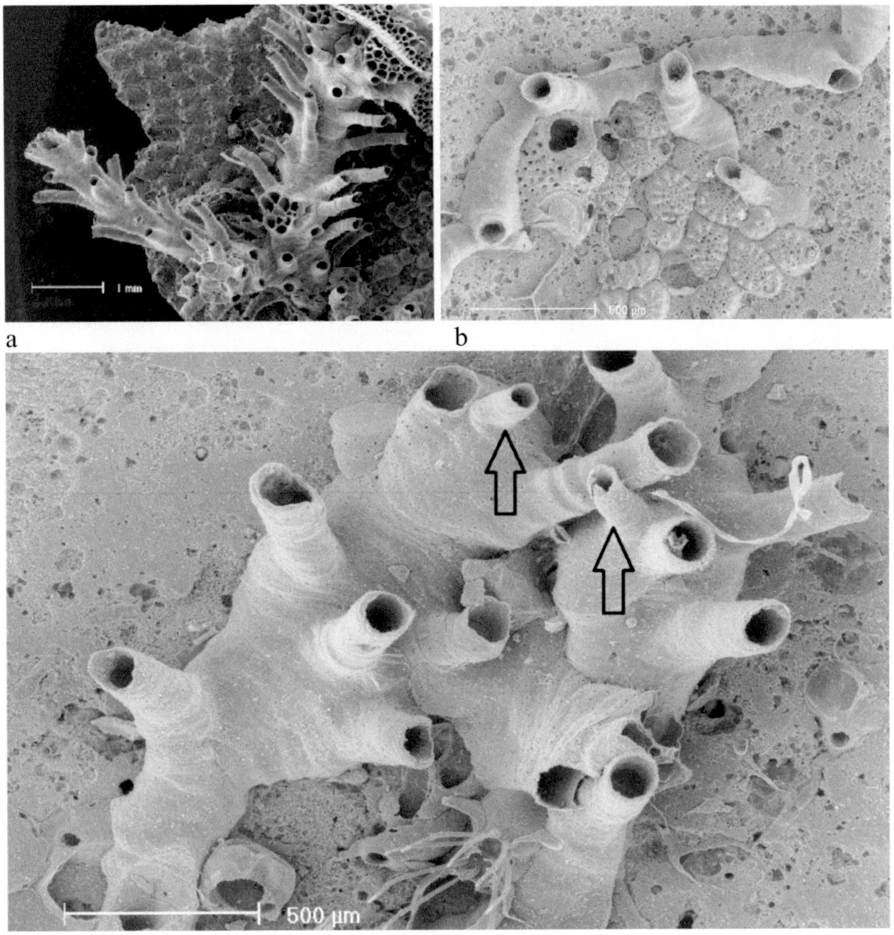

a b c

Fig. 2.24 *Entalophoroecia deflexa.* (**a**) Two erect colony parts, on beached plastic, Wenduine (Belgium). (**b–c**) Colony growing in a dead shell from the Westhinder Bank (Belgium). (**b**) Ancestrular region. (**c**) Branches with gonozooids and two ooeciostomes (*arrows*)

Family Lichenoporidae

1. (a) Basal lamina smooth, no open tubes extending to the margin. Gonozooid aperture a simple tube.................................... *Disporella hispida* (p. 49)
2. (b) Basal lamina surrounding the colony has open tubes up to the rim. Gonozooid aperture is funnel-shaped.................................... *Patinella verrucaria* (p. 49)

Genus *Disporella*

Disporella hispida (Fleming, 1828)

Description Colony is cup-shaped with broad basal lamina extending beyond the budding zone at the colony margin or dome-shaped, with less distinct basal lamina at the growth margin. Colony is usually 2–15 mm in diameter. Composite colonies have multiple centers from which zooid rows radiate. Zooids are initially in radiating rows, with each zooid sometimes connected to its predecessor and to the next zooid by struts. The peristomes themselves are not connate. The free part of the peristomes varies in length and decreases toward the edge of the colony. Proximal lip resembles a slender spine, often with one or more additional spines on each side.

The holes, created by the struts in the spaces between the zooids, initially have the same diameter as the peristomes, and they become smaller due to progressive calcification.

The colonial incubation chamber is entirely immersed in the central parts of the colony and between the zooid rows. Its roof is recognizable as a smooth lobate calcified area with tiny pores. Ooeciostome is a very short tube. Ooeciopore is round (Fig. 2.25).

Occurrence

Belgium Found on stones and shells in the RBINS collection. Dead colonies were found on boulders between the Hinder Banks in June 2005. A colony was found on the wrecks of the Birkenfels (2002 and 2021) and Ptarmigan (2021) (Zintzen 2007; van Leeuwen 2023). Often present on washed-up plastic.

The Netherlands Washed-up on cork in Scheveningen and Noordwijk (Lacourt 1949), more recently on cork (de Ruijter 2010), nylon rope (de Ruijter 2011) and on plastic (de Ruijter 2012b, 2013a, 2015a, b, 2020). *Disporella hispida* has been collected on the Cleaver Bank (van Moorsel 2003).

Germany Reported from the German North Sea (Harms 1993; Zettler et al. 2018, pers. comm. B. Kind; pers. obs. on shells from Tiefe Rinne, unpublished).

Genus *Patinella*

Patinella verrucaria (Linnaeus, 1758)

Description Colony shape and autozooidal details are similar to those in *Disporella hispida*. Basal lamina surrounds the colony, with low radiating ribs almost reaching the rim of the lamina. Ooeciostome is funnel-shaped (Fig. 2.26).

Occurrence

The Netherlands There is a report of *Patinella verrucaria* on washed-up hydroids between Zandvoort and IJmuiden and on the leg of an augustine crab *Lithodes maja*

a b c

Fig. 2.25 *Disporella hispida*. (**a**) On beached plastic, Wenduine (Belgium). (**b**) Colony area showing peristomes and short ooeciostome at the center. (**c**) Colony edge, fishing net Le Conquet (France)

at Den Helder (Lacourt 1949). This material was probably brought in from the northern North Sea by fishermen (Adema 1991).

Patinella verrucaria was collected on washed-up plastic in 2011 and 2013 near Egmond aan Zee (de Ruijter, pers. comm.).

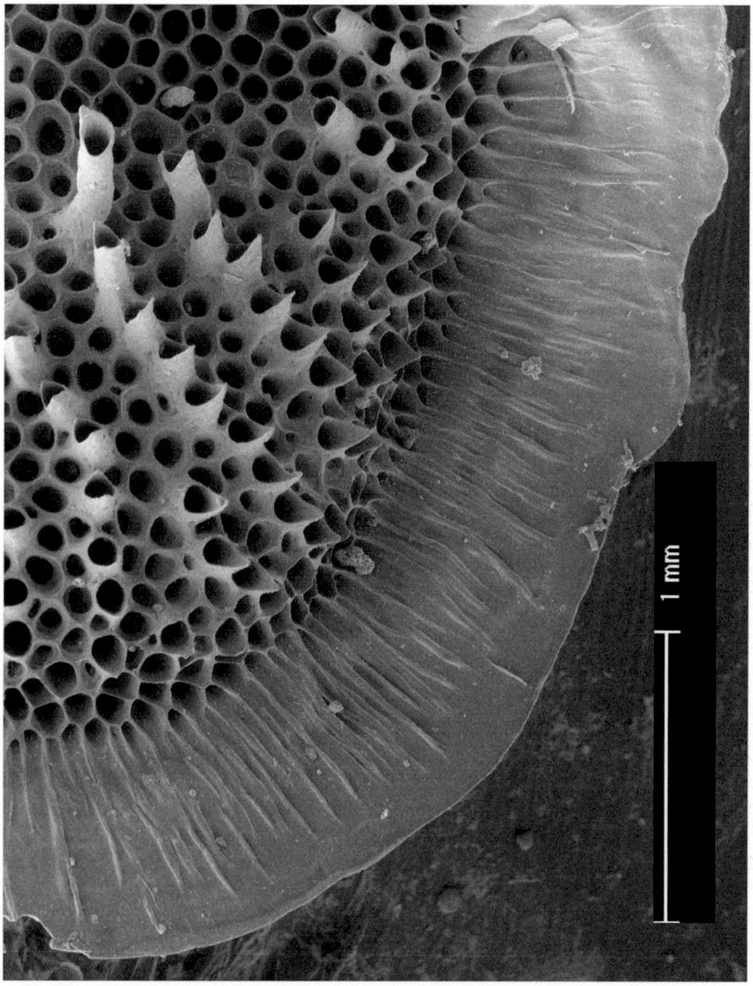

Fig. 2.26 *Patinella* sp. on plastic, harbor of St-Malo (France)

References

Adema JPHM (1991) De krabben van Nederland en België (Crustacea, Decapoda, Brachyura). Nationaal Natuurhistorisch Museum, Leiden

Armonies W, Asmus H, Buschbaum C, Lackschewitz D, Reise K, Rick J (2018) Microscopic species make the diversity: a checklist of marine flora and fauna around the Island of Sylt in the North Sea. Heligol Mar Res 72:11

Bitschofsky F (2013) Distribution over space and time in some selected epizoobiontic North Sea bryozoans. In: Ernst A, Schäfer P, Scholz J (eds) Bryozoan studies 2010. Springer, Berlin-Heidelberg, pp 1–12

Bloklander G, Leenhouts P (1948) Wat de najaarsstormen ons brachten. Het Zeepaard 8:24–25

Davoult D, Dewarumez J-M, Luczak C, Migné A (1999) Nouvelles signalisations d'espèces benthiques sur les côtes françaises de la Manche orientale et de la Mer du Nord. Cahier de Biologie Marine 40(2):121–127

De Blauwe H (2000) Riemwiervoetjes en hun begroeiing. De Strandvlo 20(3):130–137

De Blauwe H (2003) Aanvoer van zuidelijke soorten aan de Belgische kust in september 2001. De Strandvlo 23(1):5–8

De Blauwe H (2005) Bryozoa op wieren en plastic uit de herfstvloedlijn van het najaar 2004. De Strandvlo 25(1):14–16

De Blauwe H (2009) Mosdiertjes van de Zuidelijke bocht van de Noordzee: Determinatiewerk voor België en Nederland. Vlaams Instituut voor de Zee (VLIZ): Oostende

De Blauwe H, de Ruijter R (2015) *Tubulipora aperta* harmer, 1898 (Bryozoa, Cyclostomata) aangespoeld in Nederland. Het Zeepaard 75(3):105–107

de Ruijter R (2006) CS-verslag. Het Zeepaard 66(4):118–121

de Ruijter R (2010) CS-verslag. Het Zeepaard 70(1):2–10

de Ruijter R (2011) CS-verslag. Het Zeepaard 71(1):8–14

de Ruijter R (2012a) CS-verslag. Het Zeepaard 72(1):3–10

de Ruijter R (2012b) CS-verslag. Het Zeepaard 72(3–4):70–77

de Ruijter R (2013a) CS-verslag. Het Zeepaard 73(1):3–11

de Ruijter R (2013b) CS-verslag. Het Zeepaard 73(5–6):156–165

de Ruijter R (2014a) CS-verslag. Het Zeepaard 74(1):3–12

de Ruijter R (2014b) CS-verslag. Het Zeepaard 74(2):38–45

de Ruijter R (2014c) CS-verslag. Het Zeepaard 74(3):74–85

de Ruijter R (2015a) CS-verslag. Het Zeepaard 75(1):3–12

de Ruijter R (2015b) CS-verslag. Het Zeepaard 75(4/5):111–119

de Ruijter R (2018a) CS-verslag. Het Zeepaard 78(1):3–11

de Ruijter R (2018b) CS-verslag. Het Zeepaard 78(5):173–183

de Ruijter R (2019) CS-verslag. Het Zeepaard 79(3):75–86

de Ruijter R (2020) CS-verslag. Het Zeepaard 80(3):103–110

de Ruijter R (2021) CS-verslag. Het Zeepaard 81(2):55–61

de Ruijter R (2023) CS-verslag. Het Zeepaard 83(2):42–49

Faasse MA, De Blauwe H (2004) Faunistisch overzicht van de mariene mosdiertjes van Nederland (Bryozoa: Stenolaemata, Gymnolaemata). Nederlandse Faunistische Mededelingen 21:17–54

Faasse M, Van Moorsel GWNM, Tempelman D (2013) Moss animals of the Dutch part of the North Sea and coastal waters of The Netherlands (Bryozoa). NFM 41:1–14

Faasse MA, Coolen JWP, Gittenberger A, Schrieken N (2016) Nieuwe mosdiertjes van Noordzeewrakken (Bryozoa). NFM 46:43–48

Harms J (1993) Check list of species (algae, invertebrates and vertebrates) found in the vicinity of the Island of Heligoland (North Sea, German bight) – a review of recent records. Helgolander Meeresuntersuchungen 47(1):1–34

Hayward PJ, Ryland JS (1985) Cyclostome bryozoans: keys and notes for the identification of the species. Synop Br Fauna 34:1–154

Heerebout R (1970) Verspreidingsecologie van de Bryozoa in het Deltagebied, speciaal in relatie tot het brakke water. Delta Instituut voor Hydrobiologisch Onderzoek, Yerseke. Doctoraal verslag D1–1970

Lacourt AW (1949) Bryozoa of The Netherlands. Arch Néerl Zool 8(3):1–33

Lengkeek W, Didderen K, Dorenbosch M, Bouma S, Waardenburg H (2013) Bureau Waardenburg bv and Stichting De Noordzee, rapport nr. 13–226, 76 ppx. Culemborg, Netherlands: Bureau Waardenburg bv and Stichting De Noordzee, rapport nr; 2013, p 13

Loppens K (1906) Bryozoaires marins et fluviatiles de la Belgique. Annales de la Société Royale Zoologique et Malacologique de Belgique 41:286–321

Maitland RT (1851) Fauna Belgii septentrionalis: systematische beschrijving der dieren welke in Noord-Nederland of aan deszelfs kusten voorkomen, met bijvoeging derzelver synonymen, benevens der plaatsen waar dezelve gevonden zijn, gerangschikt naar Prof. J. van der

Hoeven's Handboek der Dierkunde (2de Uitgave). C.G. van der Hoek: Leyden, Nederland. XII, XXXVIII, pp 1–234

Ostrovsky AN (1998) Variability of ooeciostome shape and position in Antarctic idmidroniform bryozoans (Bryozoa: Cyclostomatida). Zool Anz 237(2):97–106

Schack CR, Gordon DP, Ryan KG (2018) Modularity is the mother of invention: a review of polymorphism in bryozoans. Biol Rev Camb Philos Soc 94(3):773–809

Taylor PD, Zaton M (2008) Taxonomy of the bryozoan genera *Oncousoecia*, *Microeciella* and *Eurystrotos* (Cyclostomata: Oncousoeciidae). J Nat Hist 42(39–40):2557–2574

van Leeuwen S (2023) De biodiversiteit op wrakken in de Noordzee, waargenomen tijdens drie duikexpedities in 2021. Het Zeepaard 83(3):99–113

van Leeuwen S, Faasse MA, Driessen F, Leeuw D, van Heugten B, Ligthart M, Heerebout G (2017) Vondsten in de visnetten van Stichting Duik de Noordzee Schoon, september 2016. Het Zeepaard 77(5):201–213

van Moorsel GWNM (2003) Ecologie van de Klaverbank. Biotasurvey 2002. ecosub, Doorn

Vanhaelen M-T, Jonckheere I, De Blauwe H (2006) Grote wierstranding aan de Belgische Westkust tijdens de zomer van 2005. De Strandvlo 26(1):5–11

Verkuil J (1998) CS-verslag. Het Zeepaard 58(1):3–9

Zettler ML, Beermann J, Dannheim J, Ebbe B, Grotjahn M, Günther C-P, Gusky M, Kind B, Kröncke I, Kuhlenkamp R, Orendt C, Rachor E, Schanz A, Schröder A, Schüler L, Witt J (2018) An annotated checklist of macrozoobenthic species in German waters of the north and Baltic seas. Helgol Mar Res 72(1):5. https://doi.org/10.1186/s10152-018-0507-5

Zintzen V (2007) Biodiversity of shipwrecks from the Southern Bight of the North Sea. PhD Thesis, University of Louvain, Department of Biology, pp 1–343

Chapter 3
Ctenostomatida

Abstract This chapter discusses the order Ctenostomatida found in the southern bend of the North Sea, between Cap Gris Nez in France and the German-Danish border. Thirty-four species have been identified. In addition to an identification key, the species are briefly described and illustrated, and the known distributional data are listed in detail. *Amathia tertia* (Winston and Hayward 2012) is here reported for the first time in Europe.

Classification

Family Alcyonidiidae

Alcyonidium mytili Dalyell, 1848
Alcyonidium albidum Alder, 1857
Alcyonidium condylocinereum Porter 2004
Alcyonidium diaphanum (Hudson, 1778)
Alcyonidium gelatinosum (Linnaeus, 1761)
Alcyonidium hirsutum (Fleming, 1828)
Alcyonidium hydrocoalitum Porter 2004
Alcyonidium mamillatum Alder, 1857
Alcyonidium parasiticum (Fleming, 1828)
Alcyonidium polyoum (Hassall, 1841)
Alcyonidium ambulans (Remane, 1936)

Family Flustrellidridae

Flustrellidra hispida (O. Fabricius, 1780)

Family Arachnidiidae

Arachnidium fibrosum Hincks, 1880
Arachnidium lacourti d'Hondt and Faasse 2006

H. De Blauwe, *Marine and Brackish Bryozoans from the Southern Bight of the North Sea*, https://doi.org/10.1007/978-3-031-92513-9_3

Family Victorellidae

*Tanganella müller*i (Kraepelin 1887)
Victorella pavida Saville-Kent, 1870

Family Walkeriidae

Walkeria uva (Linnaeus, 1758)

Family Triticellidae

Triticella flava Dalyell, 1848
Triticella pedicellata (Alder, 1857)

Family Farrellidae

Farrella repens (Farre, 1837)

Family Hypophorellidae

Hypophorella expansa Ehlers, 1876

Family Penetrantiidae

Penetrantia concharum Silén, 1946

Family Immergentiidae

Immergentia suecica Silén, 1947

Family Spathiporidae

Spathiopora sertum Fischer, 1866
Spathipora sp.

Family Vesiculariidae

Amathia spinosa (Linnaeus, 1758)
Amathia lendigera (Linnaeus, 1758)
Amathia citrina (Hincks, 1877)
Amathia cf. *gracilis* sp. 1 (Leidy, 1855)
Amathia cf. *gracilis* sp. 2 (Leidy, 1855)
Amathia imbricata (Adams, 1800)
Amathia tertia (Winston and Hayward 2012)

Family Buskiidae

Buskia nitens Alder, 1857

Unassigned ctenostomes

Panolicella nutans Jebram 1985
Anguinella palmata van Beneden, 1845

Specific Morphology

Zooids (autozooids and kenozooids) of ctenostomes have a membranous body wall as they lack a calcium carbonate exoskeleton. The zooidal orifice is not closed by an operculum.

Ctenostomes have a remarkable diversity of colony forms that range from runners and sheet-like encrusters to semi-erect or fully erect forms. They include a few solitary species and several endolithic ones that bore into calcified substrates. Autozooidal shape varies from box-shaped flat encrusters, often arranged in a quincunx, to creeping colonies with thin basal portions attached to the substrate and truly stolonal forms (Schwaha 2021a, b).

The aperture is usually slightly elevated as apertural papilla or a peristome. Some species have elongated this peristome to a peristomial tube; the basal portion may be reduced to a very thin and elongated tube (*Nolella* or *Victorella*). In this case, the thin interconnecting tubes are not true stolons as they are part of the zooid and not separated from the peristome via a pore plate—a transverse septum with pore(s) and cellular pore plug(s), which is the case in true kenozooidal stolons (Schwaha 2021a, b).

Kenozooids are present either as true stolons (Walkerioidea and Vesicularioidea (Fig. 3.1b)), spinozooids (*Flustrellidra*, Fig. 3.1c)), vicarious spacers (*Alcyonidium*),

Fig. 3.1 (**a**) *Alcyonidium condylocinereum*, encrusting colony with small erect lobes (Nieuwpoort, Belgium). (**b**) *Amathia citrina*, zooids on kenozooidal stolon. (Photo by B. Kind). (**c**) *Flustrellidra hispida* with brown spinozooids. (Photo by B. Kind)

intercalary kenozooids (*Immergentia* in Johnson et al. 2024), or rhizoids that anchor colonies to a substrate.

The polypide comprises a retractile ciliated tentacle crown (lophophore) with associated gut, muscles, and nerve ganglion.

The intertentacular organ is a ciliated tube between the adneural tentacles that functions as a conduit for alien sperm and an exit for fertilized eggs (i.e., larvae). It is present in broadcasting species and some brooders. A supraneural pore is located at the same position in brooders (Ostrovsky and Porter 2011).

Brown bodies are colored spheroids in zooids resulting from the aggregation of the residue of degenerated polypides.

The anatomy of the gut (Fig. 3.13b, *Alcyonidium mytili*) and the proportions of its parts are significant for taxonomic assignment, especially in encrusting *Alcyonidium* (see Ryland and Porter (2006) for *Alcyonidium gelatinosum* and *A. polyoum*).

Key to the Genera

1. (a) Colony forming a flat crust (Fig. 3.2a) on the substrate or erect fleshy lobes. Autozooids budding from other autozooids without any linking stolon. 2
   (b) Zooids singly or in groups on a common stolon (Fig. 3.2b), or autozooids forming branching, linear chains and may be erect, cylindrical with a narrow stolon-like proximal adnate part (Fig. 3.2c). 4
   (c) Colony free-living, interstitial on sand and shell gravel, consisting of a single autozooid. .... *Alcyonidium ambulans* (p. 78)
2. (a) Colony forming a thick, purplish-brown crust, orifices surrounded by pointed, spine-like kenozooids (Fig. 3.1c, 3.2d). On intertidal algae, especially on *Fucus serratus*. .... *Flustrellidra hispida* (p. 80)
   (b) Zooids without spine-like kenozooids. 3
3. (a) Peristomes sometimes with longitudinal stripes, colony an encrusting sheet or linear chains of zooids, frontal surface with spine-shaped appendages (Fig. 3.2e). .... *Arachnidium fibrosum* (p. 83).
   (b) Colony forming flat or nodular sheets and/or erect fleshy lobes on hard substrates, hydroid stems, or algae. .... Family Alcyonidiidae (p. 62)
4. (a) Colony a gray-brown muddy, dense, erect tuft, 5–20 cm high. Zooids cylindrical, budding from each other, orifice terminal. The colony looks more like a hydroid, easily distinguished from it by its typical tentacle crown (Fig. 3.2f). .... *Anguinella palmata* (p. 86)
   (b) Not as described. 5
5. (a) Zooids budding from a branching stolon (Fig. 3.2b). 8
   (b) Colony a chain of zooids, each with a slender proximal portion (Fig. 3.2c). 6

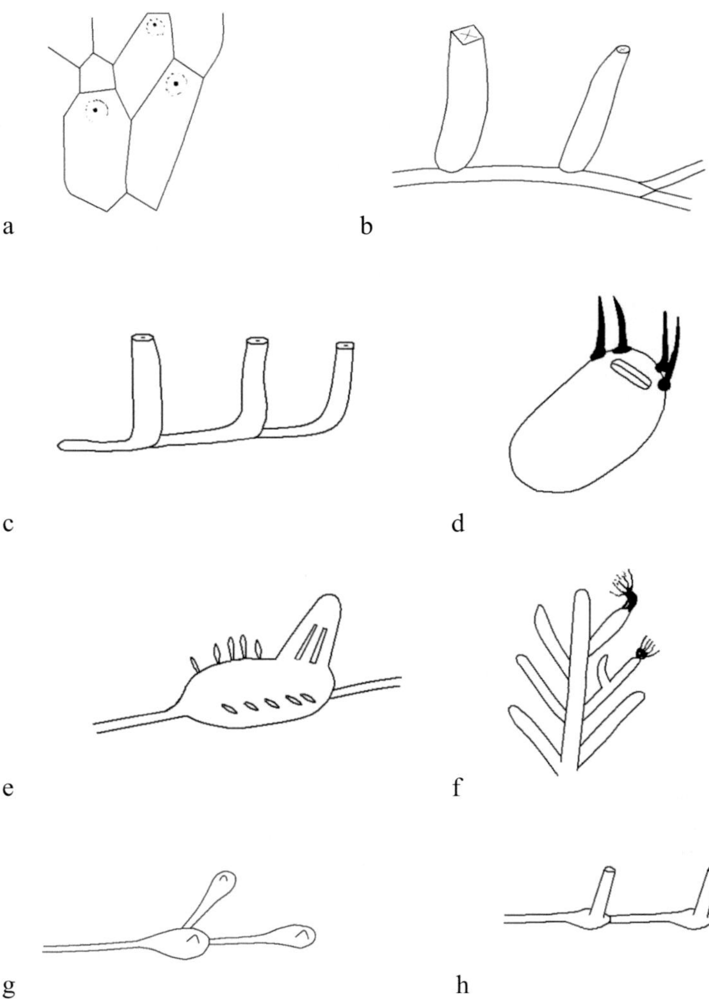

Fig. 3.2 (**a**) Zooids without linking stolon. (**b**) Zooids born on a stolon. (**c**) Zooids with a narrow stolon-like proximal adnate part. (**d**) Orifices surrounded by spine-like kenozooids. (**e**) Zooids with spine-shaped appendages. (**f**) *Anguinella palmata*. (**g**) Zooids completely adnate, drop-shaped. (**h**) Zooids with a slender proximal portion and an erect cylindrical portion

6. (a) Zooids completely adnate, drop-shaped, sometimes linked to the proceeding zooid by a filiform proximal portion. Zooid chains often closely packed and thus appearing to form a loose sheet. (Fig. 3.2g). .... *Arachnidium* (p. 81).
   (b) Colony a chain of zooids, each with a slender proximal portion and an erect cylindrical portion (peristome) (Fig. 3.2h). 7

7. (a) In fresh or brackish water only, often on stems of reed. .... Victorellidae (p. 86)
   (b) In marine conditions. .... *Panolicella nutans* (p. 84)
8. (a) Colony within the linings of the tubes of sedentary polychaetes (*Chaetopterus* and *Lanice*). Stolons and holes, made by previous zooids are common, zooids oval with a pair of spherical kenozooids (Fig. 3.3a). .... *Hypophorella expansa* (p. 96)
   (b) Not as described. 9
9. (a) Colony boring in shells and limestone. Most frequently discovered in the inner side of dead bivalve shells as rounded, oval, comma-shaped, or s-shaped holes about 0.1 mm in diameter. Stolons or cystid appendages between zooids seen by transmitted light within the shell or in translucent or eroded shell layers. 15
   (b) Not in, but on the substrate. 10
10. (a) A thin stolon creeping over the surface of hydroid, erect bryozoan, shells, or algae, with at intervals very small (0.5 mm) slipper-shaped autozooids, adnate, with only the distal portion bent away from the substrate (Fig. 3.3b). .... *Buskia nitens* (p. 113)
   (b) Not as described. 11

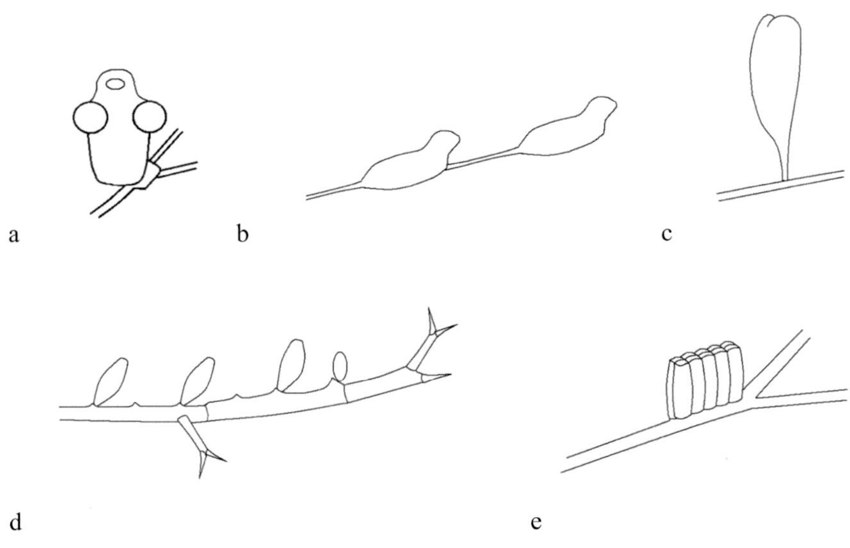

Fig. 3.3 (**a**) *Hypophorella expansa*. (**b**) *Buskia nitens*. (**c**) *Farrella repens*. (**d**) *Amathia spinosa*. (**e**) Autozooids arranged in biserial clusters; adjacent zooids are typically laterally joined

11. (a) Autozooids stalked (Fig. 3.3c). The peduncle is an erect kenozooid, separated by a septum (interior transverse wall) from the autozooid; orifice circular. .... Family Triticellidae (p. 91)
    (b) The peduncle is formed by a tapered proximal portion of the autozooid, without septum; orifice bilabiate; 11–15 tentacles. .... *Farrella repens* (p. 93)
    (c) Autozooids may be tapered proximally, but not stalked. 12
12. (a) All colony branches terminating in a short spine-like kenozooid. Oval autozooids on a socket-like protuberance of the stolon (Fig. 3.3d). Colony tufts 5–30 cm in height. .... *Amathia spinosa* (p. 102)
    (b) Branches do not terminate in a short spine-like kenozooid. 13
13. (a) Erect, stiff, brown tufts. Autozooids arranged in well-defined biserial, straight, or spiraling clusters of definitive zooid numbers; adjacent zooids are typically laterally joined (Fig. 3.3e). .... *Amathia* (p. 104)
    (b) Autozooids grouped more loosely and never laterally joined. 14
14. (a) Stolons adnate or erect, composed of cylindrical kenozooids, distinctly divided by septa (Fig. 3.4a) mostly near groups of zooids. Zooids short, with a quadrate orifice. .... *Walkeria uva* (p. 89)
    (b) Stolons without septa (Fig. 3.4b). .... *Amathia* (previously *Bowerbankia*) (p. 104)

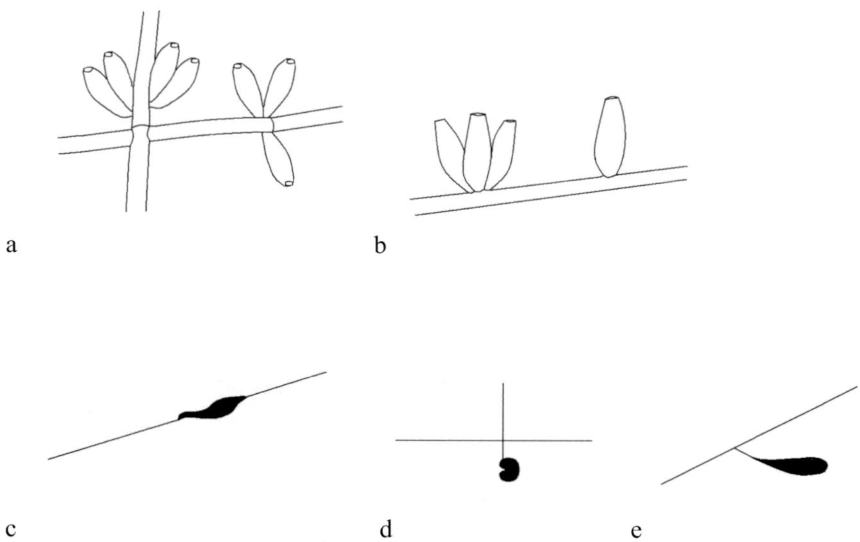

Fig. 3.4 (**a**) *Walkeria uva*. (**b**) *Amathia* (previously *Bowerbankia*). (**c**) *Immergentia suecica*. (**d**) *Penetrantia concharum*. (**e**) *Spathipora*

15. (a) Orifice on cystid appendage, S-shaped (Fig. 3.4c). .... *Immergentia suecica* (p. 98)
    (b) Autozooids pedunculate, orifice beside the stolon. 16
16. (a) Orifice kidney-shaped (Fig. 3.4d). .... *Penetrantia concharum* (p. 98)
    (b) Orifice oval or comma-shaped (Fig. 3.4e). .... *Spathipora* sp. (p. 100)

Systematic Part

Family Alcyonidiidae

Genus *Alcyonidium* (Lamouroux, 1813)

The lack of calcification and skeletal features makes it difficult to distinguish species, but recent research has clarified the true identity of some. Peristome length varies greatly in the genus. Some species like *A. gelatinosum* have little to no distinguishable apertural area, *A. albidum* has a distinct apertural papilla, and *A. mammillatum* has a large peristomial tube. The number of tentacles varies within single species. A combination of characteristics such as colony shape, color, number of tentacles, mode of reproduction, and substrate preference make a reliable determination possible. Genetic research provides a definitive answer in cases of doubt.

Some species of *Alcyonidium* release many oligolecithal zygotes via an intertentacular organ (broadcasters) that probably develop into planktotrophic cyphonautes larvae with a bivalved shell. Most species, however, brood embryos in internal brood chambers, formed after degeneration of the polypide, and release larvae (larviparous). The latter are barrel-shaped lecithotrophic larvae, with large yolk reserves. Reproduction in the genus *Alcyonidium* varies among species and does not seem to correlate with any specific colony morphology (Schwaha 2021a, b).

The presence of an intertentacular organ (ITO) (Fig. 3.5a), a temporary structure for the release of gametes, is often associated with a broadcasting reproductive

Fig. 3.5 *Alcyonidium condylocinereum*, tentacle crowns with intertentacular organ (*arrow*)

strategy but is also found in some brooders (Ostrovsky and Porter 2011; Schwaha 2021a, b).

1. (a) Autozooids with small spherical or conical kenozooids on the surface. Encrusting, sometimes with short erect lobes. On intertidal seaweeds. .... *Alcyonidium hirsutum* (p. 70)
   (b) Autozooids without small kenozooids on the surface. 2
2. (a) Colony erect, by itself, or as an encrustation around the stalk of a hydroid. 3
   (b) Colony encrusting on seaweeds, stones, mussels, or other substrate, no erect colony parts occur. 6
3. (a) Colony large (often 10–30 cm high), firm and rubbery, grayish or brownish, smooth or nodulous, branched or unbranched, growing erect from an encrusting base. Mainly sublittoral, often washed ashore after storms. 4
   (b) Colony surrounds a hydroid stem, usually only a few centimeters high, transparent, or mud-colored. 5
4. (a) Colony shape very variable, honey-colored to brown. Base of the erect colony part clearly narrowed. .... *Alcyonidium diaphanum* (p. 67)
   (b) Branches always cylindrical and grayish. .... *Alcyonidium condylocinereum* (p. 64)
5. (a) Forms a gray-brown, mud-colored crust around hydroids. When magnified, only the peristomes of the zooids are visible on a surface that contains a lot of dirt particles. .... *Alcyonidium parasiticum* (p. 76)
   (b) Colony sometimes with attached dirt, but not with a continuous dense silt layer. Colony rather transparent. Usually grows on stems of *Tubularia indivisa* or *Amathia spinosa*. .... *Alcyonidium hydrocoalitum* (p. 72)
6. (a) Zooids opening on a high peristome so that the colony surface looks bristly. .... *Alcyonidium mamillatum* (p. 73)
   (b) No tall cones, unless the polypide is partially bulged. 7
7. (a) Colony forming irregular, lobate patches or an open meshwork, often with single linear series of autozooids budded from the periphery. Autozooids randomly orientated. Spaces within the colony filled by irregularly shaped kenozooids. Subtidal on hard substrates, algae, or epizootic. .... *Alcyonidium albidum* (p. 64)
   (b) Not as described. 8
8. (a) Zooids with thick white lateral walls. .... *Alcyonidium gelatinosum* (p. 69)
   (b) Zooids without thick white lateral walls. 9
9. (a) On hard substrate. 10
   (b) On seaweed. 11
10. (a) Crust relatively thick and easily peeled off the substrate, often erect colony parts or knobby outgrowths on the colony crust, 16–18 tentacles. On stones, snail shells, and crab shells. .... *Alcyonidium condylocinereum* (p. 64)

(b) Young crust thin, difficult to peel off the substrate. Found especially on mussels, as well as on stones and metal. Usually 13–16 tentacles. .... *Alcyonidium mytili* (p. 74)
(c) 17–18 tentacles, white or pink embryos in groups of three or four. .... *Alcyonidium gelatinosum* (p. 69)

11. (a) Usually 19 tentacles, only on *Fucus serratus*, yellow or orange embryos in groups of six or seven. .... *Alcyonidium polyoum* (p. 77)
(b) 17–18 tentacles, white or pink embryos in groups of three or four............. .... *Alcyonidium gelatinosum* (p. 69).
(c) 13–16 tentacles, never incubates embryos, white eggs, an intertentacular organ is present in fertile zooids in the breeding season. .... *Alcyonidium mytili* (p. 74)

Alcyonidium albidum Alder, 1857

Description Colony forming a thin, light brown, or transparent sheet, irregularly lobate or loosely anastomosing, budding linear series of zooids on its periphery. Orientation of the zooids seems haphazard where colony lobes or linear series meet. Subsequently, the gaps are filled with kenozooids that are as large as autozooids or larger or much smaller. Slightly raised peristome in the shape of a shallow disc surrounded a horseshoe-shaped distal fold. Frontal surface translucent, with polypide and side walls clearly visible. Tentacle numbers 17–20. Broadcaster developing an ITO (Fig. 3.6).

Occurrence

The Netherlands The single colony collected is attached to an empty shell of a netted dogwhelk *Nassarius reticulatus*, which was occupied by a hermit crab, as may be judged from the presence of an attached hydrozoan *Hydractinia echinata*. The shell was found in a box core sample, taken on March 2, 2010, 30 km offshore the coast of Walcheren (51°43′06″ N, 03°06′43″ E, depth 30 m) (Faasse et al. 2013).

Germany Mentioned from the German sublittoral North Sea (Zettler et al. 2018). BioConsult Bremen (pers. comm. P. Nehmer) detected the species only in one campaign in June 2021 in a very muddy area in the northeast of the German economic zone. It occurred three times in dredge hauls (coordinates of the starting points): (1) 55°27,500′ N, 5°59,315′ E; (2) 55°22,993′ N, 6°00,476′ E; (3) 55°17,099′ N, 6°10,088′ E.

Alcyonidium condylocinereum Porter, 2004

Only described in 2004 and formerly recognized and reported, together with *A. diaphanum*, as *Alcyonidium gelatinosum*, which in fact is another species.

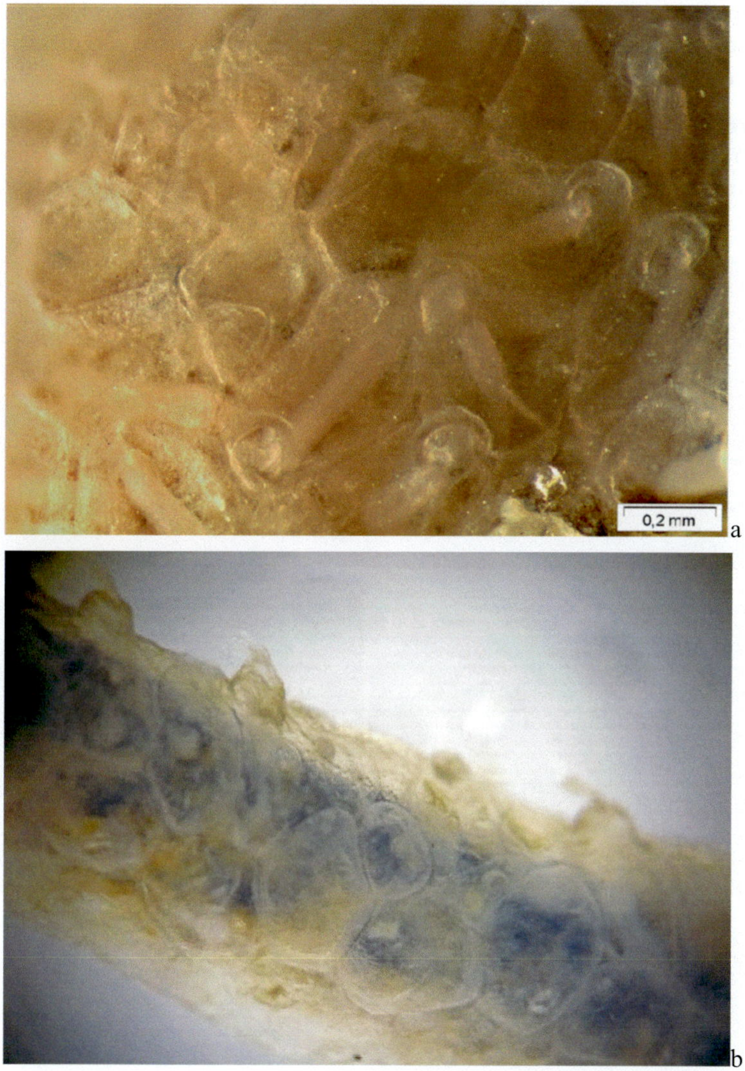

Fig. 3.6 *Alcyonidium albidum.* (**a**) Off the Walcheren coast. (Photo by D. Tempelman in Faasse et al. 2013). (**b**) Germany, collected by D. Miljutin (BioConsult)

Description Forms an extensive crust on stones, shells, or crabs at first. Later, knobby outgrowths develop on the crust that develop into erect colony parts. These colony parts branch; both the main branch and the branches are cylindrical. The surface has bumps about 0.5 cm in diameter. The crust can be easily peeled off the substrate. The crust and upright parts are gray to gray-brown. There is often a shallow gap between the zooids, which in a silty environment becomes filled with silt. As a result, the colony can have a brownish color, which leads to confusion with *A. diaphanum*. The zooids have 18–19 tentacles, usually 18. Material from the infralittoral fringe at Koksijde, however, usually had 16–17 tentacles. During the reproductive season, an intertentacular organ is present in fertile zooids and zygotes are released into the water column (broadcaster) (Fig. 3.7).

Occurrence Between 1900 and 1910, several colonies were collected on the Flemish Banks and near the Scheldt estuary (RBINS collection). Erect colony parts are found on the beach in B and NL, as well as colonies attached to strings, snails

Fig. 3.7 *Alcyonidium condylocinereum.* (**a**) Erect colony found beached at Nieuwpoort. (Belgium, Photo by F. Nuyten). (**b**) Colony on *Ensis* sp., after sand suppletion, Oostende. (Belgium, Photo by F. Kerckhof). (**c**) Encrusting colony (right), erect colony (left) on a groyne, Koksijde (Belgium). (**d**) Peeled colony. (**e**) Colony on *Hinia reticulata.* (the Netherlands, Photo by A. Wijker)

inhabited by hermit crabs (especially *Hinia*, but also *Euspira*), and bivalve shells such as *Ensis* and on crabs.

Found Living In Situ in Belgium at
Hinder Banks 2005: On *Hinia reticulata* and *Natica* inhabited by hermit crab.

Nieuwpoort 2012 on groynes: https://waarnemingen.be/photos/2907171/

Koksijde from 2003 on (De Blauwe 2004a) and 2014 on groynes: https://waarnemingen.be/photos/6039749/

Heist Feb 2013: on *Ensis directus* (own observation).

De Panne and Koksijde 2016, 2018, 2019, 2021, 2024 on permanent fishing gear/ropes: https://waarnemingen.be/photos/24443350/, https://waarnemingen.be/observation/297689640/

Found Living In Situ in the Netherlands at
Westkapelle (Faasse and De Blauwe 2004) and Veerhaven Anna Jacobapolder on pontoon 2019 and 2020:

https://waarneming.nl/observation/171126013/ https://waarneming.nl/observation/193271909/

Germany Present in German North Sea (Zettler et al. 2018). Beached at Juist (October 2024, https://observation.org/observation/329574396/).

Alcyonidium diaphanum (Hudson, 1778)

Formerly recognized and reported, together with *A. condylocinereum* and *A.polyoum*, as *Alcyonidium gelatinosum*, which in fact is another species. Porter (2004) has shown that *A. proliferans*, *A. topsenti*, and *A. gelatinosum* form *lacourti* can be synonymized with *A. diaphanum*.

Description Erect colony parts smooth, firm, and fleshy, flexible, up to 30 cm long, in estuaries up to 50 cm long. The erect parts narrow at their base. Material in the RBINS shows that the encrusting base is very limited in surface area. The erect colony parts are very variable. Six shapes are distinguished: sausage-shaped, cylindrical, shoelace-shaped, bushy, leaf-shaped, or knotted. Zooids densely packed, with openings on all sides of the colony, the polypides are perpendicular to the colony surface and have 14–16 tentacles. Multiple embryos are incubated inside fertile zooids (brooder). There is no intertentacular organ. The color is honey to brown. May be confused with the grayer species *A. condylocinereum*, which has a different tentacle number and different mode of reproduction (Fig. 3.8).

Occurrence Grows on stones and shells (*Euspira catena* and *Ostrea edulis*) and large stones (http://britishbryozoans.myspecies.info/bryozoa/alcyonidium). Sublittoral widespread throughout the North Sea; on both sides of the English Channel, although less common in the middle of the channel. Dense populations in the Thames estuary, in the bay of the Seine, and on the Dogger Bank. Its southernmost European record appears to be from Glénan (Brittany). Occurs

a

b

Fig. 3.8 *Alcyonidium diaphanum.* (**a**) Long erect colony, with to the right the very small encrusting base on a pebble. (**b**) Many colony forms are possible

from the infralittoral fringe to 100 m depth with dense populations at 10–30 m depth (Porter et al. 2002).

In the past, erect colony parts often washed up on the Belgian and Dutch coasts after autumn storms, but in recent years to a much lesser extent.

Belgium The RBINS collection shows that this species was distributed between the French border and the estuary of the Scheldt in the first half of the last century (De Blauwe 2009). Only a few colonies were collected during research on the

Hinder Banks (June 2005). Perhaps reduced by beam trawling and the eradication of oyster beds. The establishment of offshore wind farms and the reintroduction of the native oyster can have a positive impact on populations in Belgian waters. Many colonies were observed in August 2023, beached or dredged at sea.

The Netherlands One recent, unverified offshore observation in the Netherlands: https://waarneming.nl/observation/200252037/ A colony was collected on the Dogger Bank in 2016; it is not certain whether it was attached to the site or was adrift.

Germany Collected in the German North Sea (Zettler et al. 2018). A small colony beached at Düne (Heligoland) on April 21, 2024 (own. obs.).

Alcyonidium gelatinosum (Linnaeus, 1761)

This species was incorrectly considered a new species *A. reticulum* Ryland and Porter (2000) for a short time. Later, it turned out that this species had already been described by Linnaeus in 1761 under the name *Alcyonidium gelatinosum* (Ryland and Porter 2003).

Description (After Ryland and Porter 2000 as *A. reticulum***)** The colony forms a single-layer crust, opaque white, yellowish, or brown, which is easily peeled. Zooids 0.3–0.5 × 0.2–0.4 mm, rectangular or polygonal, flat or slightly convex, translucent when young, older zooids with a distinct white border, particularly during autumn/winter.

15–18 tentacles, usually 17 or 18. An intertentacular organ is never present. Polypides usually absent during the reproductive season. Embryos white or pink, with three or four per brooding zooid. Gut caecum sub-globular, less than one-half the length of the long stomach cardia.

Ancient zooids of *A. mytili* often have a white color (Fig. 3.13c), but the zooids do not have the striking white border of *A. gelatinosum* (Fig. 3.9).

Occurrence Intertidal on sheltered beaches on stones, shells, and *Fucus serratus*. Also, subtidally in Roscoff (France) on shells, around 20 m deep (pers. comm. T. Schwaha).

Belgium Absent. *A. gelatinosum* in Belgian coastal waters, dredged in the period 1973–1975 (Redant 1976) is *A. diaphanum* or *A. condylocinereum*. For Belgium, there is only one certain observation and it is on beached plastic: https://waarnemingen.be/observation/87299633/

The Netherlands Grows in situ at Schelphoekkreek, the tide pool at Topshuis on Neeltje Jans and Yerseke, with embryos in July (Faasse and De Blauwe 2004). Literature data from the Netherlands before 2004 are unreliable.

Germany Collected in German North Sea (Zettler et al. 2018) and Heligoland (pers. comm. Britta Kind).

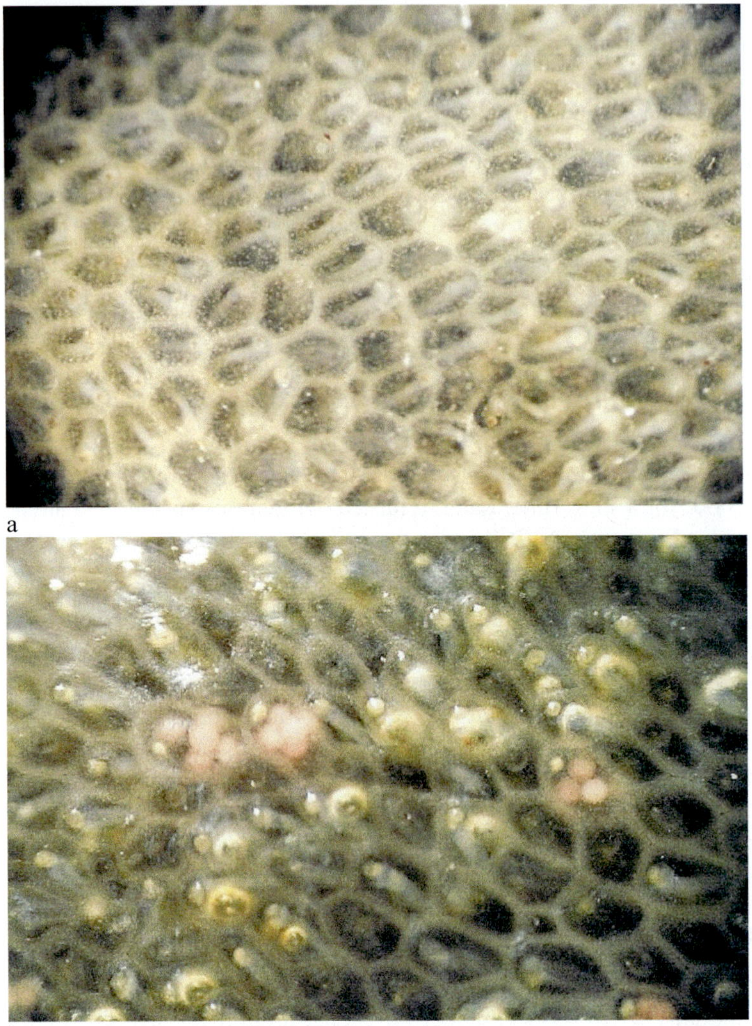

a

b

Fig. 3.9 *Alcyonidium gelatinosum.* (**a**) Living colony area showing autozooids with mostly retracted polypides. (**b**) Living colony area showing autozooids with both protruded and retracted polypides, fertile zooids with pink embryos, Sillon de Talbert (France)

Alcyonidium hirsutum (Fleming, 1828)

Description Forms thick, fleshy crusts or cylinders on seaweed. The colony surface feels velvety soft. The zooids have a frontal surface that is convex and smooth with the orifice almost in the center. This is the only species of *Alcyonidium* where kenozooids originate frontally on the zooids. The kenozooids resemble small spheres and are located on the edge of each zooid. It is as if the entire surface of the colony is covered with spheres. Internal brooder (Fig. 3.10).

a

b

Fig. 3.10 (**a**) *Alcyonidium hirsutum* (large colony) and *A. polyoum* (two small colonies). (**b**) Zooids and kenozooids (small spheres), St-Jacut-de-la-Mer (France)

Occurrence On intertidal seaweeds, especially on *Fucus serratus*, also on firm red algae.

Belgium Absent. Material available in the RBINS, which was labeled *A. hirsutum*, appears to be *A. condylocinereum* and mainly *A. diaphanum*.

The Netherlands Collected in situ near Zierikzee and Colijnsplaat (Faasse and De Blauwe 2004).

Alcyonidium hydrocoalitum Porter, 2004

Description Forms a firm gelatinous crust around the stems of the hydroids *Tubularia indivisa* and *Obelia bidentata*, possibly also around other hydroids and on the bryozoan *Amathia spinosa*. Up to 6 mm thick, smooth with irregular outline, transparent or light brown-gray. 14–18 tentacles, usually 16. During the breeding season, an intertentacular organ is present in fertile zooids (Fig. 3.11).

Fig. 3.11 *Alcyonidium hydrocoalitum*, Vlissingen. (the Netherlands, Photo by M. Faasse)

Occurrence Only a few records are known.

United Kingdom Oxwich Bay in South Wales and the Strait of Dover (Porter 2004).

The Netherlands Oosterschelde and Westerschelde (Faasse and De Blauwe 2004). Washed up in the Netherlands on Texel on *Obelia bidentata* on October 13, 2019 (de Ruijter 2020a). Collected at the Dutch Borkum Reef Grounds (Coolen et al. 2015).

Belgium In May 1905 and June 1906, respectively, one and five colonies fished on the Flemish Banks on the bryozoan *Amathia spinosa* (RBINS collection) (De Blauwe et al. 2006). Recently, a few colonies were found on *Tubularia* on shipwrecks (Zintzen 2007, Zintzen and Massin 2010), namely on the Kilmore (51°23′ N, 2°29′ E; three colonies in 2003, one in 2004, and one in 2005) and on the Birkenfels (51°38′ N, 2°32′ E; one colony in 2002).

Two in situ observations (2013 and 2016) and three on beached material at the harbor of Zeebrugge:

https://waarnemingen.be/observation/73764949/
https://waarnemingen.be/observation/115346208/ on *Obelia bidentata*.

Alcyonidium mamillatum Alder, 1857

Description Forms irregular, lobate, or star-shaped crusts on shells or cylinders around erect substrates such as hydroids. The colony has a spiny appearance. Zooids in one continuous layer, flat or somewhat convex, separated by clear grooves. Translucent light brown, darker toward the peristome. Orifice on a conspicuous cylindrical peristome, often clearly ringed in older zooids. The peristome remains evident with tentacles retracted and is perpendicular to the frontal surface. Dirt can hide the surface of zooids, but the peristome remains visible (Fig. 3.12).

Occurrence Sublittoral on hydroids, *Laminaria* holdfasts, solitary ascidians, decapods, and living or dead bivalves. Probably widely distributed in the northeastern Atlantic, as far south as the English Channel (Hayward 1985).

Belgium Records (Loppens 1906) are unreliable and cannot be verified. Loppens considers this species to be fairly common on mussels but does not mention *A. mytili*, a typical mussel epiphyte.

The Netherlands The same can be said of Lacourt (1949) and the identification key of Lacourt (1978). Lacourt reproduced the drawings of Marcus (1940), so that these drawings do not provide proof of correct Dutch identifications. Faasse and De Blauwe (2004) list a number of records and there is another recent record from 1996 on buoys in the Eemshaven (Koulman et al. 2003). A colony was collected in July 2016 on the Dogger Bank (pers. comm. Merkens Bart).

Germany Collected in German North Sea (Zettler et al. 2018) and Heligoland (pers. comm. Britta Kind).

Fig. 3.12 *Alcyonidium mamillatum.* (Germany, coll. B. Kind, Photo by H. De Blauwe in VLIZ)

Alcyonidium mytili Dalyell, 1848 (=*Alcyonidioides mytili* Dalyell, 1848)

Since we cannot confirm cyphonautes in any other *Alcyonidium* properly, also with the species in our tree (Waeschenbach et al. submitted), it still remains highly dubious to use the genus name *Alcyonidioides* (pers. comm. T. Schwaha).

Description (After Cadman and Ryland 1996) The colony forms a well-adhered crust on all kinds of substrates and is therefore difficult to peel off. Young zooids on the colony margin are very thin and completely transparent. The polypide is highly visible in the octagonal or irregular polygonal zooids. Older colony parts are thicker (up to 2 mm). Budding creates a new layer of zooids on top of the older ones, which are compressed. Their frontal surface is slightly convex and less translucent, usually brown, also white or gray, often covered with a diatom layer. During the growing season, the zooids store fat droplets, giving them a white appearance.

In the breeding season (November to March, according to Cadman and Ryland (1996); March and April in Zeebrugge, own observation), the frontal wall becomes more transparent again so that the eggs can be seen as white dots at 40x magnification. At the start of reproduction, an intertentacular organ develops (light blue and one-third of tentacle length, not on young zooids). It remains until the end of reproduction and until the polypide degenerates.

Polypide degeneration leads to the formation of one or two brown bodies that persist during successive polypide generations. For example, zooids typically contain one to five brown bodies in addition to the polypide. The polypide bears 13–18 tentacles, but usually 15 or 16 (Cadman and Ryland 1996), in the Zeebrugge marina usually 15 tentacles, and in young colonies only 13 or 14 tentacles. Extensive colonies on substrates other than mussels are difficult to distinguish from related species (Fig. 3.13).

a b c d

Fig. 3.13 *Alcyonidium mytili*. (**a**) Colony on mussel (Yerseke, The Netherlands). (**b**) Young colony on mussel at Zeebrugge (Belgium). (**c**) Older colony on mussel (Yerseke). (**d**) Polypides degenerate to brown bodies

Occurrence Reports from before 1996 are unreliable. Often, but not as common as *Conopeum reticulum*, on mussels presented for consumption. On algae, barnacles, mussels, oysters, stones, concrete, and iron in the infralittoral fringe and below.

Belgium Large colonies can be found on stones in Nieuwpoort and on mussels at Zeebrugge, Heist, and Duinbergen.

The Netherlands Faasse and De Blauwe (2004) mention a number of sites from the literature from Zeeland and the Wadden Islands. Additional: Westkapelle (own. obs.), Mok Bay on Texel (de Boois 1998), shipwreck "Klipper" (UTM 534415 5758926) (Lengkeek et al. 2013).

Germany Mentioned by Zettler et al. (2018), Ryland and Porter (2006), have positive identification of this species from Sylt, in the North Friesian Islands.

Alcyonidium parasiticum (Fleming, 1828)

Description Forms silty-colored cylinders around hydroids. The surface is covered with silt and fine sand. This gives the typical appearance and hides the details of the zooids. Colony up to 3 cm long and 5 mm thick, usually thinner. At first, single-layered with elongated zooids. Frontally emerging zooids are hexagonal or irregular and small. Zooids are delimited by clear groves. The terminal conical opening is transversely oval. The frontal surface bears many short filaments around the rim and around the opening. These filaments retain the silt. To see them, the silt has to be removed. Confusion with a dense mat of *Arachnidium fibrosum* is possible; the zooids of *A. parasiticum* are up to 0.5 mm long, while those in *A. fibrosum* are 0.6–1.3 mm (Fig. 3.14).

Occurrence In Belgium and the Netherlands (de Ruijter 2019a, 2020a) often washed ashore on thecate hydroids (*Hydrallmania falcata*, *Sertularia*) and shrub-like bryozoans (*Amathia spinosa*), usually found in the wash-up banks of wood and peat blocks.

Belgium At the RBINS, many colonies are present on hydroids, collected at the beginning of the last century in the area between the Channel and the mouth of the Scheldt (De Blauwe et al. 2006). Loppens (1906) found this species generally on hydroids, and rather rarely on the beach. Several colonies were recently found in situ in the Baai van Heist and on a groyne at Koksijde on the bryozoan *Anguinella palmata*.

The Netherlands Near Wissenkerke, on the Cleaver Bank (pers. comm. G. van Moorsel) and in 2016 on the Dogger Bank. Collected at the Dutch Borkum Reef Grounds (Coolen et al. 2015). Lengkeek et al. (2013) report the species from the shipwreck "SS Nautilus" (UTM 5807515886316).

Germany Mentioned by Zettler et al. (2018).

Fig. 3.14 *Alcyonidium parasiticum.* (**a**) On hydroid beached at Zeebrugge (Belgium). (**b**) Tentacles with ITO. (**c**) Zooids. (**d**) Zooids with filaments

Alcyonidium polyoum (Hassall, 1841)

Description Colonies never with white zooid walls. Encrusting algae, mainly *Fucus serratus*. Zooids quite long (0.5–0.8 mm). 18–21 tentacles, usually 19–20. Buff-colored to orange embryos are incubated in clusters of six to seven in the zooid. Colonies in which many embryos are present have a striking orange color. Gut caecum deep and U-shaped, almost reaching the constriction between the cardia and the esophagus. Reproductive season depending on the locality (Fig. 3.15).

Occurrence Reports from before 2000 are unreliable. Usually on algae, rarely on stones or shells.

Belgium Not found in Belgium and not present in the collection of the RBINS. Redant (1976) mentions this species in Belgian coastal waters in the period 1973–1975, but given the substrate preference of *A. polyoum*, the dredged material certainly belongs to a different species.

Fig. 3.15 *Alcyonidium polyoum.* (**a**) Colony with embryos (Yerseke). (**b**) Damgan (France). (**c**) Zooids with embryos (Yerseke, the Netherlands)

The Netherlands In the Delta area, this species was observed near Zierikzee, the Goese Sas, Wemeldinge, Burghsluis, and Yerseke.

Alcyonidium ambulans (Remane, 1936) (=*Monobryozoon ambulans* Remane, 1936)

A revised classification proposed by Waeschenbach et al. (submitted) transfers the genus *Monobryozoon* to *Alcyonidium*.

Description "Colony" consisting of a single feeding autozooid. It has usually eight or nine proximal kenozooidal appendages, termed ambulatory processes owing to their ability to move via muscular contractions that enable the animal to crawl, move, or reorientate on the sandy substrate upon which they live on. One of the appendages has a proximal expansion indicating the formation of an early, asexual bud. The largest bud may be only a little smaller than the mother zooid, from which it finally separates. Appendages are muscular, capable of movement, and perhaps adhesive to hold the colony stable, attached to sand grains. These ambulatory processes are probably all buds and appear at an early developmental stage with small buds possessing them. Autozooid sac-shaped, with bulbous proximal end; commonly 0.5–1.0 mm long with retracted polypide, but up to 2.0 mm when protruded. Orifice of autozooid terminal, polypide relatively large, with 12 tentacles. A distinct cardia is missing; the caecum is prominent and yellow to brown in living and fixed animals. Single, large eggs recorded during November in colonies collected off the Yorkshire coast (Fig. 3.16).

Occurrence

Germany *Alcyonidium ambulans* lives on fine to medium sand, and in shell gravels, with an admixture of fine silt. This species was first discovered living in *Amphioxus* sand off Heligoland (Rémane 1936, 1938) and rediscovered at the same location by Thomas Schwaha in April 2022 at 54°15′04.7″N, 7°56′48.8″E at about 40 m depth (Schwaha et al. 2024).

Its small size, transparency, and specialized habitat demand careful collecting and preparation of samples before they can be found (Gray 1971). The only other locality where it was found is in muddy shell gravel off the Yorkshire coast in 1971.

Franzenella limicola (Franzen, 1960), described from soft mud in Sweden is larger and has two to three long thin cystid appendages or buds.

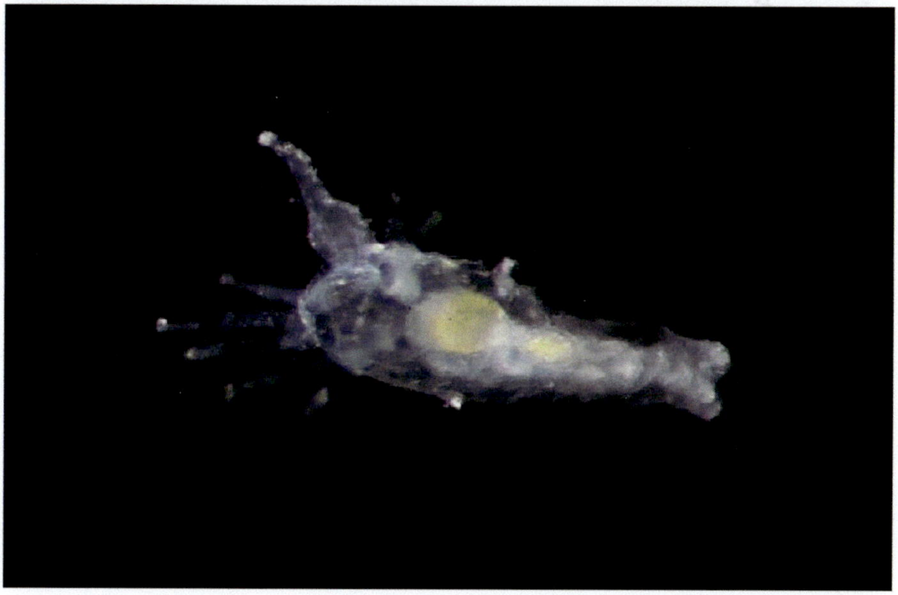

Fig. 3.16 *Alcyonidium ambulans*, Heligoland. (Germany, Photo by T. Schwaha)

Family Flustrellidridae

Genus *Flustrellidra*

Flustrellidra hispida (O. Fabricius, 1780)

Description Forming extensive, thick, lobed patches on *Fucus serratus* and fleshy cylinders on *Gigartina* and *Chondrus*, sometimes on other algae. The colony is dark reddish brown and has a furry appearance. In the Netherlands, colonies are quite thin and not very extensive. Divers especially notice the large bluish-white tentacle crowns. Autozooids large, oval to hexagonal, and convex. Frontally smooth, translucent, the polypide as a white stripe. The subterminal opening is wider than long, with proximal and distal lip. Polypide relatively large with many tentacles (27–40). Yellow-white embryos in February. From 4 to 6 typical brown spines around the distal half of each zoid (Fig. 3.17).

Fig. 3.17 *Flustrellidra hispida.* (**a**, **b**) Colony op *Fucus serratus*. (**c**) Zooids with brown spines. (**d**) Tentacles. Living colonies, Brittany (France)

Occurrence

Belgium Loppens (1906) considers this species to be very rare on algae, mainly on *Fucus serratus*. No location is given, possibly he only found them washed up on the beach. No recent growing sites are known from Belgium. Rarely does a colony wash up on plastic, probably originating from the French Channel coast (De Blauwe 2000 and 2005).

The Netherlands Lacourt (1949) found washed-up colonies on Schouwen and Scheveningen in Naturalis in Leiden.

Rarely does a colony wash up on plastic *Himanthalia* buttons (de Ruijter 2017b) or *Sargassum mucticum* (de Ruijter 2019a). Only known in the Netherlands on *Fucus serratus,* for a limited number of locations in the Oosterschelde, since 2024:

Burghsluis, pers. com. Brendan Oonk, https://waarneming.nl/observation/301436874/, https://waarneming.nl/observation/299655911/

Germany Mentioned by Zettler et al. (2018) and collected at Heligoland (pers. comm. B. Kind and own observation in 2023 and 2024).

Family Arachnidiidae

Genus *Arachnidium*

1. (a) Zooidal surface with fine filaments that hold detritus. Forming linear series or dens mats of zooids. 12–16 tentacles. In seawater. .... *Arachnidium fibrosum* (p. 83).
   (b) Zooids without fine filaments. Polypide with eight tentacles. In brackish water. .... *Arachnidium lacourti* (p. 81).

Arachnidium lacourti d'Hondt and Faasse, 2006

Description Colony attached to hard substrates. Zooids in dichotomous branching rows. A large part of the zooids can be attached to the substrate by means of small rectangular teeth. The polypide has eight tentacles. Each zooid consists of three parts: an attached elongated proximal portion of variable length, a wide distal portion, and a cylindrical peristome that is oblique and has a circular opening. According to the original description (d'Hondt and Faasse 2006), the zooid is distally free from the substrate; this is not retained as a characteristic here because the colony is detached from the edge of a barnacle. The rectangular ancestrula does not have a polypide (Fig. 3.18).

Occurrence This species is known from the Netherlands in oligohaline waters of the Western Scheldt and the Nieuwe Waterweg. Type material comes from Bath in the brackish eastern part of the Western Scheldt. Grows in the lower half of the tidal area on the undersides of boulders, on barnacles (*Balanus improvisus*), and on the Pacific oyster (*Crassostrea gigas*). In the Western Scheldt, *A. lacourti* has not been

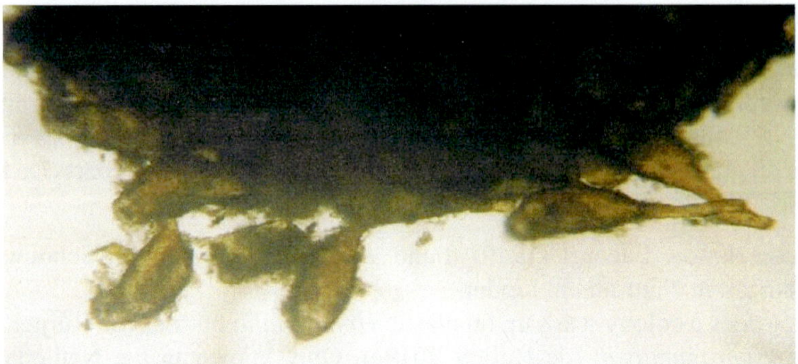

a

b

Fig. 3.18 *Arachnidium lacourti*. (**a**) Type material at the edge of a barnacle, Bath (the Netherlands). (**b**) Colony from Nieuwe Waterweg. (the Netherlands, Photo by M. Faasse)

found after 2003; in the Nieuwe Waterweg, a small colony was found near Bath in 2011 (Faasse et al. 2013). No recent records despite intensive searches by Faasse (pers. comm.)

Remark This species might in fact be a victorellid, but more material needs to be collected for study to test this attribution.

Arachnidium fibrosum Hincks, 1880

Description Zooids arranged irregularly in branching chains, sometimes forming dense mats. Zooids distally oval with a narrow tubular portion proximally, of varying length. Aperture on a striking cone-shaped peristome. Peristome often (or always?) with beige longitudinal stripes. Non-kenozooidal cystid appendages (filiform processes) develop around the edge of the zooids, often on the frontal surface as well. Sometimes a flat star-shaped widening develops from a bud, which can give rise to new buds. Branches can also split without forming a zooid at the split. This creates a network of short tubes. 12–16 tentacles, mostly 13. Dense colonies retain silt and grains of sand and between the filaments and can thus be confused with *Alcyonidium parasiticum*. The species often remains unnoticed in preserved material, especially if *Jassa* spp. (Arthropoda, Malacostraca) tubes are present. Research on living material facilitates the discovery and identification (Fig. 3.19).

Occurrence

Belgium Common subtidally, where it gives a muddy appearance to stones on the Hinder Banks and on mussels and stones on windmills (August 2018—Belwind, pers. comm. Francis Kerckhof). Infralitoral in the Baai van Heist on *Alcyonidium* sp. and *Crisularia plumosa*. Recorded on beached *Alcyonidium diaphanum* and *Abietinaria abietina* at Zeebrugge. On a buoy near Koksijde in January 2021 (pers. comm. Francis Kerckhof). *Arachnidium fibrosum* was extremely common on reef balls deployed in 2013 on the Bligh Bank and studied in 2014 (pers. comm. F. Kerckhof).

The Netherlands On empty shells near Goese Sas at 5–10 m depth (Faasse and De Blauwe 2003) and near Wissenkerke (pers. comm. Faasse M.). At the PAWF at 5, 10, and 17 m depth (Vanagt et al. 2013). Encountered in two samples taken from the foundation of platform L10-G at 10 m depth. Both colonies were attached to *Mytilus edulis*. Another colony of *A. fibrosum* was observed on a scour protection rock collected at the bottom of platform L15-A (Beukhof et al. 2016).

Germany Collected in several occasions in the German North Sea (Bryozoa workshop Heligoland 2023).

Recent publication: Schwaha, T. and De Blauwe, H. (2020). Morphology of ctenostome bryozoans: 1. *Arachnidium fibrosum*. J. Morphol. 281(12): 1598–1606. https://hdl.handle.net/10.1002/jmor.21275

a b c

Fig. 3.19 *Arachnidium fibrosum*. (**a**) Dense colony, peristomes with beige longitudinal stripes. (Germany, Photo by B. Kind). (**b**) With black filaments (transmitted light) on *Bugulina* sp. Baai van Heist (Belgium). (**c**) Lophophores with intertentacular organs

Family unassigned (pers. comm. Thomas Schwaha)

Genus *Panolicella*

Panolicella nutans Jebram, 1985

Description Scattered branching zooid chains, zooids up to 2 mm apart, connected by very slender filiform proximal zooid parts. The zooid consists of a basal portion, proximally long filamentous, and an erect cylindrical portion (up to 0.9 mm high) containing the polypide. Opaque with attached detritus. The opening is round and

a

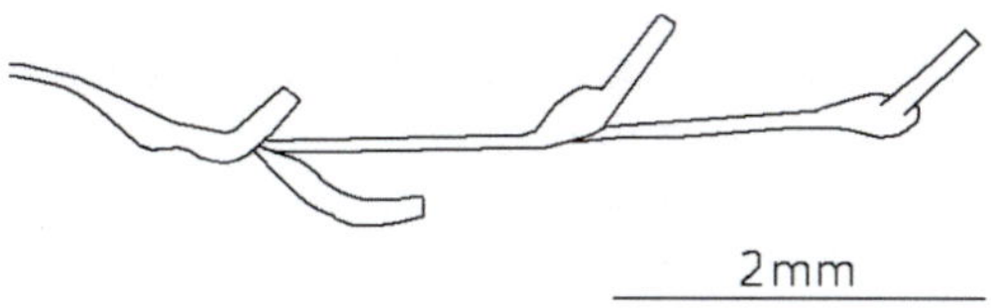

b

Fig. 3.20 *Panolicella nutans*. (**a**) Peristome (middle) between two zooids of *Amathia* cf. *gracilis*, tidal pool Goese Sas. (**b**) Part of a colony

terminal. The polypide bears 9–11 tentacles, rarely 7. Usually only discovered when a substrate containing living material is examined, the tentacles revealing its presence (Fig. 3.20).

Occurrence

Belgium In 2004, a colony was found on a stone from a channel between the Hinder Banks and in 2002 on the wreck of the Birkenfels (Zintzen 2007; Zintzen and Massin 2010). In the Netherlands found in a tide pool near the Goese Sas on a shell fragment (De Blauwe 2003), in the Westkapelse Kreek on the hydroid *Gonothyrea loveni*, and at Bommenede (Grevelingen) on a stone. Often occurs together with *Amathia* species.

Faasse and De Blauwe (2004) mentioned *Nolella pusilla* (Hincks, 1880) from the Netherlands. Reverter-Gil and Fernández-Pulpeiro (2006) concluded that the description of *Nolella pusilla* is inadequate for reliable identification. Type specimens probably no longer exist. Reverter-Gil and Fernández-Pulpeiro (2006) concluded that material from the Netherlands, mentioned under this name, belongs to the species *Panolicella nutans* Jebram, 1985, like numerous colonies collected in Galicia (Spain). (Faasse et al. 2013).

Family unassigned (pers. comm. Thomas Schwaha)

Genus *Anguinella*

Anguinella palmata van Beneden, 1845

Description Forming gray-brown, mud-colored tufts, 5–20 cm high. More reminiscent of a hydroid or of a miniature *Codium* seaweed. Several main branches have spirally implanted side branches. The zooids emerge from the side of the side branches. New zooids can spring from every zooid. The colony is attached with slender rhizoids. Zooids opaque, elongated cylindrical, and curved slightly toward the axis of the colony. Due to the adhesion of silt, only the terminal opening remains clearly visible. Polypide relatively small with 10 or 11 tentacles. d'Udekem d'Acoz (1993) reports hibernating buds or hibernacula on colonies from Belgium in February (Fig. 3.21).

Occurrence

Belgium On stones, shells, wooden poles, and groynes, low in the tidal area, especially where there is a lot of silt. Can tolerate low and fluctuating salt levels. Loppens (1906) mentions this species as common in Belgium. Found in 1990 on groynes in Ostend and Duinbergen (d'Udekem d'Acoz 1991) and on the eastern breakwater in Zeebrugge in February 1993 (d'Udekem d'Acoz 1993). Locally common along the Belgian coast (Baai van Heist, groyne Koksijde).

The Netherlands Port of Den Helder (van der Sleen 1920), the Texelstroom (van Benthem Jutting 1922), and Wadden area (Lacourt 1978). Locally common in Westkapelle, western part of the Westerschelde and the Oosterschelde. Large colonies at Borssele in February 2017 (pers. comm. Marco Faasse).

Germany Collected at Heligoland (pers. comm. Britta Kind).

Family Victorellidae

1. (a) Cardia divided into two equal-sized portions by a median sphincter. Intertentacular organ in fertile zooids. .... *Victorella pavida* (p. 89)
   (b) Cardia undivided. No an intertentacular organ. .... *Tanganella mülleri* (p. 87)

Occurrence *Victorella pavida* and *Tanganella mülleri* may be found together in areas of low and fluctuating salinity. In shallow water on submerged stones, shells, barnacles, plants (*Phragmites australis*), and wood as well as artificial substrata such as concrete.

Fig. 3.21 *Anguinella palmata*. (**a**) Living colony growing on a stone. (**b**) Zooids. Yerseke, (the Netherlands)

Genus *Tanganella*

Tanganella mülleri (Kraepelin, 1887)

Description Colony forming diffuse chains of zooids or dense tufts. Zooid proportions variable, adnate portion, and peristomial tube variable in length. Polypide with eight tentacles. Cardia undivided. However, gut morphology is difficult to observe. No intertentacular organ (Fig. 3.22).

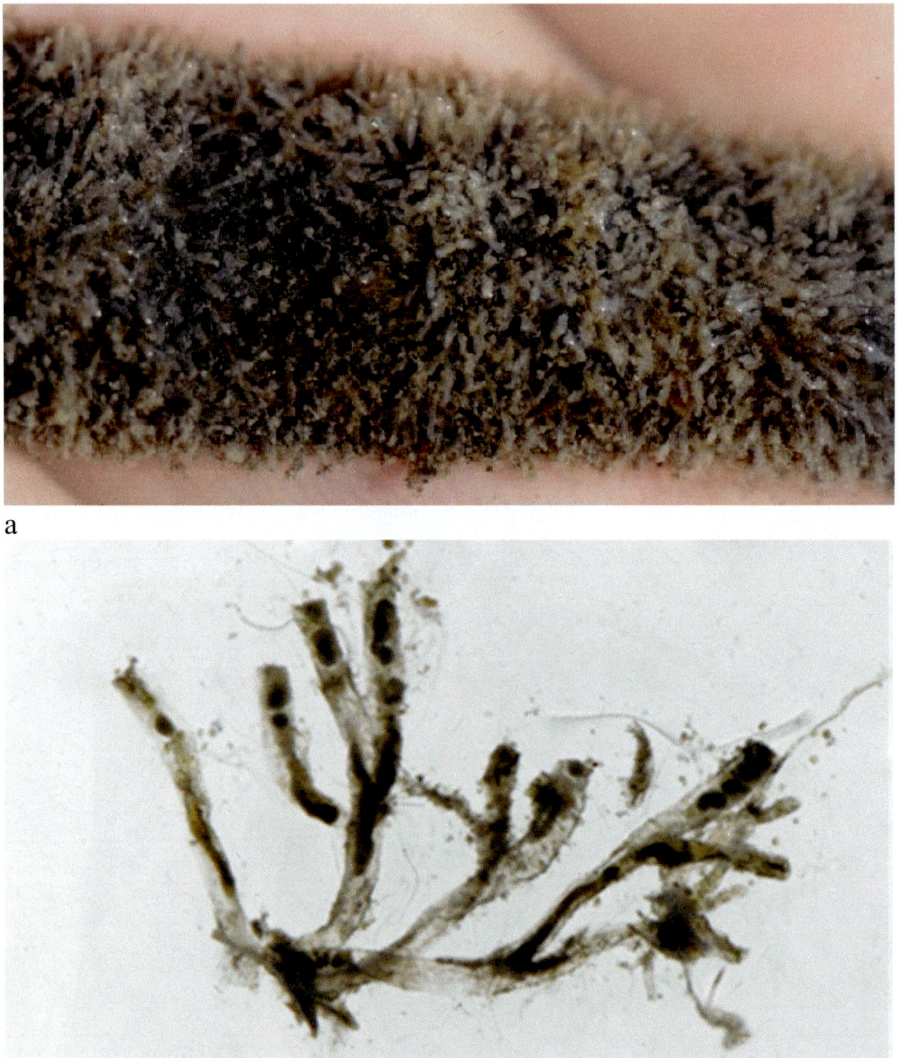

Fig. 3.22 *Tanganella mülleri*. (**a**) Colony from Goes (the Netherlands). (**b**) Part of a detached colony. (Photo by T. Schwaha)

Occurrence

Belgium Two observations at Nieuwpoort, one in the brackish marina in September 2000 and one in "het nieuw bedelf" in November 2006. Both identified by present author as *Tanganella mülleri* but the material is not preserved.

The Netherlands Collected at Goes, in July 2018 (identification by T. Schwaha).

Germany Reported by Kraepelin (1887) and Braem (1951).

Genus *Victorella*

Victorella pavida Saville-Kent, 1870

Description Forming diffuse branching chains or developing into dense autozooid clusters. In summer, colonies have the appearance of velvet. Usually, the colony is attached, sometimes parts of the colony are free from the substrate. A zooid consists of an attached basal portion and an erect cylindrical peristome 0.3–1.6 mm long. Erect zooids may be cylindrical or slightly bulbous at the base. The longest peristomes occur where the zooids grow close together. The attached part is usually very short, but sometimes long. New zooids originate distally and laterally from the basal part of each zooid or from the peristome. Polypide with eight tentacles. Zygotes released through an intertentacular organ. Produces dark brown/black hibernacula (dormant resting buds). With respect to the lifespan of a *Victorella pavida* colony, new colonies emerge from dormancy in spring when temperatures are approximately 13 °C. By November and the onset of winter, zooids begin to degenerate, and eventually, only the asexually produced dormant resting bodies (hibernacula) remain. The hibernacula germinate again in spring and the cycle begins again. Approximately 25 eggs can be produced per gravid zooid from June to September. *Victorella pavida* has short-lived planktonic larvae, which probably settle from July to September (Carter 2004; Carter and Jackson 2007) (Fig. 3.23).

Occurrence

The Netherlands *Victorella pavida* is collected in 2006 near Goes.

Germany Reported by Kraepelin (1887) and Braem (1951).

Family Walkeriidae

Genus *Walkeria*

Walkeria uva (Linnaeus, 1758)

Description At intervals, groups of autozooids develop on a branching stolon. Stolon formed by a chain of cylindrical kenozooids that are clearly separated, each kenozooid 0.5–1.5 mm long, attached to the substrate or not. Autozooids oval, translucent, small (0.4 mm long), and not stalked. The square opening is terminal. Polypide relatively large with eight tentacles, two of which are curved outward (Fig. 3.24).

Occurrence

Belgium Only known from the harbor of Zeebrugge (in 2013 and 2024).

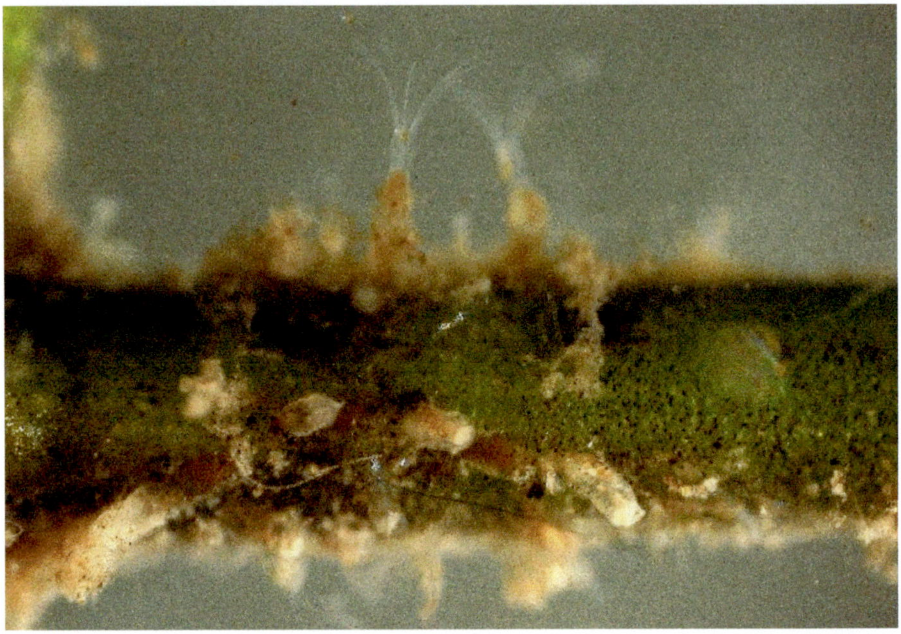

a

b

Fig. 3.23 *Victorella pavida*. (**a**) Habitus of living colony on reed (*Phragmites australis*). (**b**) Zooids detached from the reed stem to show their basal portion (Goes, the Netherlands)

The Netherlands Collected in Zeeland in 2006. Other reports of colonies with central erect stem, bearing paired pinnate branches (Photos 46 and 47 in De Blauwe 2009) possibly concern *Mimosella gracilis*.

a

b

Fig. 3.24 *Walkeria uva*. (a) Kenozooids and a few autozooids, the Netherlands. (**b**) Colony from Germany. (Photo by B. Kind)

Germany Collected in the German North Sea (Zettler et al. 2018) and at Heligoland (pers. comm. B. Kind).

Family Triticellidae

1. (a) 17–20 tentacles. Autozooid asymmetrical, angular; the border to the peduncle is abrupt and distinct, without kink or corrugation. Septum between autozooid and peduncle distinct. ……………………..…......…..... *Triticella flava* (p. 92)

(b) 12 tentacles. Autozooid fusiform, not angular; annulate proximally, grading gently into the peduncle, septum indistinct. .... *Triticella pedicellata* (p. 92)

Genus *Triticella*

Triticella flava Dalyell, 1848

Description The zooids resemble thick hairs and are easily overlooked. Stalked zooids arise from a network of slender stolons. The stolons consist of a linear series of kenozooids, ending at a septate junction, where chains of short flattened kenozooids are budded. From there, very slender peduncles grow vertically and support a single distal autozooid. Distinct septum between peduncle and autozooid. Zooid length (0.4–1.2 mm) and peduncle length (0.2–2.8 mm) very variable (Fig. 3.25).

Occurrence Noticed in samples from the Frisian front since 2006, near Terschelling and north of Den Helder. Found in the period 2008–2012 at 12 stations, where at least 37% of the *Upogebia* acted as hosts for *Triticella flava* (Tempelman et al. 2013). The colonies are found on the claws and antennae of *Upogebia deltaura* and *Upogebia stellata*.

Recently, *T. flava* was found in the North Sea (German EEZ and Dutch NCP) on the carapace of males of *Goneplax rhomboides* (Neumann et al. 2010) and on *Upogebia deltaura* claws and antennae and *Liocarcinus holsatus* (pers. comm. Lisa Schüler).

Triticella pedicellata (Alder, 1857)

Description Zooids arising from a slender kenozooidal stolon that is closely attached to the substrate. Kenozooids grow vertically to form the peduncle of a slender, fusiform autozooid. Septum only seen in young zooids, the proximal portion of autozooids with transverse corrugations and wrinkles. Orifice terminal, not bilobate. Lophophore with 10–12 tentacles. Length of zooid between 0.4 and 0.8 mm, length of peduncle between 0.64 and 2.4 mm (Hayward 1985) (Fig. 3.26a).

Occurrence A rarely reported species or species complex.

Germany Collected by Britta Kind on a decapod antenna from the Tiefe Rinne near Heligoland in July 2022.

Additional pictures (Fig. 3.26b–d) from a colony collected by Lisa Schüler in the Baltic Sea (54.43 N, 11.41 E) on April 29, 2024, on *Neptunea antiqua*. Figure 3.26d: length of zooid + peduncle 2.6 mm. *T. pedicellata* is found mostly on gastropod shells or on the hydroid *Dicoryne conferta*, pers. comm. L. Schüler.

Fig. 3.25 *Triticella flava*. (**a**) On claw of *Upogebia deltaura* from oyster grounds. (the Netherlands, Photo by D. Tempelman). (**b**, **c**) On antenna from the German North Sea. (coll. by L. Schüler, Photo by H. De Blauwe)

Family Farrellidae

Genus *Farrella*

Farrella repens (Farre, 1837)

Description Kenozooids form a slender creeping stolon. Erect cylindrical zooids at the distal end of each kenozooid. The stolon branches off at this point, forming groups of zooids every 1–2 mm. Zooids transparent, cylindrical, or club-shaped,

a b c d

Fig. 3.26 *Triticella pedicellata*. (**a**) Colony from Tiefe Rinne. (Germany, Photo by B. Kind). (**b–d**) Colony from the Baltic Sea. (Germany, coll. L. Schüler, Photos by H. De Blauwe)

and the proximal part suddenly narrows and resembles a stem. The terminal opening is conspicuously bilobed. Polypide with 11–15 tentacles. The length of the zooids and of the narrow proximal portion is variable. Unlike *Triticella*, the polypide sometimes degenerates and forms a residual "brown body." The bilobed opening is the most striking distinction from *Triticella* (Fig. 3.27).

Occurrence Tolerant of variable salinities.

Fig. 3.27 *Farrella repens*, a few zooids, bilobed terminal opening, Zeebrugge (Belgium)

Belgium Loppens (1906) considered this species to be very common along the Belgian coast. Common sublittorally.

The Netherlands Lacourt (1949) mentions various sites from Zeeland to Friesland.

Common sublittorally on stones, shells, hydroids, and crabs (Faasse and De Blauwe 2004).

Germany Britta Kind (unpublished) collected *Farrella repens* on Hydrozoa in 2014 in the northeast harbor of Heligoland. Detlef Henning (Bioconsult) and Lisa Schüler (IfAÖ) have stated that this species is quite common in coastal waters of the German North Sea and Wadden Sea (pers. comm. Britta Kind).

Family Hypophorellidae

Genus *Hypophorella*

Hypophorella expansa Ehlers, 1876

Description The colony lives within the linings of organic polychaete tubes. The colony is composed of elongated stolonal kenozooids with a distal capsule-like expansion. An autozooid is laterally attached to the capsule. Single autozooids are attached in an alternating right–left succession on subsequent stolons. The most obvious apomorphic features of *Hypophorella* are space balloons and a gnawing apparatus. The former are two fronto-lateral spherical structures on autozooids, which provide space inside the tube. The latter perforates layers of the polychaete tube wall and consists of two rows of cuticular teeth that, together with the entire vestibular wall, are introverted during the retraction process (Pröts et al. 2019). Polypide with 12–14 tentacles.

Remove the outer layer of the *Chaetopterus* tubes. The inner layers are studied under transmitted light under a stereomicroscope. Usually, the zooids have disappeared and a small borehole remains, usually surrounded by a brown spot. The outer layers of the tube may contain older colonies, with only stolons and holes where the zoids have disappeared. The stolons are very fine but still easily recognizable. Zooids may still be present in fresh material.

For searching in *Lanice* tubes, the samples should be stained with Bengal rose making the *Hypophorella* colonies clearly visible (Fig. 3.28).

Occurrence Lives in tubes of *Chaetopterus variopedatus* and *Lanice conchilega*.

Belgium Tubes of *Chaetopterus variopedatus* washed up on the Belgian coast often containing remnants or traces of colonies (De Blauwe 2001). Present on the Hinder Banks in tubes of *Chaetopterus variopedatus*.

The Netherlands Found in tubes of *Lanice conchilega* (pers. comm. M. Faasse 2017).

Germany Collected at Heligoland by Britta Kind (pers. comm.).

Family Penetrantiidae

Genus *Penetrantia*

Characteristic are the kidney-shaped openings that lie next to the main stolon. The *Penetrantia concharum* complex comprises two cryptic recent species in Europe (pers. comm. Decker and Schwaha). A fossil species, *Penetrantia soulei* Pohowsky, 1978, is known from the Plio-Pleistocene of the Kaloot (Zeeland).

In *Penetrantia soulei* additional stolons arise on the zooids and gonozooids and immediately split into three branches, which grow to the surface of the substrate.

a b c

Fig. 3.28 *Hypophorella expansa*. (**a**) Worm tube of *Chaetopterus variopedatus*, washed ashore. (**b**) Prepared membrane with stolons and drill holes. (**c**) Stolons and two zooids

Gonozooid barely longer than half zooid length with a nearly spherical brood chamber in its distal part (Decker et al. 2023). Zooidal opening kidney-shaped. This genus and the next two, which also drill into shells, have already been reported in De Blauwe (2004b).

Penetrantia concharum Silén, 1946

Description The colony bore into empty or living shells. Occurs in many shell species, bivalves and gastropods, rarely in scaphopods, and also in the plates of barnacles. Zooidal opening kidney-shaped. The drill holes are interconnected by fine thread-like stolons just below the shell surface. Zooids cylindrical, slender, proximally pointed, and perpendicular to the shell surface. Opening terminal, closed by a tan, D-shaped operculum. The stolons consist of many cylindrical kenozooids of variable length. Tubulets, as small tubes on the stolons, occur in regular intervals in *Penetrantia* and most other borings species. These might be themselves also kenozooids. They are interpreted as a kind of snorkel that determines how far the stolon is away from the surface of the substrate. The zooids are connected to the stolons by means of a short peduncle. Polypide with ten tentacles. The surface openings have a characteristic kidney shape. The drill holes have a constant size, approximately 0.1 mm in diameter. Brood chamber develops in the distal half of the gonozooid (Fig. 3.29).

Occurrence

Belgium Borings are very common in shell valves on the Flemish Banks.

The Netherlands Collected in shells at Vlieland, Egmond and Dishoek (de Ruijter 2012, 2016, 2019b). Collected on the Dogger Bank in 2016.

Germany First reported by Jebram (1972) from Heligoland. Common on dredged shells near Heligoland (pers. comm. Britta Kind, Sebastian Decker, own obs.).

Family Immergentiidae

Genus *Immergentia*

Immergentia suecica Silén, 1947

Description The colony forms an irregular branching network. Zooids not stalked, interconnected by cystid appendages emanating from the borehole, not true kenozooidal stolons. Cystid anastomoses frequently present, which also differentiates immergentids from *Penetrantia* and *Spathipora*. Side branches of the colony are opposite to each other. Opening like a swollen S or its mirror image. The zooids are not perfectly perpendicular to the substrate, and the proximal (lower) part of the zooid is oriented proximally.

Recent species of *Immergentia* brood embryos within their tentacle sheaths (Schwaha 2021a, b; Johnson et al. 2024) (Fig. 3.30).

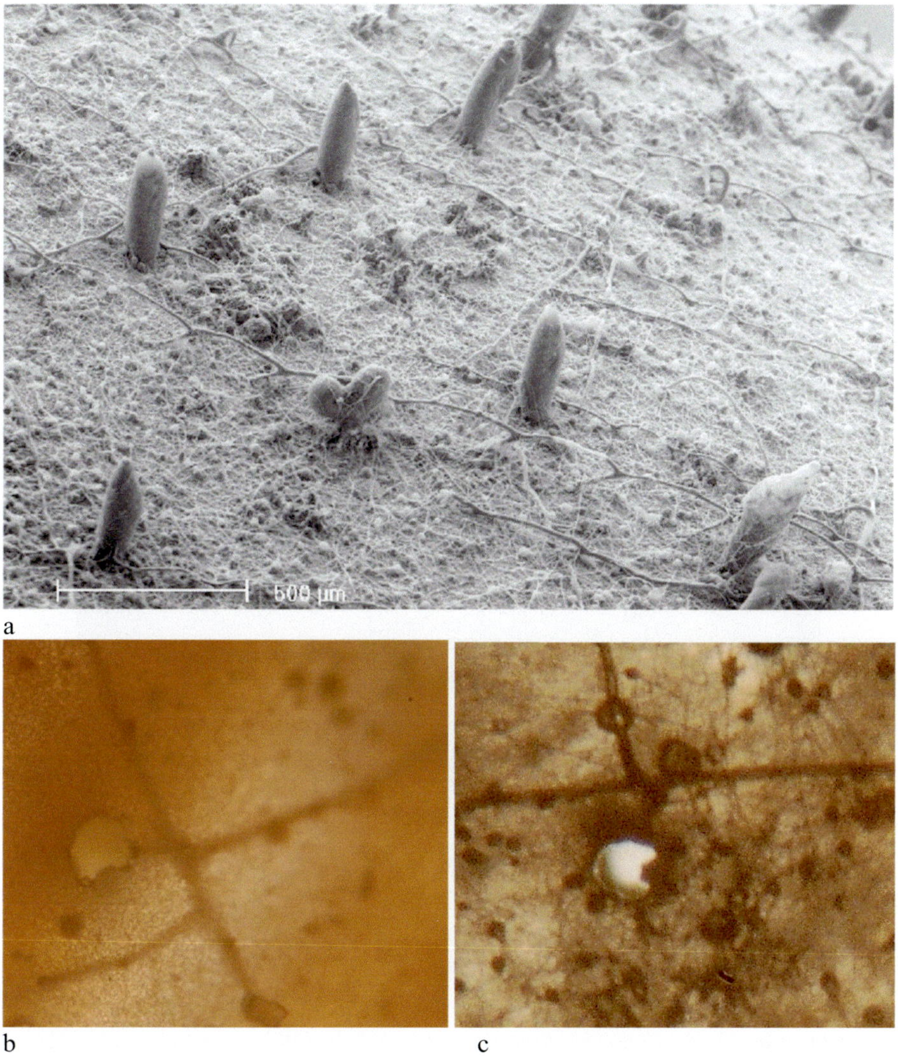

Fig. 3.29 *Penetrantia concharum.* (**a**) SEM image of cast in polyester, showing stolons, zooids, and gonozooid with brood chamber (left of center). Westhinder Bank (Belgium). (**b**, **c**) Borings, Kwinte Bank (Belgium)

Occurrence

Belgium Borings occur in shell valves on the Flemish Banks. The least common borer of the three included here.

The Netherlands Collected on the Dogger Bank in 2016.

Germany First reported by Jebram (1972) from Heligoland, later also by Britta Kind (pers. comm.).

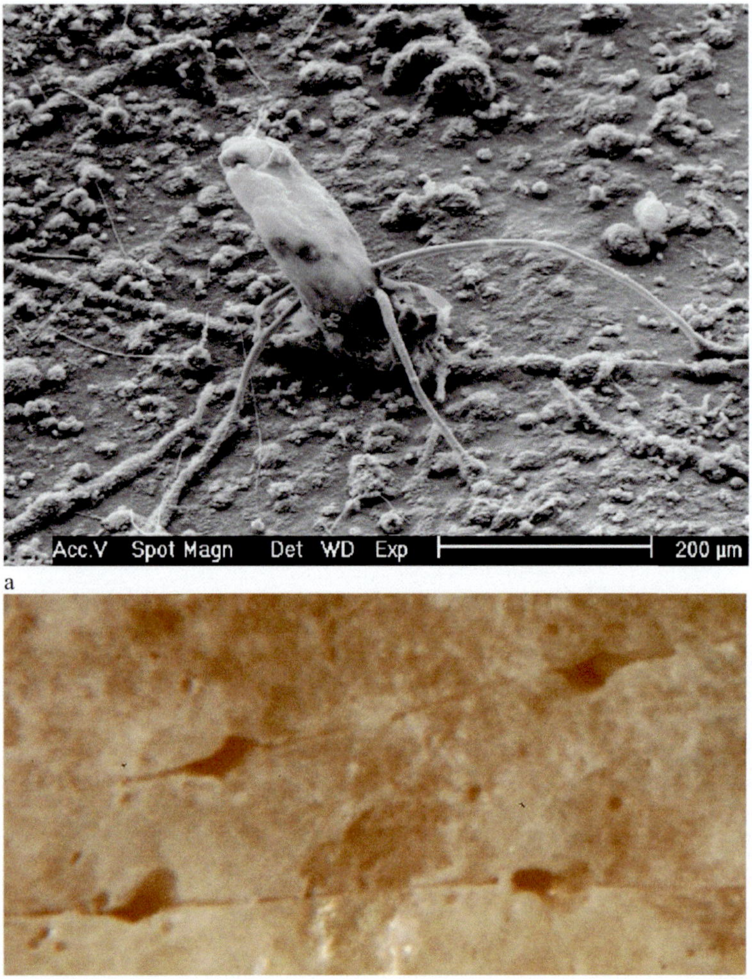

a

b

Fig. 3.30 *Immergentia suecica*. (**a**) SEM image of cast in polyester, showing zooid and cystid anastomoses. (**b**) Borings. Kwinte Bank (Belgium)

Family Spathiporidae

Genus *Spathipora* (Fischer, 1866)

Spathipora borings are common in shells (ancient or fossil). For identification, it is necessary to make polyester casts of the cavities, unless the shells are translucent enough. Species differ in the site where the peduncle is inserted along the zooid.

Most of the shells in which the borings were found have a fossil appearance, so there is a good chance that the *Spathipora* made the borings long ago.

Recent species of *Spathipora* brood embryos within their tentacle sheaths similar to other vesicularioidean ctenostomes (Schwaha 2021a, b).

Spathipora sertum Fischer, 1866

Diagnosis The peduncle enters the zooid at its tapered proximal end, as in Fig. 3.31c. The proximal part does not cross the stolon basally. Frontal vane with tubulets as in other species.

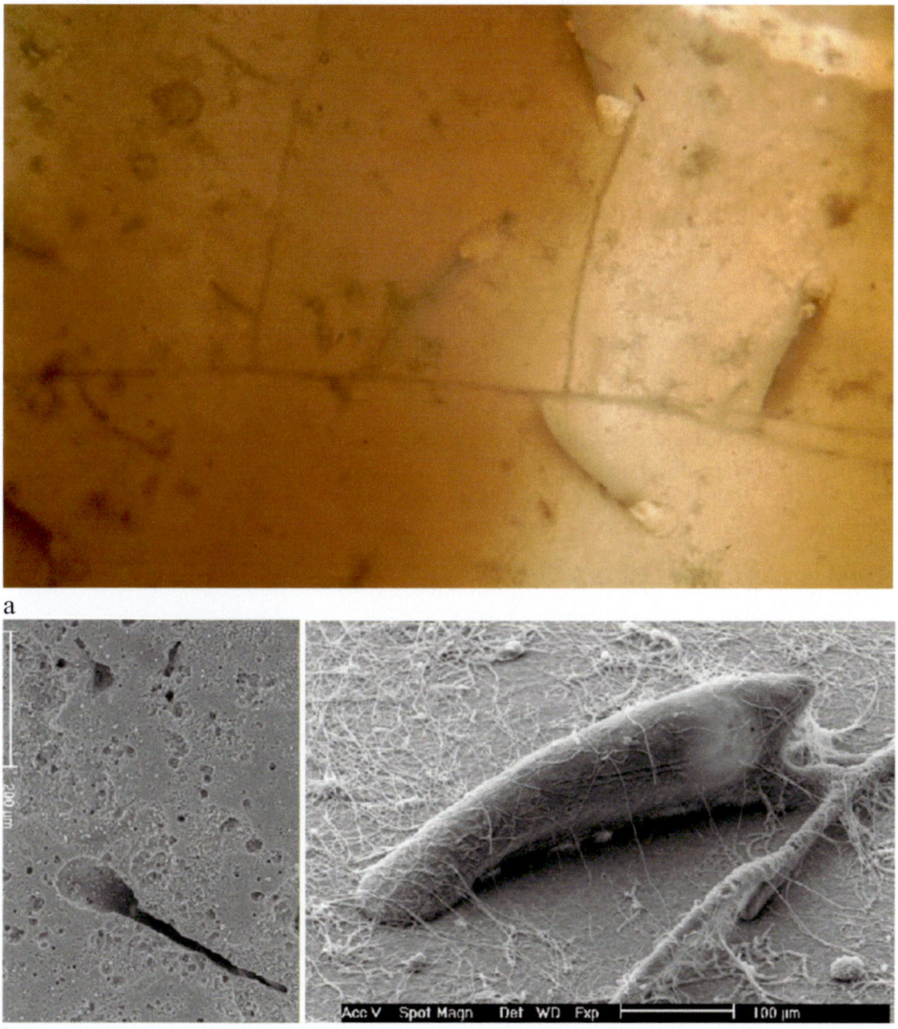

Fig. 3.31 *Spathipora sertum*. (**a**, **b**) Borings in shell. (**c**) Cast in polyester. Kwinte Bank (Belgium)

Spathipora sp.

Description Boring Bryozoa with peduncle inserted near or proximal to midlength of zooid. Zooids tilted horizontally, varying in species, and commonly bearing a frontal vane with tubulets. Aperture and openings associated with the vane typically form a comma-shaped pattern at the surface of the substrate (Fig. 3.31).

Occurrence

Belgium Borings are common in shell valves on the Flemish Banks. A live colony was found by Marco Faasse (pers. comm. 2020) after staining with Bengal pink when breaking away a piece of the last turn of a *Hinia reticulata*, collected 0.5–1 km off the coast of De Panne.

The Netherlands Borings recorded from Schiermonnikoog (Holsteijn 1998), Vlieland (de Ruijter 2012), Egmond (de Ruijter 2016), and 60 km west of Walcheren (Heerebout 1970).

Germany First reported by Jebram (1972) from Heligoland later also by Britta Kind (pers. comm.).

Family Vesiculariidae

Genus *Amathia*

The genera *Bowerbankia* and *Zoobotryon* have been placed in synonymy with *Amathia* by Waeschenbach et al. (2015). The revised classification proposed by Waeschenbach et al. (submitted) transfers the genus *Vesicularia* to *Amathia*.

Amathia spinosa (Linnaeus, 1758)

Vesicularia spinosa (Linnaeus, 1758)

Description Forms erect light gray-brown dense clumps, 5–30 cm long. The main branch is sturdy and branches continuously at intervals. The branches composed of kenozooids are separated by transverse walls. The side branches branch in the same way. All growing tips end in a pointed stolon. Zooids on a row along the branches, especially on the youngest branches, not permanent. Oval, translucent, narrowing proximally but not stalked, on a protrusion of the branch. Polypide with eight tentacles (Fig. 3.32).

Occurrence

Belgium Loppens (1906) mentions this species as fairly common on rocks, oysters, and seaweed. At the beginning of the last century, Gilson only encountered this species in the vicinity of the Hinder Banks. The absence in samples taken near the coast is remarkable. It is apparently not a species of shallow water.

Fig. 3.32 *Amathia spinosa*. (**a**) Colony washed up at Zeebrugge (Belgium). (**b**) Zooids on kenozooidal branch with spiky end, washed up at Oostende (Belgium)

Research in 2005 showed that *Amathia spinosa* is still present in the channels between the Hinder Banks (De Blauwe et al. 2006). On April 30, 2006, a colony was found attached to a groyne in Koksijde. In 2003, a colony was collected on the wreck of the Bourasque (Zintzen 2007; Zintzen and Massin 2010). Often washes up along the Belgian coast in autumn and winter.

The Netherlands *Amathia spinosa* was found during the monitoring program on the Oyster Grounds. The loose colonies leave some doubt as to whether they lived at the exact box core locations (MWTL 2006: 54°49′24″N, 003°22′00″E, depth 42 m, 2009: 53°30′00″N, 03°00′00″E, depth 36 m) (Faasse et al. 2013).

There are older reports of which it is uncertain whether the colonies were washed ashore or established (Lacourt 1949).

Washed up on the shores of Katwijk and Westkapelle (Maitland 1851) and Noordhollands Duinreservaat in 2020: https://waarneming.nl/observation/184498187/

Other *Amathia*: diagnosis (After Waeschenbach et al. 2015)

Colony forming a creeping network or erect branches of stolons. Stolon branching irregularly or regularly, dichotomic, trichotomic, or tetratomic. Autozooids are cylindrical and saccular, forming individual or clusters of zooids in intervals; zooids budded directly from the main stolon, without secondary appendages. Lophophore with 8–12 tentacles, and gut with a distinct gizzard. Embryos brooded internally in tentacle sheath (Schwaha et al. 2019).

1. (a) Erect, stiff, brown tufts. Autozooids arranged in straight comb-like rows, typically, adjacent zooids are laterally joined. .... *Amathia lendigera* (p. 110)
   (b) Autozooids group more loosely and are never laterally joined. 2
2. (a) Zooids in groups, each group in a semi-spiral around an erect stolon. Zooids shorter than 0.5 mm, polypide bright yellow. .... *Amathia citrina* (p. 104)
   (b) Zooids not spirally implanted on the stolon. 3
3. (a) Polypide with eight tentacles. 4
   (b) Polypide with ten tentacles. ....(Fig. 3.33) 5
4. (a) Embryos pink. .... *Amathia* cf. *gracilis* 1 (p. 107)
   (b) Embryos yellow. .... *Amathia* cf. *gracilis* 2 (p. 107)
5. (a) Embryos pink. .... *Amathia tertia* (p. 112)
   (b) Embryos yellow. .... *Amathia imbricata* (p. 108)

Amathia citrina (Hincks, 1877)

Bowerbankia citrina (Hincks, 1877)

Description At first adherent, with later spreading erect tufts up to 5 cm high. Stolons cylindrical, as wide or wider than the zooids. Internodes 1.5–2 mm long. Zooids in separate groups in a semi-spiral on the stolon, just before a bifurcation. Active feeding colonies are yellowish. Inactive, degenerate colonies are to be confused with *A. pustulosa*, a similar species with colorless polypide, which was

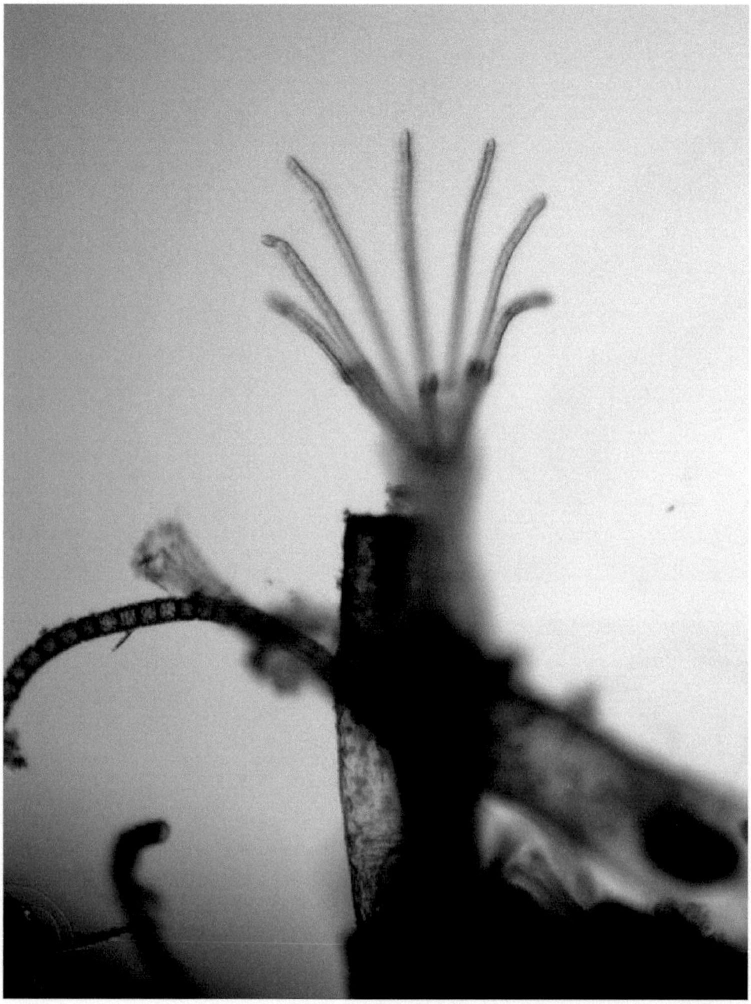

Fig. 3.33 *Amathia tertia*, lophophore with ten tentacles (Cuxhaven, 2023)

recently found in Boulogne (Northern France) by Marco Faasse (pers. comm.). Zooids cylindrical and short, 0.4 mm long, truncated distally. Degenerate zooids shorter and distally pointed or rounded. Zooids colorless, translucent, polypide bright lemon yellow, with eight tentacles (Fig. 3.34).

Occurrence Found in Zeeland since 1997 on, near Zierikzee, Strijenham, and Yerseke, with presence from July to early October (Faasse and De Blauwe 2004).

Fig. 3.34 *Amathia citrina*. (**a**) Living colony. (**b**) Detail showing yellow tentacles. (**c**) Spiral position of the zoids. Yerseke (the Netherlands)

Amathia gracilis (Leidy, 1855)

Bowerbankia gracilis **Leidy, 1855**

Amathia gracilis is described from the USA. Results (Waeschenbach et al. 2015) from BI phylogenetic reconstruction and the presence of a 20% difference between cox1 sequences suggest that the European and US specimens are clearly separate species. The identification of the European material still needs to be established. Based on embryo color, morphometrics, and allozyme work, there are likely to be two British species referred to as *B. gracilis*: one with yellow embryos and one with pink embryos. Colonies that harbor yellow vs. pink embryos exhibit slight morphological differences, specifically zooid and stolon dimensions. Embryo color needs to be observed in living material.

Amathia cf. *gracilis* sp. 1: Pink Embryos

Description Colony adherent, diffuse, or with dense clumps of zooids. Stolons may be free from the substrate. Stolon very slender, half as broad as the zooids, translucent at first, later opaque brown. Autozooids budding from sides of the stolon, scattered or in dense groups, truncate distally, often with a caudal projection (short proximal tail-shaped kenozooid). Zooids 0.6–1.2 mm long. Polypide with eight tentacles. Embryos pink.

Often forms dense mats. The largest zooids are active feeding zooids; smaller ones often breed embryos or have a degenerating polypide. Often two or three "brown bodies" are present in the zooids (Fig. 3.35).

Occurrence On the lower beach and shallow sublittoral on rocks, shells, algae, and often on hydroids. Tolerates low and variable salinity levels.

Belgium Loppens (1906) generally found this species in brackish water, but probably underestimated the numbers due to confusion with *A. imbricata*. Very common on ship hulls and submerged structures in marinas, in creeks, and on groynes.

The Netherlands Lacourt (1949) faced the same problem. Very common on ship hulls and submerged structures in marinas, in creeks, and on groynes in Zeeland. Collected on the Dogger Bank in 2016.

Germany Collected at Heligoland (pers. comm. B. Kind).

Amathia cf. *gracilis* sp. 2: Yellow Embryos

Description Occasionally, *Amathia* cf. *gracilis* (eight tentacles, stolon thinner than zooids) except that the embryos are yellow instead of pink. Waeschenbach et al. (2015) call European *A. gracilis* with yellow embryos *A.* cf. *gracilis*. It is genetically different from the true *A. gracilis* from Rhode Island (Fig. 3.36).

Fig. 3.35 *Amathia* cf. *gracilis* sp. 1—pink embryos. (**a**) Part of a colony, Zeebrugge (Belgium). (**b**) Zooids from Burghsluis (the Netherlands)

Occurrence The author found such colonies in Zeebrugge and Arcachon (France), they have also been found in Swansea, Wales (Hayward 1985). According to J.S. Ryland (pers. comm.), this may be an undescribed species. Further morphological, DNA, and ecological research into *Amathia* is desirable.

Amathia imbricata (Adams, 1800)

Bowerbankia imbricata (Adams, 1798)

Description Colony adherent or erect, dense feathery tufts to over 5 cm in length. Stolon often wider than the zooids. Zooids in groups on the stolons. Zooids rounded cylindrical, proximally somewhat swollen, ending quadrangular distally. Zooids quite large: 0.8–1.5 mm long. Zooids with a degenerate polypide are

Fig. 3.36 *Amathia* cf. *gracilis* sp. 2—yellow embryos. Stolon thinner than the zooids, embryos yellow

shorter and swollen with rounded distal ends. Polypide with ten tentacles. Embryos deep yellow (Fig. 3.37).

Occurrence In the tidal area and in shallow coastal waters, on brown algae, on rocks and wooden structures, and on other Bryozoa and Hydrozoa. In sheltered places, luxuriant erect tufts may occur. Especially in turbid waters of estuaries and on sheltered beaches. Colonies of *Amathia gracilis* (with eight tentacles) and *A. imbricata* (with ten tentacles) sometimes grow together.

Belgium Loppens (1906) mentions this species as fairly common in the sea and in brackish water, but was unable to distinguish it well with *Amathia gracilis*, which he called var. *A. imbricata* var. *caudata*, this is apparent from his statement of the number of tentacles.

The Netherlands Occurs in the Oosterschelde, in the Canal of Goes in the canal through Zuid-Beveland (Faasse and De Blauwe 2004). Lengkeek et al. (2013) report *A. imbricata* from the shipwreck "HMS Scott" (UTM 564929 5805707).

Germany Collected at Heligoland (pers. comm. B. Kind).

Fig. 3.37 *Amathia imbricata*. Ten tentacles, stolon relatively wide, St-Lunaire (France)

Amathia lendigera (Linnaeus, 1758)

Description Forms stiff, spreading, light brown tangles, up to 4 cm in diameter. Twigs firm, cylindrical, branching dichotomous, and each internode is a single kenozooid. Zooid groups consist of five to eight pairs. Autozooids arranged close together in double, straight, comb-like rows on the distal half of each internode, just before the bifurcation. Zooids cylindrical, truncated, and with terminal opening. Polypide with eight tentacles. Fully grown zooids have the same length; at the growing points, the zooids decrease in length toward the end of the internode, so that the zooid groups there have a triangular shape. Internode typically 2 mm long (Fig. 3.38).

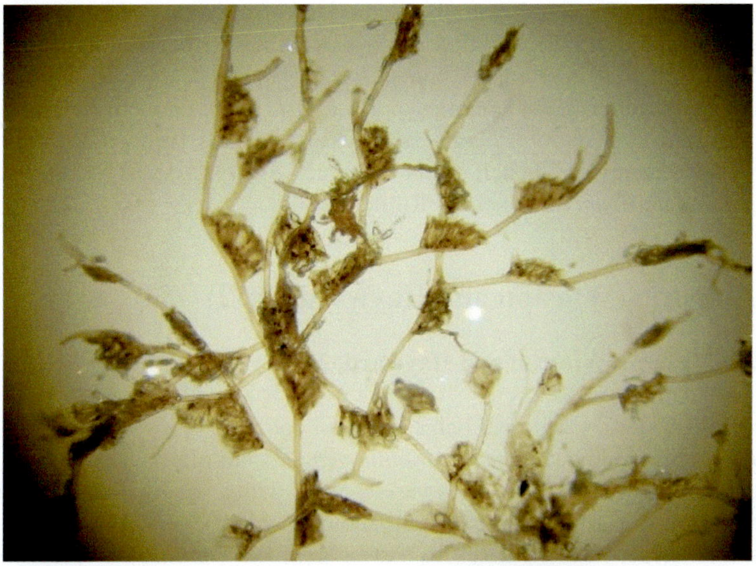

a

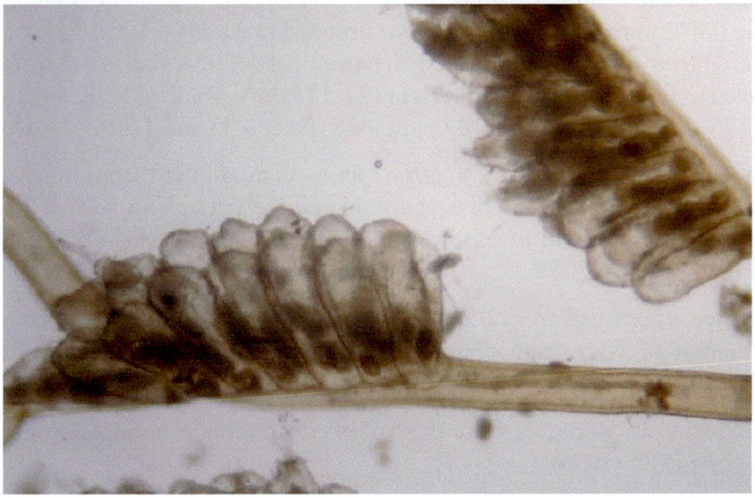

b

Fig. 3.38 *Amathia lendigera*. (**a**) Colony beached at Petten. (the Netherlands, Photo by R. de Ruijter). (**b**) Groups of zooids, Damgan (France)

Occurrence

Belgium Loppens (1906) found this species rather rarely on hydroids and algae, mainly on *Halidrys siliquosa*. Washed up on a lobster trap in Zeebrugge in December 1999, on a fishing net in Ostend in November 2000, and in a water bottle in De Haan in December 2011.

The Netherlands Reported from the Netherlands from Scheveningen on washed-up algae (Maitland, 1851). Washed up in November 2016 on a plastic basket in Petten (de Ruijter 2017a) and in January 2018 on Texel (Sytske Dijksen, pers. comm.). In addition to many colonies of *Tricellaria inopinata* and *Cryptosula pallasiana*, a colony of *Amathia lendigera* (de Ruijter 2018) grew on a valve of *Mya arenaria* that was under a bunch of bladderwrack *Fucus vesiculosus*. Two colonies *on Sargassum muticum* on Texel (de Ruijter 2020b).

Amathia tertia (Winston and Hayward, 2012)

Bowerbankia tertia Winston and Hayward, 2012

Description Colonies whiteish-beige to tan in color, creeping over algae or other substrata, similar in appearance to *Amathia gracilis*, but often less transparent, and more sediment covered. Autozooids vase-shaped, with a square distal orifice, tubular mid-section, and rounded bases. Autozooids budded separately off stolons; stolon diameter wider in comparison to that of *Amathia gracilis*. Polypides with ten tentacles, tentacle length averaging 0.36 mm, lophophore diameter 0.35 mm, mouth diameter 0.042 mm. Orange-red embryos brooded in zooids.

Material from Cuxhaven, collected on November 4, 2023, shows a narrow cauda, ten tentacles (Fig. 3.31), pink embryos (119 × 148 μm), stolon (64, 78, 84 μm wide), zooids 175–186 μm wide, 700–1072 μm long (Figs. 3.33 and 3.39).

Discussion Living colonies of *A. tertia* can be distinguished from *A. gracilis* by tentacle number (10 vs. 8). Preserved colonies are best distinguished by stolon width (mean = 0.060 mm in *A. gracilis* vs. 0.077 mm in *A. tertia*) (Winston and Hayward 2012).

Occurrence Described from the USA, capable of living in estuarine conditions, more tolerant of warm temperatures than *A. imbricata*, but not as tolerant of extremely low salinities as *A. gracilis*.

The Netherlands Reported from the harbor of Rotterdam in September 2024 by Gittenberger et al. (2025).

Germany A colony of *Amathia tertia* with ten tentacles and pink embryos was collected in the marina of Cuxhaven in November 2023 by the author. A 12S sequence of this colony corresponds 100% to a sequence of *Amathia tertia* from 2016 placed on Genbank (Waeschenbach et al. 2012).

https://www.ncbi.nlm.nih.gov/nuccore/JN681104.1 (pers. comm. Arjan Gittenberger, GIMARIS).

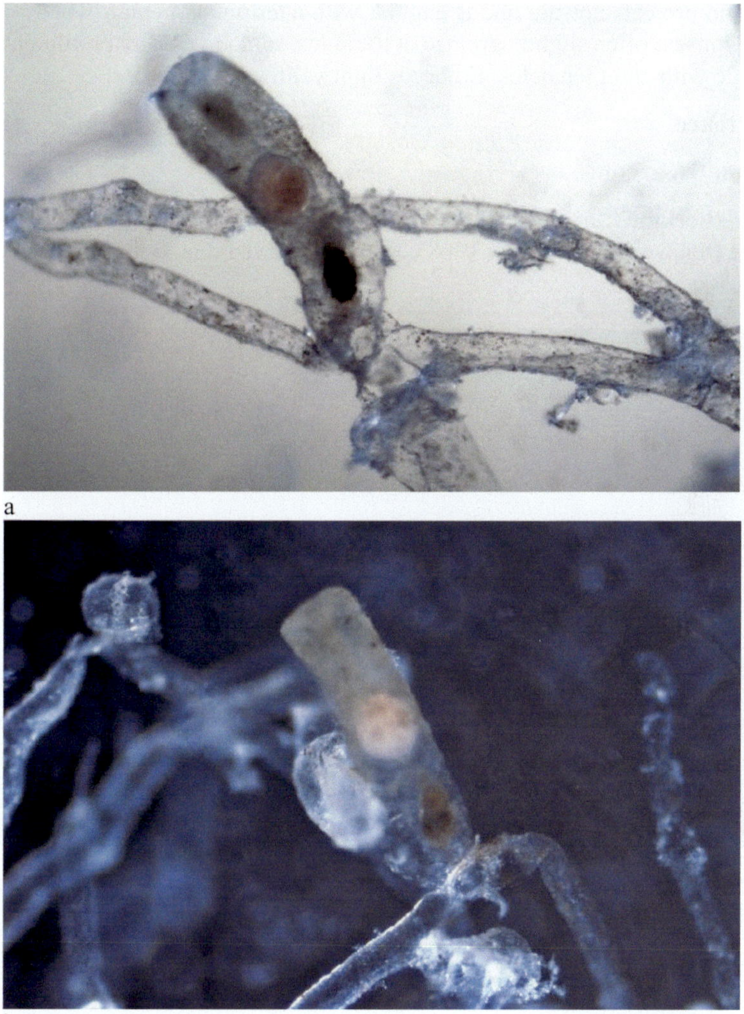

a

b

Fig. 3.39 *Amathia tertia.* (**a**, **b**) Cuxhaven (Germany)

Family Buskiidae

Genus *Buskia*

Buskia nitens Alder, 1857

Description Stolons branch inconspicuously over the substrate. They are very thin, only 0.05 mm thick, translucent, almost invisible, and closely attached to the substrate. Zooids ovate, 0.3–0.4 mm long, for the most part attached to the stolon. The distal end projects upward and is angular with a terminal opening. The frontal surface is convex, often slightly grained at the sides, sometimes with prominent bulges. Polypide with eight tentacles. Embryos light yellow (Fig. 3.40).

Occurrence

Belgium No records.

The Netherlands In Zeeland at Wemeldinge, Yerseke, and Westkapelle (Faasse and De Blauwe 2004; Jebram 1968). Rarely observed due to its small size. Grows

a

b

Fig. 3.40 *Buskia nitens*. (**a**, **b**) Part of colony on dead oyster shell, Yerseke (the Netherlands) on hydroids such as *Halecium*, *Hydrallmania*, erect bryozoans, and rarely on rocks and small algae. The colony shown is growing on a Pacific oyster.

Germany Collected at Heligoland (pers. comm. B. Kind).

References

Beukhof ED, Coolen JWP, van der Weide BE, Cuperus J, De Blauwe H, Lust J (2016) Records of five bryozoan species from offshore gas platforms rare for the Dutch North Sea. Marine Biodiversity Records 9:91

Braem F (1951) Über *Victorella* und einige ihrer nächsten Verwandten, sowie über die Bryozoenfauna des Ryck bei Greifswald. Zoologica 102:1–59

Cadman PS, Ryland JS (1996) Redescription of *Alcyonidium mytili* Dalyell 1848 (Bryozoa, Ctenostomatida). Zool J Linnean Soc 116:437–450

Carter MC (2004) The biology and genetic diversity of the trembling sea mat *Victorella pavida* (Bryozoa: Ctenostomata) from Swanpool, Falmouth. M. Res Thesis, University of Plymouth

Carter MC, Jackson A (2007) *Victorella pavida* Trembling Sea mat. In: Tyler-Walters H, Hiscock K (eds) Marine life information network: biology and sensitivity key information reviews. Marine Biological Association of the United Kingdom, Plymouth

Coolen JWP, Bos OG et al (2015) Reefs, sand and reef-like sand: a comparison of the benthic biodiversity of habitats in the Dutch Borkum reef grounds. J Sea Res 2015(103):84–92

d'Hondt J-L, Faasse MA (2006) A new *Arachnidium* species, *Arachnidium lacourti* spec. nov. (Bryozoa: Ctenostomona) from the coast of The Netherlands. Zoölogische Mededelingen Leiden 80–1(5):87–90

d'Udekem d'Acoz C (1991) Quelques données sur *Anguinella palmata* van Beneden, 1845 et sa faune associée en Belgique (Bryozoa, Ctenostoma: Nolellidae). De Strandvlo 11(2):57

d'Udekem d'Acoz C (1993) Existence d'hibernacles chez *Anguinella palmata* van Beneden, 1845 (Bryozoa, Ctenostomata, Nolellidae). De Strandvlo 13(4):146–148

De Blauwe H (2000) Riemwiervoetjes en hun begroeiing. De Strandvlo 20(3):130–137

De Blauwe H (2001) *Hypophorella expansa* Ehlers, 1876 een algemeen mosdiertje in aangespoelde kokers van de perkamentkokerworm *Chaetopterus variopedatus*. De Strandvlo 21(3):102–104

De Blauwe H (2003) Eerste melding van het mosdiertje *Nolella pusilla* (Hincks 1880) voor Nederland (Bryozoa: Ctenostomata). Het Zeepaard 63(1):27–30

De Blauwe H (2004a) De zeevingers *Alcyonidium condylocinereum* Porter, 2004 en *Alcyonidium diaphanum* (Hudson, 1778) (Bryozoa: Ctenostomatida) aan de Belgische kust. De Strandvlo 24(2):74–77

De Blauwe H (2004b) Schelpen anders bekeken. De Strandvlo 24(3/4):113–120

De Blauwe H (2005) Bryozoa op wieren en plastic uit de herfstvloedlijn van het najaar 2004. De Strandvlo 25(1):14–16

De Blauwe H (2009) Mosdiertjes van de Zuidelijke bocht van de Noordzee: Determinatiewerk voor België en Nederland. Vlaams Instituut voor de Zee (VLIZ), Oostende, p 445

De Blauwe H, Kerckhof F, Houziaux JS (2006) Een opmerkelijke diversiteit aan Bryozoa (mosdiertjes) op de Hinderbanken: een tussentijds verslag. De Strandvlo 26(4):125–134

de Boois I (1998) Verslag van excursie op Texel. Het Zeepaard 58(5):141–143

de Ruijter R (2012) CS-verslag 2012 (4). Het Zeepaard 72(5–6):116–121

de Ruijter R (2016) CS-verslag. Het Zeepaard 76(1):3–11

de Ruijter R (2017a) CS-verslag. Het Zeepaard 77(1):3–11

de Ruijter R (2017b) CS-verslag. Het Zeepaard 77(5):182–194

de Ruijter R (2018) CS-verslag. Het Zeepaard 78(5):173–183

de Ruijter R (2019a) CS-verslag. Het Zeepaard 79(2):43–55

de Ruijter R (2019b) CS-verslag. Het Zeepaard 79(3):75–86

de Ruijter R (2020a) CS-verslag. Het Zeepaard 80(1):5–13

de Ruijter R (2020b) CS-verslag. Het Zeepaard 80(5):199–209

Faasse MA, De Blauwe H (2003) Het mosdiertje *Arachnidium fibrosum* (Hincks, 1880) nieuw voor België en Nederland. De Strandvlo 23(2):47–49

Faasse MA, De Blauwe H (2004) Faunistisch overzicht van de mariene mosdiertjes van Nederland (Bryozoa: Stenolaemata, Gymnolaemata). Nederlandse Faunistische Mededelingen 21:17–54

Faasse M, Van Moorsel GWNM, Tempelman D (2013) Moss animals of the Dutch part of the North Sea and coastal waters of The Netherlands (Bryozoa). Nederlandse Faunistische Mededelingen 41:1–14

Franzèn A (1960) *Monobryozoon limicola* n. sp. a ctenostomatous bryozoan from the detritus layer on soft sediment. Zool Bidr Uppsala 33:135–147

Gittenberger A, Rensing M and Keeler-Perez H (2025) Marine alien species detection network of the Netherlands: 2024 results. Issued by the Office for Risk Assessment and Research, the Netherlands Food and Consumer Product Safety Authority of the Ministry of Agriculture, Nature and Food Quality. GiMaRIS report 2025_05: 52 pp

Gray JS (1971) Occurrence of the aberrant bryozoan *Monobryozoon ambulans* Remane, off the Yorkshire Coast. J Nat Hist 5(1):113–117. https://doi.org/10.1080/00222937100770051

Hayward PJ (1985) Ctenostome bryozoans. Synopses Br Fauna 33:1–169

Heerebout GR (1970) Verspreidingsecologie van de Bryozoa in het Deltagebied, speciaal in relatie tot het brakke water. Delta Instituut voor Hydrobiologisch Onderzoek, Yerseke

Holsteijn H (1998) Een dagje op het strand – Strandwandelverslag, Schiermonnikoog 21-3-1998. Het Zeepaard 58(3):65–68

Jebram D (1968) Zur Bryozoen-Fauna der Niederlande. Neth J Sea Res 4(1):86–94

Jebram D (1985) *Panolicella nutans* gen. et sp. n. (Bryozoa, Ctenostomata), its description, development and laboratory cultivation. Zool Scr 14(1):11–18

Johnson MJ, Saadi AJ, Kuklinski P, Smith AM, López-Gappa J, Schwaha T (2024) Digging into boring bryozoans: new characters and new species of Immergentiidae. Org Divers Evol 24:217. https://doi.org/10.1007/s13127-024-00645-y

Koulman A, de Bruin J, Begeman J (2003) Boeien in de Eemshaven: een eigen biotoop. Het Zeepaard 63(2):34–38

Kraepelin K (1887) Die deutschen Süsswasser-Bryozoen. Eine Monographie. I. Anatomisch-systematischer Teil. Abh Naturw Hamburg 10(9):1–168

Lacourt AW (1949) Bryozoa of The Netherlands. Arch Néerl Zool 8(3):1–33

Lacourt AW (1978) De Nederlandse mariene mosdiertjes: Bryozoa (Ectoprocta, Gymnolaemata). Wetenschappelijke Mededelingen van de Koninklijke Nederlandse Natuurhistorische Vereniging, 129. KNNV/Strandwerkgemeenschap: Hoogwoud, Nederland, p 1–21

Lengkeek W, Didderen K, Dorenbosch M, Bouma S, Waardenburg H (2013) Bureau Waardenburg bv and Stichting De Noordzee, rapport nr. 13–226, 76 ppx. Culemborg, Netherlands: Bureau Waardenburg bv and Stichting De Noordzee, rapport nr; 2013, p 1–13

Loppens K (1906) Bryozoaires marins et fluviatiles de la Belgique. Annales de la Société Royale Zoologique et Malacologique de Belgique 41:286–321

Maitland RT (1851) Fauna Belgii septentrionalis: systematische beschrijving der dieren welke in Noord-Nederland of aan deszelfs kusten voorkomen, met bijvoeging derzelver synonymen, benevens der plaatsen waar dezelve gevonden zijn, gerangschikt naar Prof. J. van der Hoeven's Handboek der Dierkunde (2de Uitgave). C.G. van der Hoek: Leyden, Nederland. XII, XXXVIII, p 1–234

Marcus E (1940) Mosdyr (Bryozoa eller Polyzoa). Danmark Fauna 46:1–401

Neumann H, Ehrich S, Kröncke I (2010) Establishment of the angular crab *Goneplax rhomboides* (Linnaeus 1758) (Crustacea, Decapoda, Brachyura) in the southern North Sea. Aquat Invasions 5(Suppl. 1):S27–S30

Ostrovsky AN, Porter JS (2011) Pattern of occurrence of supraneural coelomopores and intertentacular organs in Gymnolaemata (Bryozoa) and its evolutionary implications. Zoomorphology 130:1–15

Porter JS (2004) Morphological and genetic characteristics of erect subtidal species of *Alcyonidium* (Ctenostomata: Bryozoa). J Mar Biol Assoc UK 84(1):243–252

Porter JS, Ellis JR, Hayward PJ, Rogers SI, Callaway R (2002) Geographic variation in the abundance and morphology of the bryozoan *Alcyonidium diaphanum* (Ctenostomata; Alcyonidiidae) in UK coastal waters. J Mar Biol Assoc UK 82:529–535

Pröts P, Wanninger A, Schwaha T (2019) Life in a tube: morphology of the ctenostome bryozoan, *Hypophorella expansa*. Zool Lett 5:28. https://doi.org/10.1186/s40851-019-0142-2

Redant F (1976) Epibenthos. In: Nihoul JCJ and De Coninck L (ed) Project Sea final report: 7. Inventory of fauna and flora. Project Sea final report, 7, p 167–202

Rémane A (1936) *Monobryozoon ambulans* n. gen., n. sp., ein eigenartiges Bryozoon des Meeressandes. Zool Anz 113:161–167

Rémane A (1938) Remane A. Ergänzende Mitteilungen über *Monobryozoon ambulans* Remane. Kieler Meeresforsch 2:356–359

Reverter O, Fernandez E (2006) Does *Nolella pusilla* (Bryozoa: Ctenostomatida) actually exist? Cahier Biologie Maritime 47:237–242

Ryland JS, Porter JS (2000) *Alcyonidium reticulum* sp. nov., a common intertidal bryozoan from south-west Britain. J Mar Biol Assoc UK 80:563–564

Ryland JS, Porter JS (2003) The identity of *Alcyonidium gelatinosum* (Linnaeus 1761) (Bryozoa: Ctenostomatida). J Nat Hist 37:2179–2189

Ryland JS, Porter JS (2006) The identification, distribution and biology of encrusting species of *Alcyonidium* (Bryozoa: Ctenostomatida) around the coasts of Ireland. Biol Environ 106(1):19–33

Schwaha T (2021a) Handbook of zoology, phylum Bryozoa. De Gruyter

Schwaha T (2021b) Morphology of ctenostome bryozoans: 3. Elzerina, Flustrellidra, Bockiella. J Morphol 282(5):633–651. https://doi.org/10.1002/jmor.21334

Schwaha T, De Blauwe H (2020) Morphology of ctenostome bryozoans: 1. *Arachnidium fibrosum*. J Morphol 281(12):1598–1606

Schwaha T, Moosbrugger M, Walzl M, Ostrovsky AN (2019) First ultrastructural evidence of placental nutrition in a ctenostome bryozoan: example of *Amathia verticillata*. Zoomorphology 138:221–232

Schwaha T, Decker SH, Baranyi C, Saadi AJ (2024) Rediscovering the unusual, solitary bryozoan *Monobryozoon ambulans* Remane, 1936: first molecular and new morphological data clarify its phylogenetic position. Front Zool 21(1):5. https://doi.org/10.1186/s12983-024-00527-1

Tempelman D, van Moorsel GWNM, Faasse MA (2013) De molkreeften *Upogebia deltaura* en *U. stellata* en geassocieerde soorten in de Noordzee (Decapoda: Upogebiidae). Nederlandse Faunistische Mededelingen 41:15–29

van Benthem Jutting (1922) Bryozoën. In: Redeke HC (red.), Flora en fauna der Zuiderzee, monografie van een brakwatergebied. Nederlandsche Dierkundige Vereeniging, Den Helder, p 411–417

van der Sleen WGN (1920) Lijst der aan de Nederlandsche kust aangetroffen Nederlandsche evertebraten. Tijdschrift der Nederlandsche Dierkundige Vereeniging serie 2, 18: XXXIII–XXXIX

Vanagt T, Van de Moortel L and Faasse M (2013) Development of hard substrate fauna in the Princess Amalia wind farm. ecoast, Oostende [ecoast report 2011036]

Waeschenbach A, Taylor PD, Littlewood DT (2012) A molecular phylogeny of bryozoans. Mol Phylogenet Evol 62(2):718–735

Waeschenbach A, Vieira LM, Reverter-Gil O, Souto-Derungs J, Nascimento KB and Fehlauer-Ale KH (2015) A phylogeny of Vesiculariidae (Bryozoa, Ctenostomata) supports synonymization of three genera and reveals possible cryptic diversity Zoologica Scripta (online). https://doi.org/10.1111/zsc.12130

Waeschenbach A, Zhou Z, Schwaha T, Okamura B, Gordon DP, Wood TS, Ostrovsky AN, Todd JA, Decker SH, Johnson M, Al-Ghanem MM, De Blauwe H, Florence WK, Graham R, Hall A, Hartikainen H et al (Submitted) Molecular Phylogenetics and Evolution A genome skimming phylogeny of ctenostome bryozoans

Winston JE, Hayward PJ (2012) The marine bryozoans of the northeast coast of the United States: Maine to Virginia, vol 11. Virginia Museum of Natural History Memoir, pp 1–18

Zettler ML, Beermann J, Dannheim J, Ebbe B, Grotjahn M et al (2018) An annotated checklist of macrozoobenthic species in German waters of the north and Baltic seas. Helgol Mar Res 72(1):5. https://doi.org/10.1186/s10152-018-0507-5

Zintzen V (2007) Biodiversity of shipwrecks from the Southern Bight of the North Sea. PhD Thesis, University of Louvain, Department of Biology, p 343

Zintzen V, Massin C (2010) Artificial hard substrata from the Belgian part of the North Sea and their influence on the distributional range of species. Belg J Zool 140(1):20–29

Chapter 4
Cheilostomatida

Abstract This chapter discusses the order Cheilostomatida found in the southern bend of the North Sea, between Cap Gris-Nez in France and the German-Danish border. Overall, 160 species have been identified, 58 with opesia, 19 with cribrimorph costae, and 83 with a calcified frontal shield. New non-indigenous species, such as *Conopeum chesapeakensis*, *Electra* cf. *asiatica*, *Juxtacribrilina mutabilis*, *Metroperiella* cf. *biformis*, and *Schizoporella* cf. *variabilis*, have become established, and *Celleporaria brunnea* reached our beaches once, on plastic. Three new species of the superfamily Cribrilinoidea are described, namely *Cribrilina belgica* nov. sp., *Distanseścharella spirorbicola* nov. sp., and *Puellina monostachys* nov. sp. The species complex *Microporella* "*ciliata*" from the area comprises five new species: *Microporella annae* nov. sp., *Microporella dirupoi* nov. sp., *Microporella helgolandensis* nov. sp., *Microporella kindi* nov. sp., and *Microporella vanoutryvei* nov. sp. *Porella chadleyi* nov. sp. is also newly described. In addition to an identification key and illustrations, all other species are accompanied by a brief description, and the known distributional data are listed in detail.

Classification

Suborder Membraniporina

Superfamily Membraniporoidea

Family Membraniporidae

Membranipora membranacea (Linnaeus, 1767)
Biflustra tenuis (Desor, 1848)
Biflustra grandicella (Canu and Bassler, 1929)
Jellyella eburnea (Hincks, 1891)
Jellyella tuberculata (Bosc, 1802)

H. De Blauwe, *Marine and Brackish Bryozoans from the Southern Bight of the North Sea*, https://doi.org/10.1007/978-3-031-92513-9_4

Family Electridae

Electra monostachys (Busk, 1854)
Electra pilosa (Linnaeus, 1761)
Electra cf. *asiatica* Grischenko, Dick and Mawatari, 2007
Aspidelectra melolontha (Landsborough, 1852)
Conopeum chesapeakensis (Banta, Perez and Santagata, 1995)
Conopeum reticulum (Linnaeus, 1767)
Conopeum seurati (Canu, 1928)
Einhornia crustulenta (Pallas, 1766)
Pyripora catenularia (Fleming, 1828)

Suborder Aeteina

Superfamily Aeteoidea

Family Aeteidae

Aetea anguina (Linnaeus, 1758)
Aetea sica (Couch, 1844)
Aetea truncata (Landsborough, 1852)

Suborder Scrupariina

Superfamily Scruparioidea

Family Scrupariidae

Scruparia ambigua (d'Orbigny, 1841)
Scruparia chelata (Linnaeus, 1758)

Family Eucrateidae

Eucratea loricata (Linnaeus, 1758)

Suborder Flustrina

Superfamily Calloporoidea

Family Calloporidae

Callopora craticula (Alder, 1856)
Callopora discreta (Hincks, 1862)
Callopora dumerilii (Audouin, 1826)
Callopora lineata (Linnaeus, 1767)
Callopora rylandi Bobin and Prenant, 1965
Cauloramphus spinifer (Johnston, 1832)
Ammatophora nodulosa (Hincks, 1877)
Amphiblestrum auritum (Hincks, 1877)
Amphiblestrum flemingii (Busk, 1854)
Crassimarginatella solidula (Hincks, 1860)
Parellisina curvirostris (Hincks, 1862)
Tegella unicornis (Fleming, 1828)

Superfamily Flustroidea

Family Flustridae

Flustra foliacea (Linnaeus, 1758)
Chartella papyracea (Ellis and Solander, 1786)
Hincksina flustroides (Hincks, 1877)
Securiflustra securifrons (Pallas, 1766)

Superfamily Buguloidea

Family Bugulidae

Bugula neritina (Linnaeus, 1758)
Bicellariella ciliata (Linnaeus, 1758)
Bugulina avicularia (Linnaeus, 1758)
Bugulina flabellata (Thompson in Gray, 1848)
Bugulina fulva (Ryland, 1960)
Bugulina simplex (Hincks, 1886)
Bugulina stolonifera (Ryland, 1960)
Bugulina turbinata (Alder, 1857)
Crisularia plumosa (Pallas, 1766)

Family Beaniidae

Beania mirabilis Johnston, 1840

Family Candidae

Caberea boryi (Audouin, 1826)
Cradoscrupocellaria ellisi (Vieira and Spencer Jones, 2012)
Cradoscrupocellaria reptans (Linnaeus, 1767)
Scrupocellaria scrupea Busk, 1852
Scrupocellaria scruposa (Linnaeus, 1758)
Tricellaria inopinata d'Hondt and Occhipinti Ambrogi, 1985
Tricellaria ternata (Ellis and Solander, 1786)

Superfamily Microporoidea

Family Setosellidae

Setosella vulnerata (Busk, 1860).

Family Microporidae

Steraechmella buski Lagaaij, 1952

Superfamily Cellarioidea

Family Cellariidae

Cellaria fistulosa (Linnaeus, 1758)
Cellaria salicornioides Lamouroux, 1816
Cellaria sinuosa (Hassall, 1840)

Superfamily Cribrilinoidea

Family Cribrilinidae

Cribrilina cryptooecium Norman, 1903
Cribrilina punctata (Hassall, 1841)
Cribrilina belgica nov. sp.
Collarina gautieri Harmelin, 2019
Cribrilaria arrecta Bishop and Househam, 1987
Cribrilaria bifida (d'Hondt, 1970)
Cribrilaria directa (Bishop and Househam, 1987)
Cribrilaria innominata (Couch, 1844)
Cribrilaria venusta (Canu and Bassler, 1925)
Distansescharella seguenzai Cipolla, 1921
Distansescharella spirorbicola nov. sp.
Figularia figularis (Johnston, 1847)
Glabrilaria corbula Bishop and Househam, 1987
Juxtacribrilina mutabilis (Ito, Onishi and Dick, 2015)
Membraniporella nitida (Johnston, 1838)
Puellina modica Bishop and Househam, 1987
Puellina praecox Bishop and Househam, 1987
"Puellina" monostachys nov. sp.
"Puellina" nana Reverter-Gil and Fernandez Pulpeiro, 1996

Superfamily Hippothooidea

Family Hippothoidae

Hippothoa divaricata Lamouroux, 1821
Hippothoa flagellum Manzoni, 1870
Celleporella hyalina (Linnaeus, 1767)
Plesiothoa gigerium (Ryland and Gordon, 1977)

Family Chorizoporidae

Chorizopora brongniartii s.l.

Family Trypostegidae

Trypostega venusta (Norman, 1864)

Family Haplopomidae

Haplopoma graniferum (Johnston, 1847)
Haplopoma impressum (Audouin, 1826)

Superfamily Adeonoidea

Family Adeonidae

Reptadeonella insidiosa (Jullien, 1903)
Reptadeonella violacea (Johnston, 1847)

Superfamily Umbonuloidea

Family Bryocryptellidae

Palmiskenea skenei (Ellis and Solander, 1786)
Porella concinna (Busk, 1854)
Porella chadleyi nov. sp.

Family Romancheinidae

Escharella gilsoni De Blauwe, 2006
Escharella immersa (Fleming, 1828)
Escharella labiosa (Busk, 1856)
Escharella variolosa (Johnston, 1838)
Escharella ventricosa (Hassall, 1842)
Escharoides bishopi De Blauwe, 2006
Escharoides coccinea (Abildgaard, 1806)
Escharoides mamillata (Wood, 1844)
Neolagenipora collaris (Norman, 1867)

Family Umbonulidae

Oshurkovia littoralis (Hastings, 1944)
Umbonula ovicellata (Hastings, 1944)

Superfamily Smittinoidea

Family Smittinidae

Prenantia cheilostoma (Manzoni, 1869)
Smittina affinis (Hincks, 1862)
Smittina crystallina (Norman, 1867)
Smittina landsborovii (Johnston, 1847)
Smittoidea prolifica Osburn, 1952
Smittoidea reticulata (MacGillivray, 1842)
Parasmittina trispinosa (Johnston, 1838)
Phylactella labrosa (Busk, 1854)

Family Bitectiporidae

Hippoporina pertusa (Esper, 1796)
Metroperiella cf. *biformis* (Zhang and Liu, 1995)
Pentapora foliacea (Ellis and Solander, 1786)
Schizomavella (Calvetomavella) biancae Pica and Berning, 2022
Schizomavella auriculata (Hassall, 1842)
Schizomavella cristata (Hincks, 1879)
Schizomavella cornuta (Heller, 1867)
Schizomavella hastata (Hincks, 1862)
Schizomavella linearis (Hassall, 1841)
Schizomavella teresae Reverter-Gil and Fernandez Pulpeiro, 1996

Family Watersiporidae

Watersipora subatra (Ortmann, 1890)
Terwasipora complanata (Norman, 1864)

Superfamily Schizoporelloidea

Family Schizoporellidae

Schizoporella hesperia Hayward and Ryland, 1995 or *S. magnifica* Hincks, 1886
Schizoporella japonica Ortmann, 1890
Schizoporella unicornis (Johnston in Wood, 1844)
Schizoporella cf. *variabilis* (Leidy, 1855)
Schizoporella sp.
Schizobrachiella verrilli (Maturo and Schopf, 1968)

Family Fenestrulinidae

Fenestrulina delicia (Winston, Hayward, and Craig, 2000)
Fenestrulina sp.

Family Escharinidae

Escharina haywardi (Zabala, Maluquer, and Harmelin, 1993)
Escharina johnstoni (Quelch, 1884)
Escharina vulgaris (Moll, 1803)
Herentia hyndmanni (Johnston, 1847)
Phaeostachys spinifera (Johnston, 1847)

Family Pacificincolidae

Pacificincola perforata (Okada and Mawatari, 1937)
Primavelans glabricollaris Min, Chae, Yang, Noh, Lee, and Seo, 2021

Family Cryptosulidae

Cryptosula pallasiana (Moll, 1803)

Family Cheiloporinidae

Hagiosynodos latus (Busk, 1856)

Family Microporellidae

Microporella annae nov. sp.
Microporella dirupoi nov. sp.
Microporella helgolandensis nov. sp.
Microporella kindi nov. sp.
Microporella vanoutryvei nov. sp.

Superfamily Celleporoidea

Family Celleporidae

Cellepora pumicosa (Pallas, 1766)
Buffonellaria muriella Berning and Kuklinski, 2008
Celleporaria brunnea (Hincks, 1884)
Celleporina cf. *pygmaea* (Norman, 1868)
Celleporina decipiens Hayward, 1976
Celleporina hassallii (Johnston, 1847)
Celleporina sp.
Lagenipora lepralioides (Norman, 1868)
Omalosecosa ramulosa (Linnaeus, 1767)
Turbicellepora avicularis (Hincks, 1860)
Turbicellepora magnicostata (Barroso, 1919)

Family Hippoporidridae

Hippoporidra lusitania Taylor and Cook, 1981

Family Phidoloporidae

Rhynchozoon bispinosum (Johnston, 1847)
Schizotheca divisa (Norman, 1864)
Schizotheca fissa (Busk, 1856)

Specific Morphology

Cheilostome bryozoans are those with a box-like calcified exoskeleton (Fig. 4.1). In Chap. 1, we divided the Cheilostomatida into:

- Cheilostomatida where the frontal skeletal surface of the zooid has a large opening (an opesia) that is covered by a membranous frontal wall (Fig. 4.1a). The distal part of this membrane includes the operculum. The membrane can be underlain partly by a proximal calcified shelf (cryptocyst) (Fig. 4.1b). Overall, the opesia is a larger opening than the aperture closed by the operculum. This group used to comprise a heterogeneous taxonomic group known as Anasca (i.e., without an ascus).
- Cheilostomatida where the frontal skeletal surface is a shield of fused flattened spines (costae) (Fig. 4.1c, d).
- Cheilostomatida in which the entire frontal skeletal surface is calcified, except for the opening closed by the operculum (Fig. 4.1e, f). This kind of frontal shield may be exposed (gymnocystal) or covered by a membranous frontal wall underlain by a coelomic space. The latter comprises two major variants: umbonuloid and lepralioid.

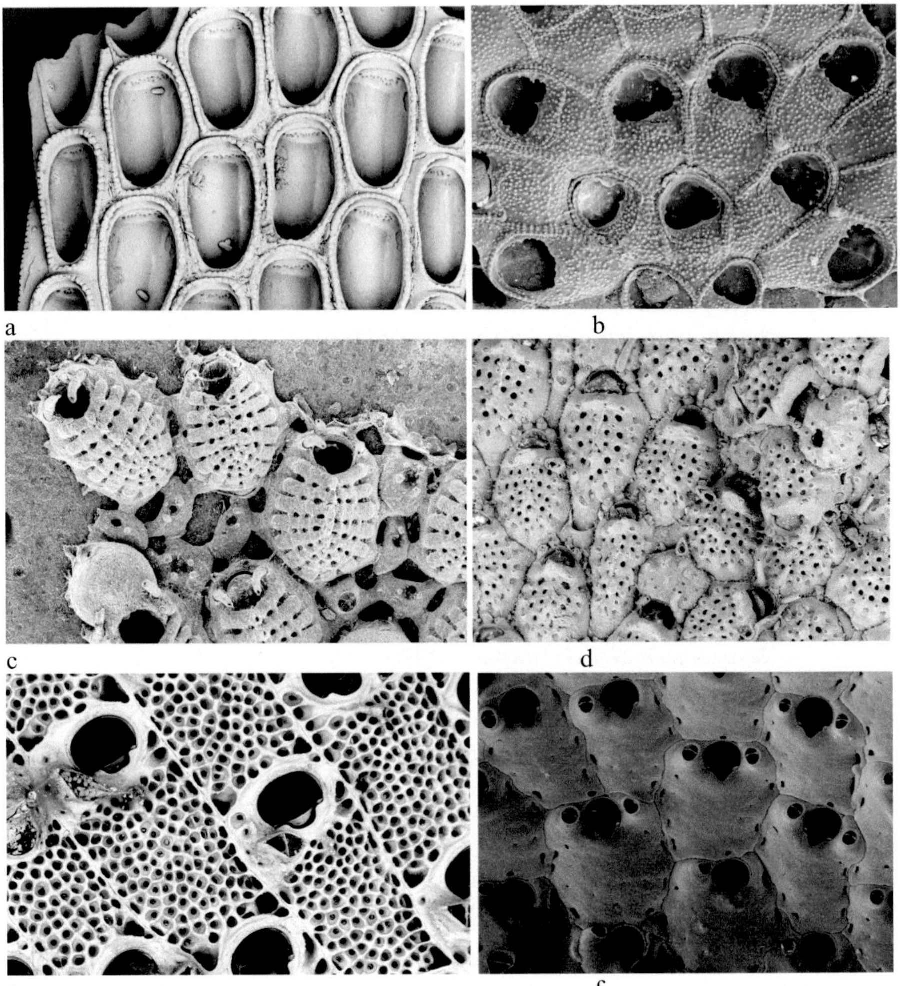

Fig. 4.1 (**a**) *Conopeum chesapeakensis*, large opesia. (**b**) *Biflustra tenuis*, opesia less than half the zooid length, proximal half of membrane underlain by cryptocystal calcification. (**c**) *Distanseschareila seguenzai* with clearly visible costae and intercostal pores. (**d**) *Collarina gautieri* with less obvious costae and intercostal and other pores. (**e**) *Schizoporella japonica* with the frontal surface evenly pseudoporous. (**f**) *Buffonellaria muriella* with areolae

In a gymnocystal shield, the frontal calcified surface is not covered by a membrane.

A lepralioid shield is formed as an internal partition (cryptocyst), overlain by a hypostegal coelom and underlain by the ascus and main coelomic chamber, respectively. Pores with communication cells connect to the main coelomic cavity (Fig. 4.1e).

Umbonuloid shields develop as an overarching fold above the primary membranous frontal wall, extending distally from the zooid proximal margin. The space between the fold and the frontal membrane functions analogously as an ascus. Pores at the zooid margins connect the hypostegal and main coeloms (Fig. 4.1f).

When describing the frontal skeletal surface, we often talk about gymnocyst and cryptocyst. These terms require a word of explanation. Cryptocystal calcification is present as a granular inner ring or narrow shelf around the opesia (Fig. 4.1a), or as a mainly proximally extended granular shelf (Fig. 4.1b), always beneath the membranous frontal wall. Gymnocystal calcification is generally smooth, and surrounds the opesia and cryptocyst; it is often more extensive proximally on the zooid.

A semicircular cuticular flap (**operculum**) hinged to the membranous frontal wall, closes the orifice. Muscles operate the elevation and closure of the operculum. As transverse parietal muscles inside the body cavity pull the frontal membrane inward, internal pressure rises above that of the surrounding seawater, longitudinal parietal muscles relax, and the lophophore everts smoothly. The lophophore retracts very swiftly by retractor muscles (Fig. 4.2).

Lateral walls between neighboring zooids have one to several small mural pore chambers. A mural pore chamber comprises a septulum (pore plate) with one or more communication pore(s) (Fig. 4.3a). In many species, larger, basal pore chambers occur in the angles of lateral/transverse and basal walls (Fig. 4.3b).

Heterozooids are individuals that do not have a protrusible tentacle crown and are unable to feed. They include kenozooids, which always lack an operculum, and avicularia (Fig. 4.4a–f), which are special heterozooids that occur only in Cheilostomatida. A specialized form of avicularium is known as a vibraculum (Fig. 4.4g).

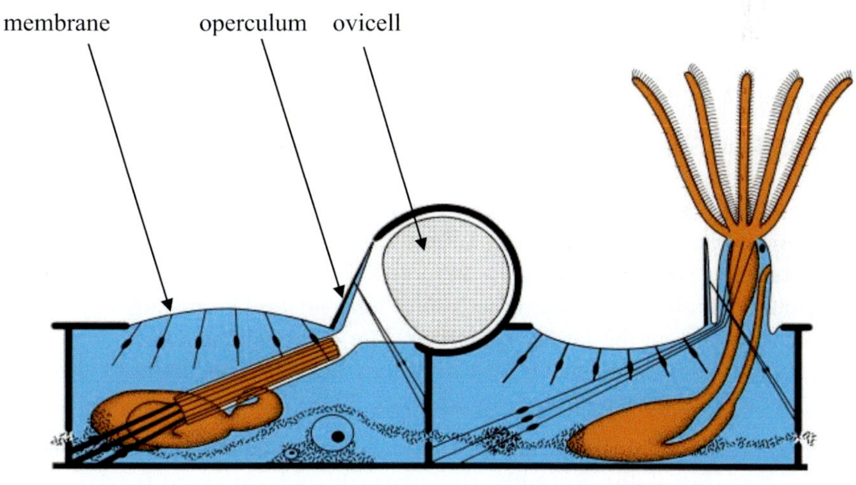

Fig. 4.2 Schematic image of the anatomy of two zooids with opesiae, showing gut and lophophore (orange, extended, and retracted), skeleton, muscles, funicular system, communication pores, ovary (left zooid), testis (right zooid), and an ovicell with developing embryo. (© Copyright Dr. Claus Nielsen (University of Copenhagen))

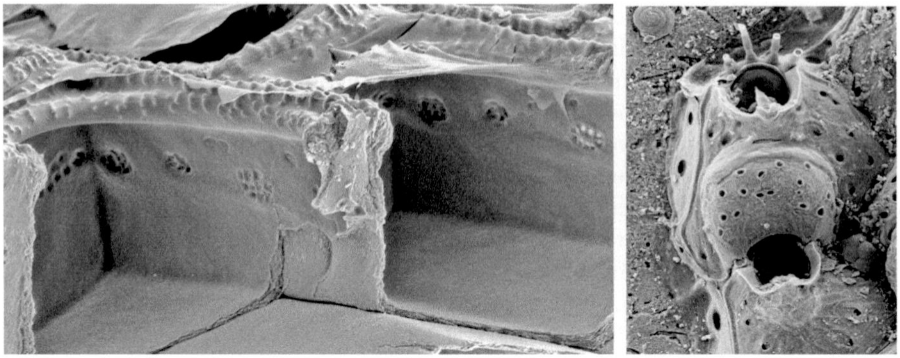

a b

Fig. 4.3 (**a**) *Biflustra grandicella*: lateral view of broken colony; lateral zooid walls with multiporous pore plates and line(s) of single communication pores in distal walls. (**b**) Two zooids of *Schizomavella cristata* showing pore chambers on the left side. Ovicellate autozooids with subimmersed ovicell whose pseudoporous ooecium is surrounded by the secondary calcification

The main function of avicularia and vibracula is thought to be defense, especially the deterrence or removal of unwanted organisms or substances that are a nuisance to the colony. In avicularia, the tentacle crown is reduced to a vestige. In kenozooids, the polypide is completely absent.

Avicularia are sessile (Fig. 4.4a) or pedunculated (Fig. 4.4b, c). The mandible can be compared to the operculum of autozooids. It is a kind of lid or flap that closes an opening and can only move in one plane. The rostrum (⟵) is the distal part of the avicularium closed by the mandible (⇨) (Fig. 4.4d). In avicularia of the "bird's head" type, the rostrum and mandible resemble the upper and lower jaw of a bird's beak (Fig. 4.4b).

Four types of avicularia are distinguished, three of which are defined by their budding position in the colony and the fourth by morphology:

- **Adventitious**: Smaller than the autozooids and develop on the lateral, frontal, or basal walls of an autozooid (Fig. 4.4a).
- **Interzooidal**: Smaller than, and squeezed in between, autozooids (Fig. 4.4e).
- **Vicarious**: Avicularia "replace" an autozooid in the budding sequence and have about the same size (Fig. 4.4f).
- **Vibracula**: Modified avicularia that bear a long bristle-like mandible; they are found principally in the family Candidae (Fig. 4.4g). Vibracula are defined by their unique morphology that may also differ from other avicularia by the rotational ability of their mandibles (Schack et al. 2018).

Other heterozooids are kenozooids. They form root-like structures and certain kinds of spines (Fig. 4.4h). Root-like structures (rhizoids) are used for the attachment of the colony, whereas spines may be used for protection. Other kenozooids are space fillers and can be common in the basal parts of robustly calcified erect colonies.

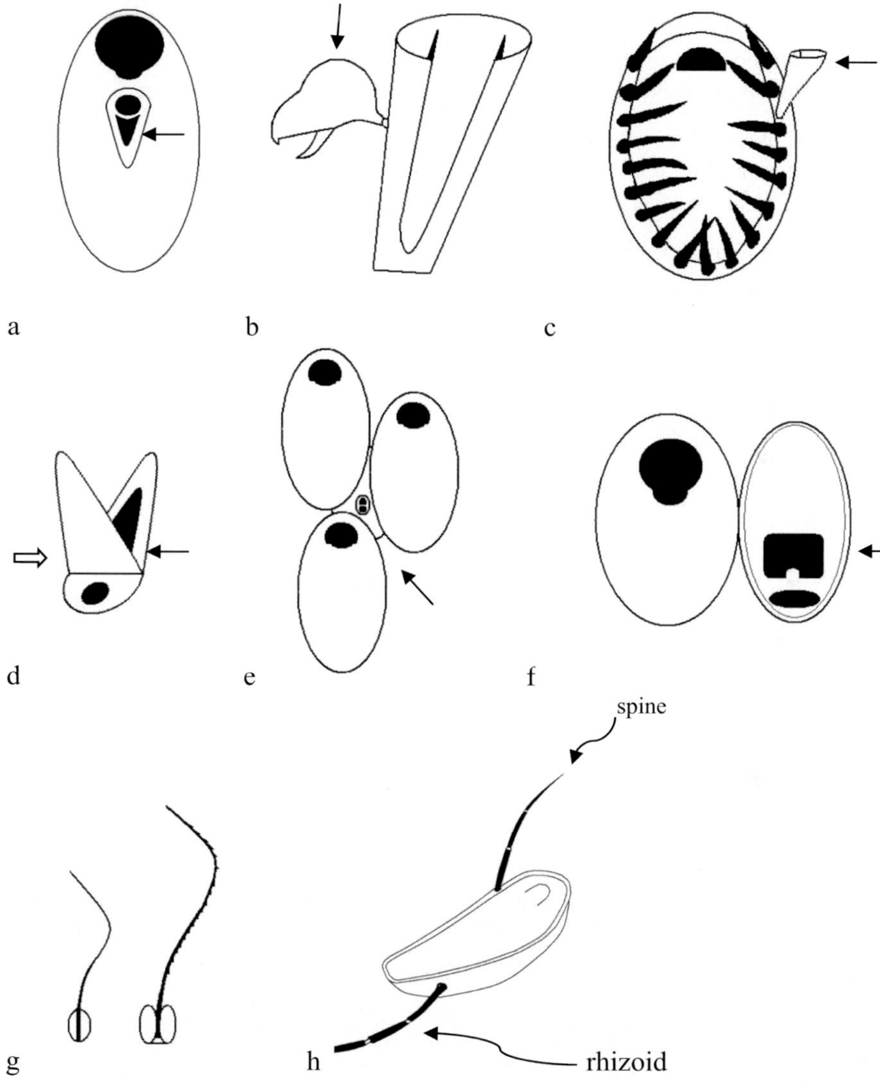

Fig. 4.4 Avicularia. (**a**) Sessile. (**b**) Pedunculated bird's head type. (**c**) Pedunculated. (**d**) Mandible and rostrum. (**e**) Interzooidal avicularium. (**f**) Vicarious avicularium. (**g**) Vibracula. (**h**) Rhizoid and spine

Not all spines are kenozooidal, only those that have a tiny septulum at the point of attachment. Some are connected to the autozooid by an uncalcified joint that makes the spine flexible instead of rigid. Other spines are body-wall projections of the parent zooid. Spines are hollow and may break off easily. Spines prevent predators such as sea slugs from moving across the colony. They are usually present at the colony edge on young autozooids. On the basis of the remaining stumps (Fig. 4.4f), one can count the original number of spines per autozooid. Side lighting is particularly useful because the spines stand out with an exposed side and a shadow side.

For a detailed classification of spines (and other polymorphs), see Schack et al. (2018). The commonest locations of spines illustrated in this book are:

- Opesial: Around the opesia (Fig. 4.5a)
- Oral: Surrounding the orifice laterally and distally (Fig. 4.5b, f)
- Distal: Distal to the orifice (Fig. 4.5c)

A **scutum** (Fig. 4.5e) is a broadened spine that curves over the frontal surface. It can be club-shaped, oval, or branched (bifid or cervicorn). Scuta are characteristic of many species in the family Candidae.

Ovicells (Figs. 4.2 and 4.5g), including their skeletal component (ooecia), are brood chambers with their incubation cavity external to the main zooidal body cavity (Ostrovsky 2013).

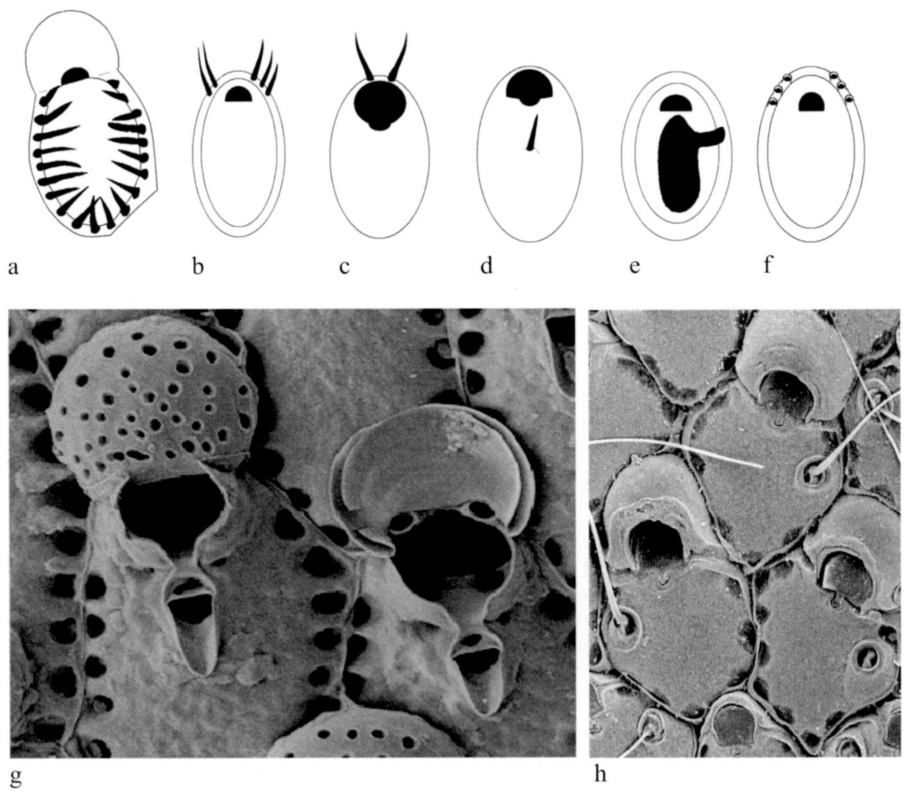

Fig. 4.5 **(a–f)** Spines, showing variations in placement and size. **(g)** *Smittoidea reticulata*: left, fully formed prominent ovicell; right, developing ooecium; its fold resting on and budded from the daughter zooid. **(h)** *Herentia hyndmanni*, operculum closes the opening of the subimmersed ovicell

Two major types of ovicells occur:

- The ooecium is produced by a distal autozooid (type A) or a distal kenozooid (visible, type B, or non-visible, type C, from the above). In all these types, the brood-chamber floor is formed by the daughter zooid, and the ovicell opening is either closed (cleithral) or not closed (acleithral) by the zooidal operculum. Depending on the immersion of the brood cavity, prominent (hyperstomial) (Fig. 4.5g, h), subimmersed, immersed, and endozooidal ovicells are distinguished.
- The ooecium is a kenozooid budded by the maternal zooid; the brood cavity is deeply immersed in the maternal zooid (e.g., *Cauloramphus*).

Cribrimorph zooids have a **costate shield**, made up of **hollow non-jointed spines or costae**. The gaps between the costae are **lacunae** (true pores), and the round pores leading into the interior of the costae are a type of window termed **pelmatidia** (or **pelmata** if large). The pair of distal costae forms an **apertural bar**.

Key to the Genera of Cheilostomatida

1. (a) The colony consists of uniserial chains of autozooids (Fig. 4.6a). 2
   (b) Colony biserial or multiserial (Fig. 4.6b). 7

Uniserial Chains of Autozooids

2 (1) (a) Branched chains of fusiform zooids lying on the substrate, each with an upright distal portion resembling a tube lamp (Fig. 4.6c). .... *Aetea* (p. 152)
(b) Not as described. 3

3 (2) (a) Frontal surface with an uncalcified oval zone (opesia) (Fig. 4.6d). 4
(b) Frontal surface completely calcified except for the orifice (Fig. 4.6e). 6
(c) Cribrimorph shield (Fig. 4.8f). .... *Puellina monostachys* nov. sp. (p. 264)

4 (3) (a) Colony consists of erect, branched chains of horn-shaped autozooids, on an adnate zooid chain (Fig. 4.6f) or a slender adnate stolon (Fig. 4.6g). .... *Scruparia* (p. 156)
(b) Colony attached to the substratum with rhizoids. Autozooids with a tubular slender proximal portion and an upright coal scoop-shaped distal portion surrounded by spines (Fig. 4.6h). .... *Beania mirabilis* (p. 224)
(c) Colony completely adnate (Fig. 4.6i). 5

5 (4) (a) Branched chains of thickly calcified, teardrop-shaped autozooids. The oval frontal membrane occupies almost half of the total zooid length (Fig. 4.6j). .... *Pyripora catenularia* (p. 184)

(b) Similar but thinner calcified autozooids, mostly with spines unless damaged. The autozooids make an oval scar in the substrate (Fig. 4.6k).. *Electra monostachys* (p. 176)

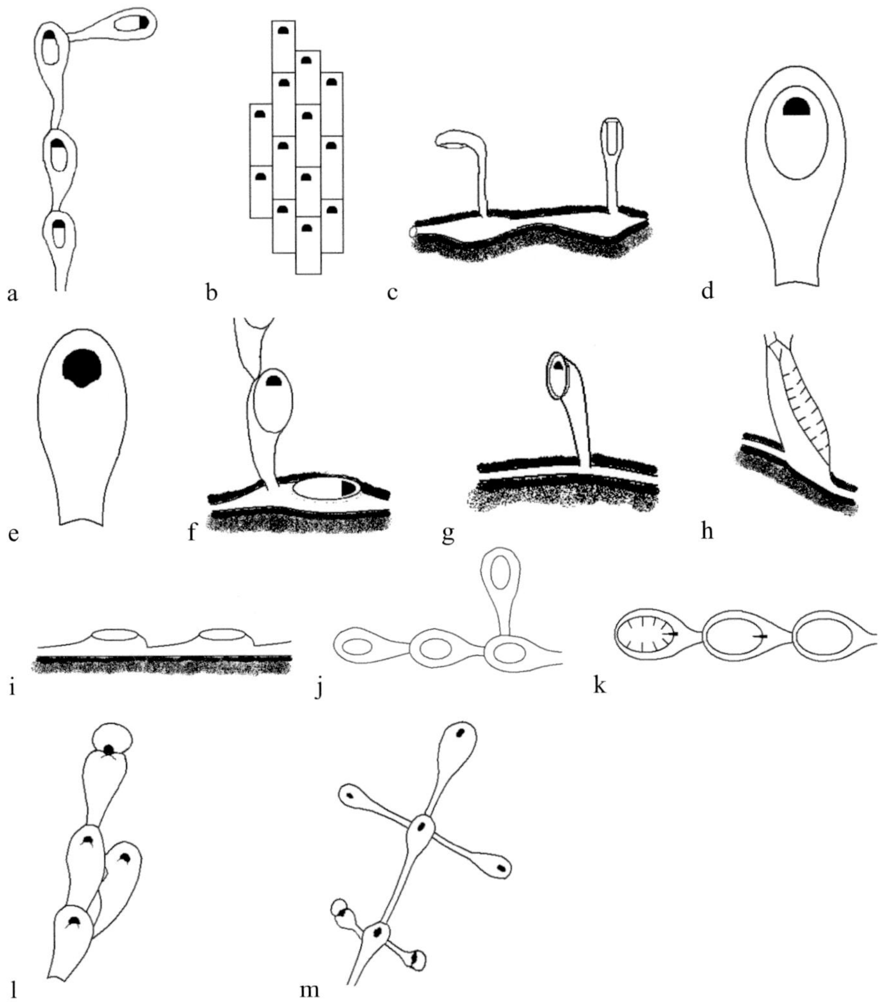

Fig. 4.6 Uniserial versus multiserial colonies: (**a**) Uniserial colony. (**b**) Multiserial colony. (**c**) *Aetea*. (**d**) Autozooids with opesia. (**e**) Autozooids with a completely calcified shield. (**f**) Adnate zooid chain. (**g**) Adnate stolon. (**h**) *Beania mirabilis*. (**i**) Completely adnate uniserial colony. (**j**) *Pyripora catenularia*. (**k**) *Electra monostachys*. (**l**) *Plesiothoa gigerium*. (**m**) *Hippothoa*

6 (3) (a) Autozooids elongate oval, in irregularly branched chains, adjacent autozooids sometimes connected by short tubes (Fig. 4.6l). .... *Plesiothoa gigerium* (p. 278)
(b) Autozooids fusiform with a long thread-like proximal part. The chains branch in a cross-shaped pattern (Fig. 4.6m). .... *Hippothoa* (p. 273)

Bi- or Multiserial Colonies

7 (1) (a) Colony erect. 8
(b) Colony encrusting, as lying sheets or as clumps, or a honeycomb-like structure of erect connected plates. 18

Erect Colonies

8 (7) (a) Colony flexible and branched. Bush-shaped, fan-shaped, or like a twig. (Fig. 4.7a, b). 9
(b) Colony rigid, branched like deer antlers. .... *Omalosecosa ramulosa* (p. 395)
(c) Colony rigid, branched flattened fronds. .... *Palmiskenea skenei* (p. 287)

9 (8) (a) Colony stiff, the white branches consist of slender stem segments, connected by yellow-brown cuticular joints. Autozooids calcified, covered by a glistening frontal membrane, orifices around the branch segments (Fig. 4.7d). .... *Cellaria* (p. 241)
(b) Colony consists of one or two layers of autozooids with orifices on one or two sides. 10

10 (9) (a) Colony consists of one layer of autozooids, with orifices on one side (Fig. 4.7e). 11
(b) Colony two layers of autozooids, with orifices on both sides (Fig. 4.7f). 16

11 (10) (a) Avicularia present (Fig. 4.7h, i). 12
(b) Avicularia absent (Fig. 4.7g). .... *Bugula neritina* (p. 213)

12 (11) (a) Avicularia with a broad base connected to the autozooid (Fig. 4.7i). 13
(b) Avicularia like a stalked bird's head (Fig. 4.7h). 15

13 (12) (a) Avicularia and vibracula present (Fig. 4.7j, k). 14
(b) Vibracula absent. Large, lateral avicularia, as wide as an autozooid (Fig. 4.7i). .... *Tricellaria* (p. 234)

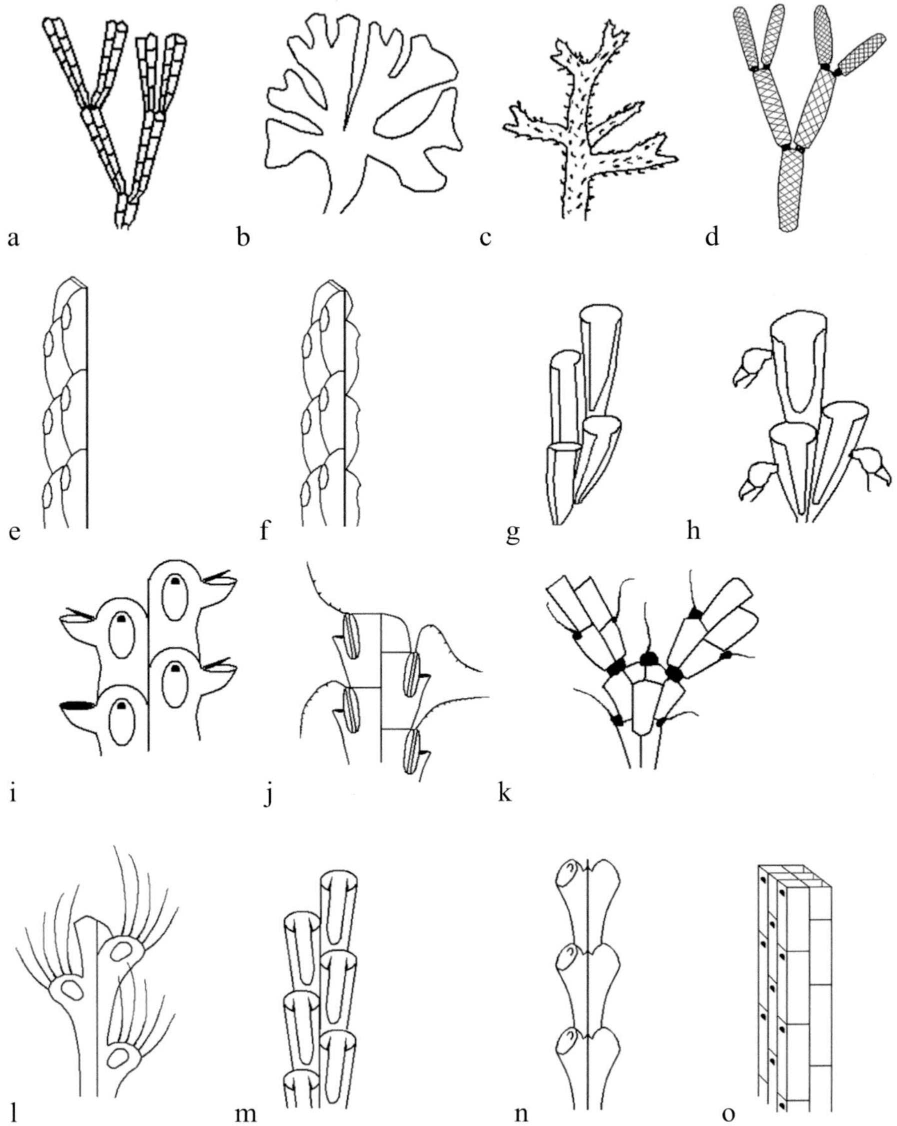

Fig. 4.7 Colony shape of erect growing colonies: (**a**, **b**) Flexible and branched colony. (**c**) Rigid colony. (**d**) *Cellaria*. (**e**) One layer of zooids. (**f**) Two layers of zooids growing back to back. (**g**) *Bugula neritina*. (**h**) Avicularia of bird's head type. (**i**) Large lateral avicularia. (**j**) *Caberea*, vibracula with toothed setae. (**k**) Small vibracula with toothless setae. (**l**) *Bicellariella ciliata*. (**m**) *Bugula*, *Bugulina*, and *Crisularia*. (**n**) *Eucratea loricata*. (**o**) Colony with erect bilaminar fronds

14 (13) (a) Basal surface with vibracula with toothed setae (Fig. 4.7j). .... *Caberea* (p. 225)
(b) Small vibracula with toothless setae, lateral and in ramifications (Fig. 4.7k). ....*Cradoscrupocellaria* and *Scrupocellaria* (p. 228)

15 (12) (a) Colony branches biserial. Autozooids with four to nine long, curved spines, at the distal end of an oval opesia (Fig. 4.7l). .... *Bicellariella ciliata* (p. 224)
(b) Colony branches biserial or multiserial, opesia elongated. Proximal ends of autozooids, in basal view, clearly bifurcated. Spines only in distal corners of the autozooids (Fig. 4.7m). .... *Bugula, Bugulina, and Crisularia* (p. 208)

16 (10) (a) Colony branches are slender, consisting of elongated autozooids in back-to-back pairs, with an oval opesia occupying about half the zooid length. The colony resembles a miniature poplar (Fig. 4.7n). .... *Eucratea loricata* (p. 159)
(b) Colony consists of erect bilaminar, branching fronds (Fig. 4.7o). 17

17 (16) (a) Autozooids urn-shaped, proximal gymnocyst with pseudopores, opesia surrounded by spines, proximal spine constant (Fig. 4.8a). .... *Electra pilosa* (p. 179)

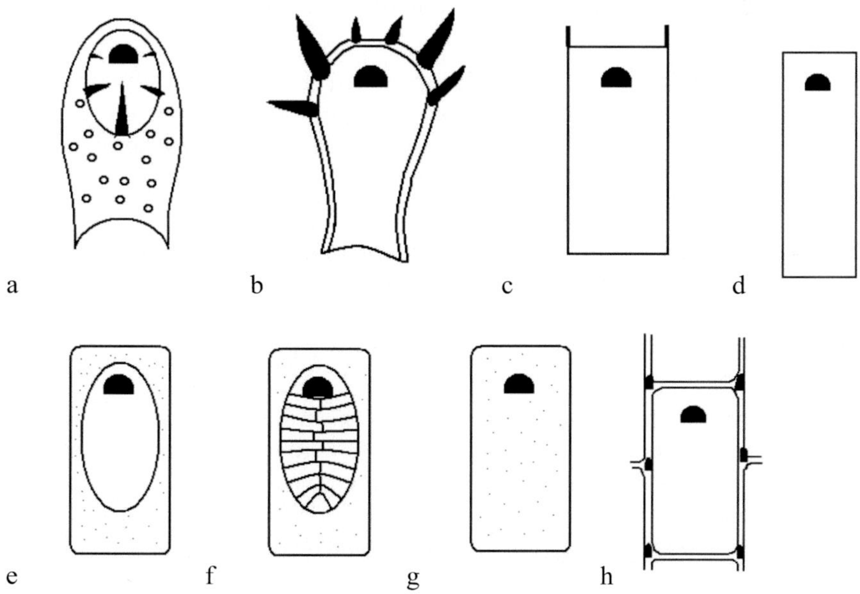

Fig. 4.8 (**a**) *Electra pilosa*. (**b**) *Flustra foliacea*. (**c**) *Chartella papyracea*. (**d**) *Securiflustra securifrons*. (**e**) Frontal surface with an opesia. (**f**) Opesia covered by a shield of fused flat spines. (**g**) Frontal surface completely calcified. (**h**) *Membranipora membranacea*

(b) Autozooids tongue-shaped, distal spines (Fig. 4.8b). .... *Flustra foliacea* (p. 187)

(c) Autozooids rectangular, with a pair of short distal spines (Fig. 4.8c). Colony papery with rounded fronds. .... *Chartella papyracea* (p. 188)

(d) Autozooids elongated rectangular, without spines (Fig. 4.8d). Colony papery with wedge-shaped fronds. .... *Securiflustra securifrons* (p. 189)

(e) In brackish water. Side branches flattened tubular, comprising four or more rows of zooids, widening into flat bilaminar fronds. No spines, ovicells, or avicularia. .... *Conopeum chesapeakensis* (p. 171)

Encrusting Colonies

18 (7) (a) Frontal surface with an opesia (uncalcified zone larger than the operculum) (Fig. 4.8e). 19

(b) Opesia covered by a shield of fused flat spines (Fig. 4.8f). 72

(c) Frontal surface completely calcified, except for the orifice and possibly large pores (Fig. 4.8g). A membrane may lie over the calcified frontal surface. 38

Encrusting Colonies with an Opesia in the Frontal Shield

19 (18) (a) Colony attached to the cephalopod shell of *Spirula*. Crudely constructed gymnocystal spines originate through the folding of the proximal gymnocyst.. *Jellyella eburnea* (p. 169)

(b) Not on *Spirula*. Gymnocyst not folded to crude spines. 20

20 (19) (a) Avicularia absent. (Note: the common *Conopeum reticulum* has kenozooids that can be mistaken for avicularia by beginners.) 21

(b) Avicularia present, sometimes only a few per colony. 31

21 (20) (a) Autozooids elongated rectangular, the frontal surface is completely membranous except possibly a few rounded proximal tubercles (Fig. 4.8h). Forms lacy colonies on brown algae. Also on green algae on plastic or pebbles. .... *Membranipora membranacea* (p. 161)

(b) Not like that, or if very similar, then not on brown algae. 22

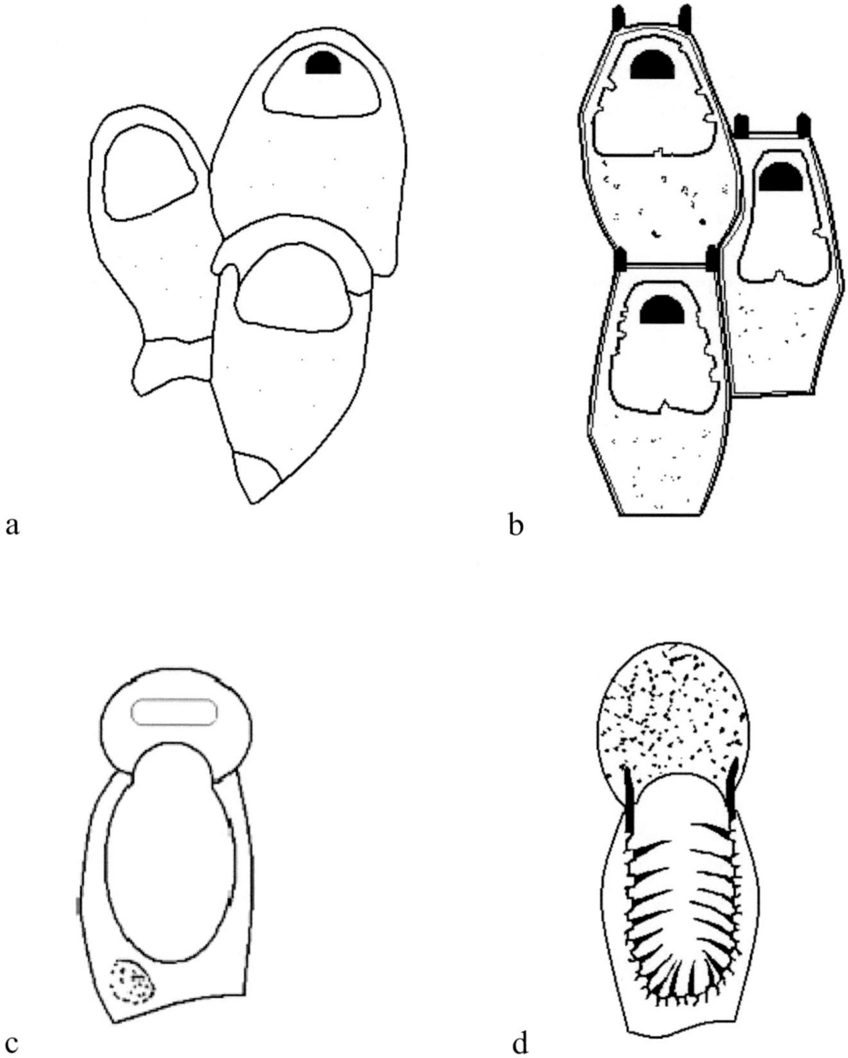

Fig. 4.9 (**a**) *Ammatophora nodulosa*. (**b**) *Biflustra*. (**c**) *Crassimarginatella solidula*. (**d**) *Callopora discreta*

22 (21) (a) A large cryptocystal plate lays under the frontal membrane, it reduces the opesia especially proximally (Fig. 4.9a, b). 23
(b) Opesia not reduced by a cryptocystal calcareous layer under the frontal membrane (Fig. 4.9c, d). 27

23 (22) (a) Ovicells do occur. Opesia semielliptical, semicircular, or trifoliate. 24
(b) Ovicells never occur. Opesia oval or rounded rectangular, never trifoliate. Ancestrula is a twin. 25

24 (23) (a) Opesia semicircular, half of the zooid length. Nodular kenozooids occur between the autozooids (Fig. 4.9a). .... *Ammatophora nodulosa* (p. 208)
(b) Opesia smaller, with a straight or convex proximal edge. Kenozooids with a reduced oval opesia. .... *Steraechmella buski* (p. 238)

25 (23) (a) On oceanic flotsam: for example, plastic or goose barnacles. Resembling *Membranipora membranacea*, with longer proximal tubercles. .... *Jellyella tuberculata* (p. 167)
(b) One or more layers of zooids on shells or pebbles or forming a honeycomb structure of plates of zooids, consisting of two layers growing back to back. Cryptocyst underlying frontal membrane mainly present proximally. At the edge of the oval opesia, minuscule spines may point toward the center of the opesia (Fig. 4.9b). 26

26 (25) (a) Forming a layer or thick crust of several layers on shells or pebbles. Cryptocystal ledge surrounding the opesia is mainly well developed proximally and laterally, its edge with the opesia is finely serrated. .... *Biflustra tenuis* (p. 163)
(b) Forming large honeycomb structures resembling *Pentapora foliacea*. Spines on the edge of the cryptocystal plate pointing to the center of the opesia are rare. .... *Biflustra grandicella* (p. 164)

27 (22) (a) Ovicells may be present. 28
(b) Ovicells never present. 29

28 (27) (a) Opesia elliptical, with a narrow calcareous margin and one or more low, rounded tubercles. Up to four spines in young zooids. Ovicell smooth, with a membranous frontal area (Fig. 4.9c). .... *Crassimarginatella solidula* (p. 200)
(b) Opesia surrounded by 17 to 22 spines with thickened bases. Ovicells uniformly granular (Fig. 4.9d). ....*Callopora discreta* (p. 194)

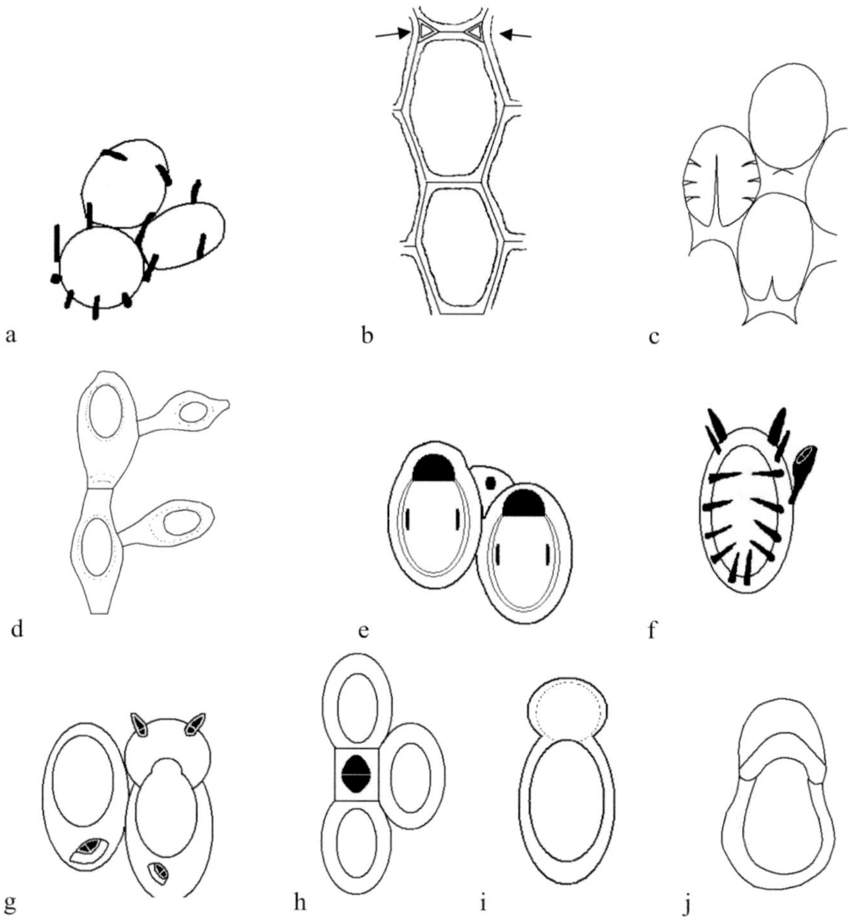

Fig. 4.10 (**a**) Young colony of *Flustra foliacea*. (**b**) *Conopeum reticulum* with kenozooids (arrows). (**c**) *Electra/Einhornia*. (**d**) *Pyripora catenularia*. (**e**) *Setosella vulnerata*. (**f**) *Cauloramphus spinifer*. (**g**) Avicularia on the autozooids and ovicells. (**h**) Avicularium in between the autozooids, its base on the substratum. (**i**) *Callopora*. (**j**) *Amphiblestrum*

29 (27) (a) Ancestrula with eight spines, zooids without gymnocyst and with two distal, cylindrical, vertical spines (Fig. 4.10a); young colony of............ .... *Flustra foliacea* (p. 187)
(b) Spines pointed, or very thin, or very long or absent. 30

30 (29) (a) Frontal view as in Fig. 4.10b. Sometimes small kenozooids (arrows) between the autozooids, often absent. Short slender lateral spines present under sheltered conditions. ……................ *Conopeum* (p. 170)

(b) Frontal view as in Fig. 4.10c. The proximal calcareous layer occupies one-sixth to one-half of the frontal surface and may have pores. If spines are present, the one at the proximal end of the opesia is the most prominent. .... *Electra/Einhornia* (p. 174)
(c) Frontal view as in Fig. 4.10d. Autozooids in one or more series, sometimes as a non-contiguous crust in dense colonies. Looks like thickly calcified *Electra monostachys* (p. 176). .... *Pyripora catenularia* (p. 184)

31 (20) (a) Avicularia small, interzooidal, with long thread-like mandible. Proximal to the D-shaped opening are two small opesiules in the calcium layer below the frontal membrane (Fig. 4.10e). .... *Setosella vulnerata* (p. 238)
(b) Not as described. 32

32 (31) (a) Opesia bordered by numerous thin, pointed spines. Avicularia pedunculate, club-shaped, with terminal mandibles, resembles a thicker spine between the other spines (Fig. 4.10f). Not present on all autozooids. Ovicells cap-like, inconspicuous. .... *Cauloramphus spinifer* (p. 200)
(b) Avicularia not pedunculate. 33

33 (32) (a) Avicularia on the zooids, their base is not on the substrate (Fig. 4.10g). 34
(b) Avicularia in between the zooids, their base is on the substrate (Fig. 4.10h)... 35

34 (33) (a) Ovicell without a frontal ridge (Fig. 4.10i) or with a transverse membranous area bounded by a sinuous ridge; 1 to 4 spines at the distal end, or up to 20 around the oval opesia. Basal pore chambers present. .... *Callopora* (p. 193)
(b) Ovicell with a frontal ridge (Fig. 4.10j). Up to four spines at the distal end. Opesia three-lobed or oval. Basal pore chambers present. .... *Amphiblestrum* (p. 205)
(c) Ovicell with a frontal ridge. Pore chambers absent. Lateral walls with multiparous septula. .... *Tegella unicornis* (p. 203)

35 (33) (a) Avicularium with semielliptic mandible (Fig. 4.11a). 36
(b) Avicularium with curved, pointed mandible (Fig. 4.11b). .... *Parellisina curvirostris* (p. 201)

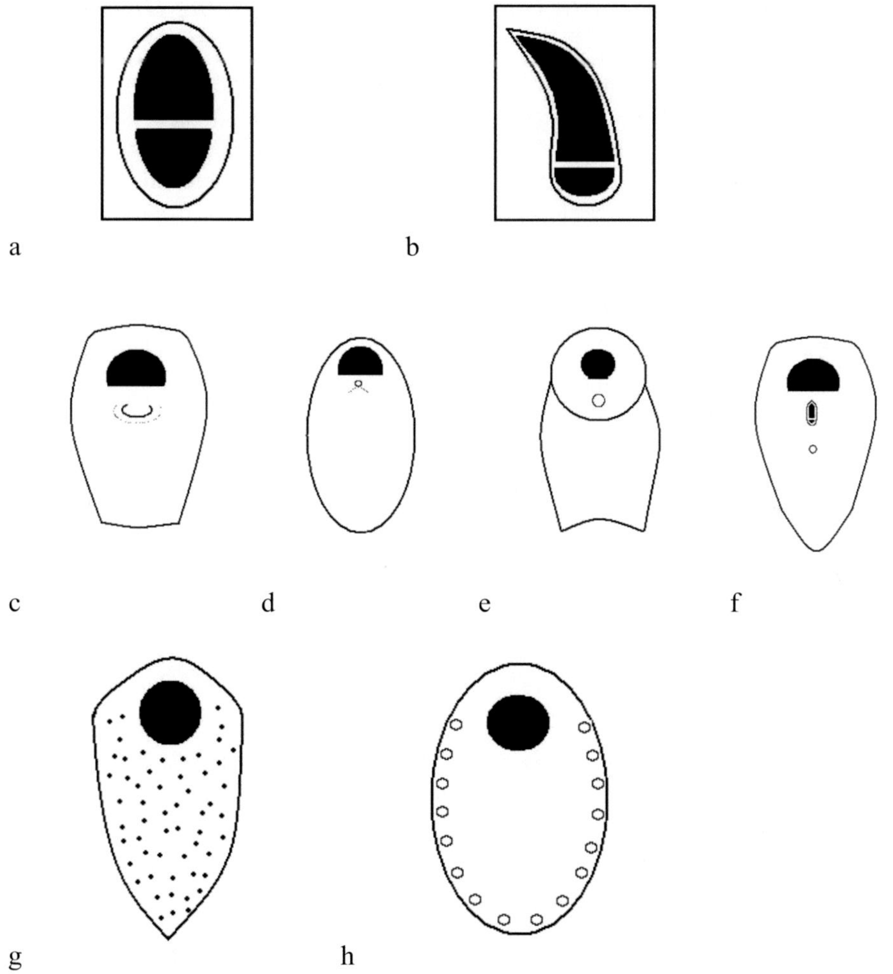

Fig. 4.11 (**a**) Semielliptic mandible. (**b**) *Parellisina curvirostris*, with curved, pointed mandible. (**c**) Ascopore crescentic. (**d**) *Haplopoma*. (**e**) Pacificincolidae. (**f**) *Reptadeonella violacea*. (**g**) Frontal shield evenly perforated. (**h**) Only areolae

36 (35) (a) Avicularium as large as an autozooid, rare. Autozooids spineless except in young zooids. Autozooids with one or two rounded knobs on the proximal frontal calcification. .... *Crassimarginatella solidula* (p. 200)
(b) Avicularia much smaller than zooid, frequent. Zooid bears spines. 37

37 (36) (a) Frontal membrane bordered by usually 10 to 14 short stout spines. Ovicells endozooidal. .... *Hincksina* (p. 191)
(b) Frontal membrane bordered by only two distal spines. .... *Flustra foliacea* (p. 187)

Encrusting Colonies with Calcified Shield, Except the Orifice

38 (18) (a) Zooids with a small hole proximal to the orificial aperture (ascopore, pore opening into an ascus, see Fig. 4.11c–e). If necessary, tilt the colony to inspect the perpendicular side of the peristome (not a spiramen, a central pore opening into the space between the overarching umbonuloid shield and the primary membranous frontal wall as in *Reptadeonella violacea* Fig. 4.11f) 39
(b) No ascopore proximal to the orifice. 42

39 (38) (a) Autozooids with an avicularium with a slender rostrum and mandible, proximolateral to the ascopore. .... *Microporella* (p. 374)
(b) Autozooids without avicularia. 40

40 (39) (a) Ascopore crescentic (Fig. 4.11c). .... *Fenestrulina* (p. 351)
(b) Ascopore round or oval (Fig. 4.11d, e). 41

41 (40) (a) Ascopore in the peristome around the orifice (Fig. 4.11e). ....
.... Pacificincolidae (p. 363)
(b) Ascopore on the distal side of a small bump proximal to the orifice (Fig. 4.11d). .... *Haplopoma* (p. 283)

42 (38) (a) Frontal shield evenly perforated (Fig. 4.11g). 43
(b) Only marginal pores (areolae) (Fig. 4.11h) or not perforated. ....
.... 53

43 (42) (a) Avicularia absent. 44
(b) Avicularia present, but sometimes not on all zooids. 48
(c) Avicularia very rare, at the distal end of the autozooid at either side of the orifice pointing outward. ....
.... *Schizobrachiella verrilli* (p. 350)

44 (43) (a) Orifice with a lyrula (Fig. 4.12a, b). 45
(b) Orifice without lyrula (Fig. 4.12c, d). 46

45 (44) (a) Orifice with a broad, rectangular lyrula and a low, thickened peristome with a deep medioproximal notch (Fig. 4.12e). ....
.... *Prenantia* and *Smittina* (p. 309)
(b) Orifice with narrow, pointed lyrula and a broad, deep, flared peristome, especially prominent proximally (Fig. 4.12f). ....
.... *Phylactella labrosa* (p. 318)

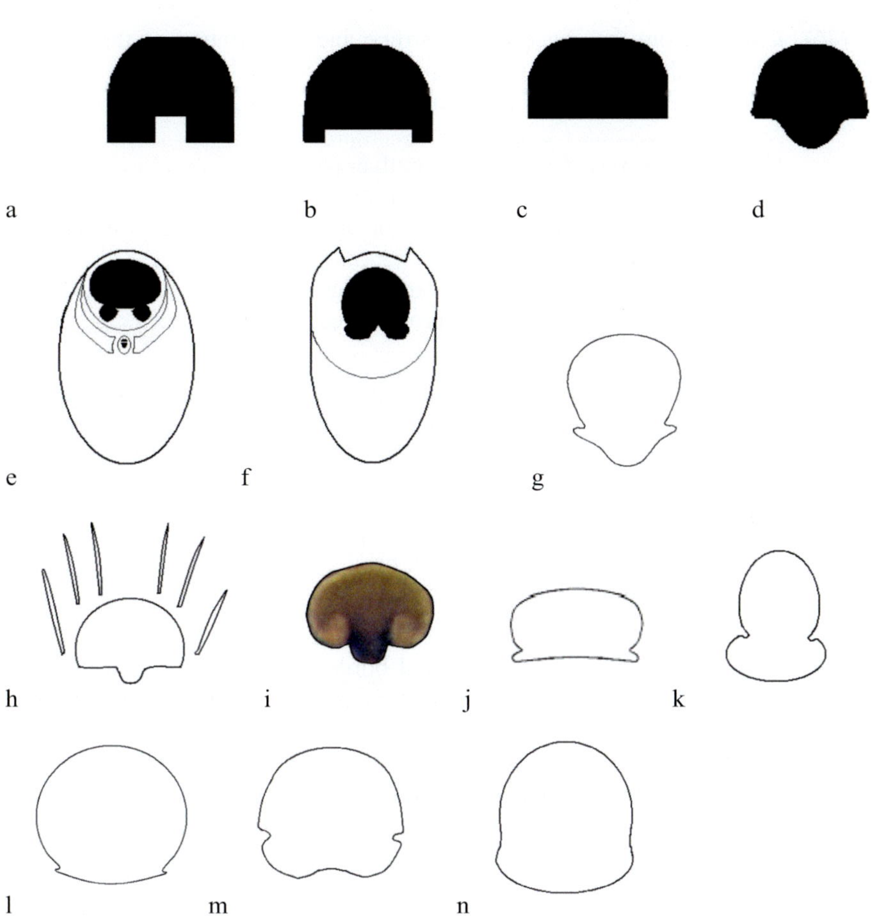

Fig. 4.12 (**a**, **b**) Orifice with a lyrula. (**c**) orifice D-shaped, with a straight proximal border. (**d**) Orifice with sinus. (**e**) *Prenantia* and *Smittina*. (**f**) *Phylactella labrosa*. (**g**) *Trypostega venusta*. (**h**) *Phaeostachys spinifera*. (**i**) Operculum of *Watersipora subatra*. (**j**) *Terwasipora complanata*. (**k**) *Hagiosynodes latus*. (**l**) *Hippoporina pertusa*/*Schizobrachiella verrilli*. (**m**) *Pentapora foliacea*. (**n**) *Cryptosula pallasiana*

46 (44) (a) Orifice appears keyhole-shaped. Dwarf zooids with small round openings scattered between the autozooids (Fig. 4.12g). .... *Trypostega venusta* (p. 280)

(b) Orifice surrounded by thick spines (Fig. 4.12h). .... *Phaeostachys spinifera* (p. 361)

(c) Orifice oval with proximal sinus, operculum dark with a paler dot on both sides of the sinus (Fig. 4.12i) .... *Watersipora subatra* (p. 337)

(d) Orifice not so. 47

47 (46) Choose from the shape of the orifice and the location of the condyles from the drawings in Fig. 4.12, p. 143:
(4.12j) *Terwasipora complanata*, wider than long .... (p. 338)
(4.12k) *Hagiosynodes latus*, longer than wide .... (p. 372)
(4.12l) *Hippoporina pertusa*, almost round .... (p. 320)
Schizobrachiella verrilli, same orifice surrounded by a peristome proximally and laterally. .... (p. 350)
(4.12m) *Pentapora foliacea*, proximal edge convex .... (p. 324)
(4.12n) *Cryptosula pallasiana*, proximal edge concave .... (p. 370)

48 (43) (a) Orifice with prominent lyrula and a medial suboral avicularium (Fig. 4.12e). .... *Smittina* (p. 309)
(b) Orifice with sinus (Fig. 4.12d) or bell-shaped (Fig. 4.12l, m). 49

49 (48) (a) Orifice bell-shaped (Fig. 4.12l, m). 50
(b) Orifice with sinus (Fig. 4.12d). 51

50 (49) (a) Opening with convex proximal border (Fig. 4.12m). Ovicells large, not always present. Develops an erect colony that resembles a honeycomb. .... *Pentapora foliacea* (p. 324)
(b) Opening with concave proximal border (Fig. 4.12n). Ovicells absent. Forms flat adherent or irregular erect crusts. .... *Cryptosula pallasiana* (p. 370)

51 (49) (a) Orifice bordered by thick spines. Avicularia distant from the opening, obliquely directed proximally. Often, a frontal umbo that bears a stout chitinous spine. Ovicell without pseudopores (Fig. 4.13a). .... *Phaeostachys spinifera* (p. 361)
(b) Few or no oral spines. Avicularia associated with the orifice. Ovicells with frontal pseudopores or areolae. 52

52 (51) (a) Avicularia proximal to the orifice. Ovicell with frontal pits (Fig. 4.13b). Frontal shield with pseudopores and peripheral areolar pores. .... *Schizomavella* (p. 324)
(b) Avicularia lateral to the orifice, often on both sides. Ovicell with frontal or marginal pits, sometimes added by radial ridges (Fig. 4.13c). Frontal shield with pseudopores and peripheral areolar pores. .... *Schizoporella* (p. 341)
(c) Avicularia proximal to the orifice and proximally directed. Frontal shield and ovicell with many evenly spaced pseudopores, frontal shield without peripheral areolar pores. Orifice wide and shallow sinuate. .... *Metroperiella* (p. 322)

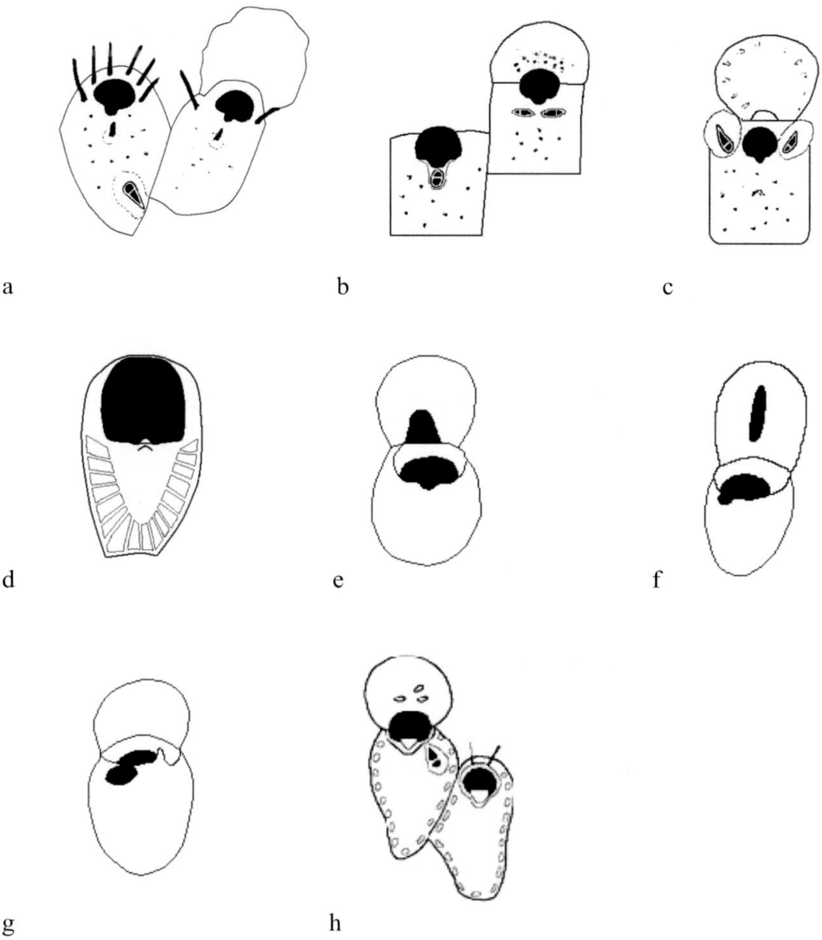

Fig. 4.13 (**a**) *Phaeostachys spinifera*. (**b**) *Schizomavella linearis*. (**c**) *Schizoporella*. (**d**) *Oshurkovia/Umbonula*. (**e**, **f**) *Schizotheca*. (**g**) *Rhynchozoon bispinosum*. (**h**) *Parasmittina trispinosa*

53 (42) (a) Orifice large. Lateral pores large, rectangular, and separated by ribs Avicularia, if present, perpendicular to the frontal plane, median suboral (Fig. 4.13d). .... *Oshurkovia/Umbonula* (p. 306)
(b) Not as described. ……....…........……….....…………………... 54

54 (53) (a) Ovicells with a frontal fissure (cuticular area) (Fig. 4.13e, f). ………
……..............………………………….....………… *Schizotheca* (p. 402)
(b) Ovicells without a slit. …………….........................……………... 55

55 (54) (a) Colony porcelain white. Orifice with a knobbed or notched peristome (Fig. 4.13g), with a conical umbo or a suboral avicularium. Frontal adventitious avicularia sporadic. .... *Rhynchozoon bispinosum* (p. 402)
(b) Not as described. 56

56 (55) (a) Avicularia lateral suboral, not present on all autozooids, the rostrum directed to the peristome. Ovicells with scattered pores (Fig. 4.13h). ... . *Parasmittina trispinosa* (p. 318)
(b) Not as described. 57

57 (56) (a) Avicularia absent. 58
(b) Avicularia on all or on few zooids, sometimes hidden in the peristome. 61

58 (57) (a) Orifice with well-developed lyrula and bordered with four to eight spines (Fig. 4.14a). .... *Escharella* (p. 293)
(b) Orifice with inside on the proximal side a slender, forked tooth, resembling a lyrula (Fig. 4.14b). Very small colonies on shells off the coast.. *Escharoides bishopi* (p. 300)
(c) Orifice without lyrula. 59

59 (58) (a) Zooids slightly calcified, sausage-shaped, close together, but with small open spaces between them. Ovicells associated with dwarf female zooids (Fig. 4.14c). .... *Celleporella hyalina* (p. 277)
(b) Zooids thickly calcified, no spaces between the zooids. Ovicells developing from normal autozooids, with peristome. 60

60 (59) (a) Orifice with high raised rim. Ovicells and frontal surface with one or more conical umbones (Fig. 4.14d). .... *Escharina johnstoni* (p. 356)
(b) Orifice bordered by a deep, protruding peristome. Ovicells with scattered pseudopores (Fig. 4.14e). .... *Neolagenipora collaris* (p. 304)
(c) Orifice with tubular peristome. Ovicells with a membranous frontal area (removed in dry specimens) (Fig. 4.14f). .... *Lagenipora lepralioides* (p. 394)

61 (57) (a) Orifice D-shaped. Zooids connected by short tubes with small spaces between them; each autozooid with a small avicularium immediately distal to the orifice or to the ovicell (Fig. 4.14g). .... *Chorizopora brongniartii* (p. 281)
(b) Not as described. 62

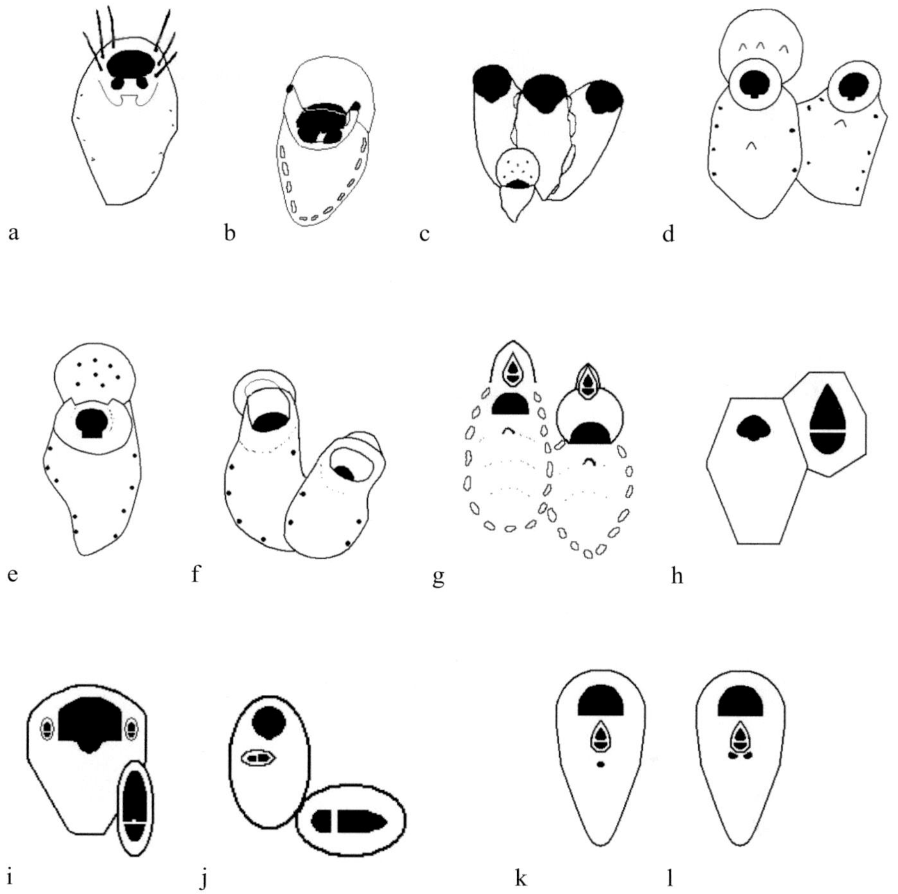

Fig. 4.14 (**a**) *Escharella*. (**b**) *Escharoides bishopi*. (**c**) *Celleporella hyalina*. (**d**) *Escharina johnstoni*. (**e**) *Neolagenipora collaris*. (**f**) *Lagenipora lepralioides*. (**g**) *Chorizopora brongniartii*. (**h–j**) Enlarged avicularia present. (**k**) *Reptadeonella violacea*. (**l**) *Reptadeonella insidiosa*

62 (61) (a) Avicularia uniform, on the zoid, at the orifice or along the zooid margin. 63
(b) Small and large avicularia present at the same time (Fig. 4.14h–j). 69

63 (62) (a) Zooids with a distally directed avicularium in front of the orifice and a large single (Fig. 4.14k) or twinned (Fig. 4.14l) pore in the center. ... . *Reptadeonella* (p. 285)
(b) No large central pore(s). 64

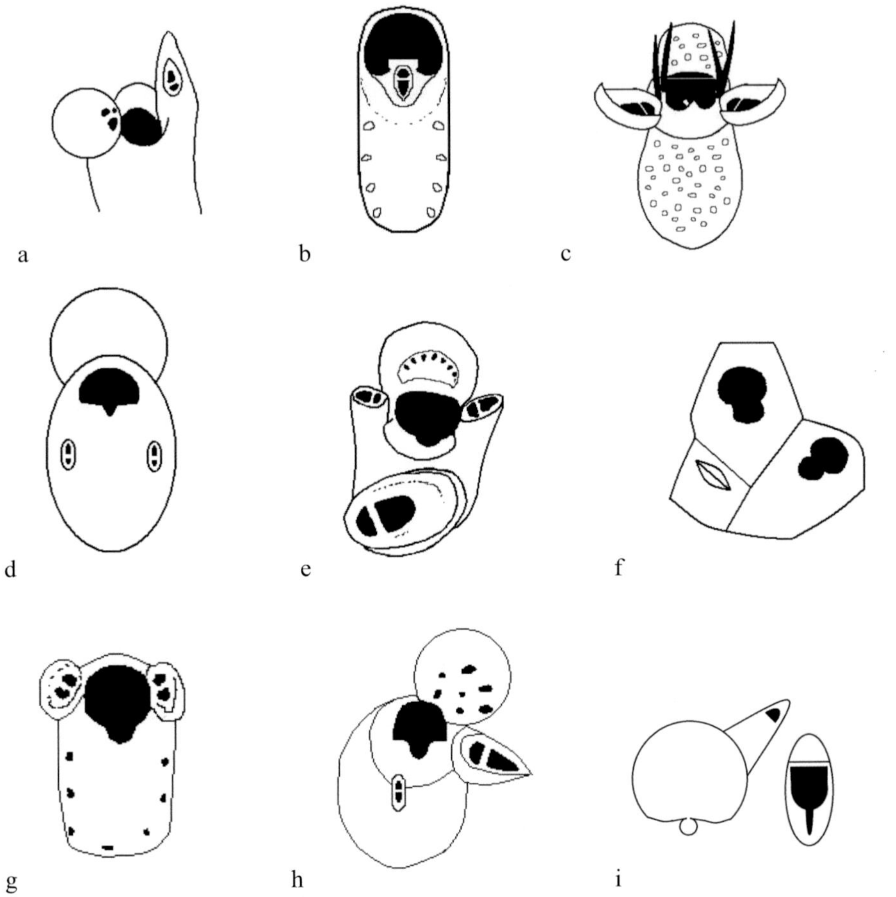

Fig. 4.15 (**a**) *Cellepora*. (**b**) orifice with lyrula. (**c**) *Escharoides*, with suboral tooth. (**d**) *Escharina* and *Herentia*. (**e**) *Celleporina*, with two cylindrical avicularia. (**f**) *Hippoporidra lusitania*. (**g**) *Buffonellaria*. (**h**) *Turbicellepora*. (**i**) *Celleporaria brunnea*

64 (63) (a) Colony pumice-like. Orifice oval, often deeply immersed in a peristome. A spiny nodule on the edge of the peristome with a lateral avicularium (Fig. 4.15a). .... *Cellepora* (p. 384)
(b) Not as described. 65

65 (64) (a) Orifice with lyrula (Fig. 4.15b) or with suboral tooth (Fig. 4.15c). 66
(b) Orifice without lyrula or suboral tooth, but with sinus. 68

66 (65) (a) Orifice large, with a small tooth proximally on the inside. Avicularia usually on either side of the orifice (Fig. 4.15c). .... *Escharoides* (p. 299)
(b) Orifice of normal size with lyrula.. 67

67 (66) (a) Ovicells with frontal pseudopores. .... *Smittoidea* (p. 314)
(b) Ovicells either with up to six pseudopores or without them. .... *Porella* (p. 290)

68 (65) (a) Orifice with prominent sinus; peristome thin and protruding or absent. Avicularia single or paired, distal to the orifice, or proximal, close to the zooid border. Ovicells without pseudopores or pits (Fig. 4.15d). Colony a flat sheet. .... *Escharina* and *Herentia* (p. 355)
(b) Colony encrusting lenticular, mound-like, nodular, or cylindrical. Orifice with sinus. Peristome usually bears two conspicuous, cylindrical avicularia with small rounded mandibles. Additional interzooidal avicularia may occur (Fig. 4.15e). Ovicells with a membranous frontal area. .... *Celleporina* (p. 390)

69 (62) (a) Forms knotty colonies on snail shells. Orifice keyhole-shaped (Fig. 4.15f). Avicularia present. .... *Hippoporidra lusitania* (p. 399)
(b) Orifice with sinus, not keyhole-shaped. 70

70 (69) (a) Orifice without peristome and with sinus. Avicularia with swollen cystid on either side of the orifice (Fig. 4.15g). Colony a flat sheet. .... *Buffonellaria* (p. 387)
(b) Orifice with peristome and sinus. Colony encrusting lenticular, mound-like, nodular, or cylindrical. 71

71 (70) (a) Single avicularium at the orifice, extra large avicularia are common. Ovicells with scattered pseudopores (Fig. 4.15h). .... *Turbicellepora* (p. 396)
(b) A small avicularium on top of a cylindrical tube on either side of the opening. Extra avicularia may occur. Ovicells with a membranous frontal area (Fig. 4.15e). .... *Celleporina* (p. 390)
(c) Opercula and mandibles dark brown, sinus rounded, avicularia in peristome, large vicarious avicularia with spade-like mandible common (Fig. 4.15i). .... *Celleporaria brunnea* (p. 387)

Encrusting Colonies with a Shield of Fused Flattened Spines

72 (18) (a) Spines of the shield with slit-shaped interstices (type rib) (Fig. 4.16a–c). 73
(b) Only small pores between the spines (type shield) (Fig. 4.16d–h). 75
(c) Colony pink. Frontal shield of rib type or shield type, eventually together with an intermediate type, within the same colony. .... *Juxtacribrilina mutabilis* (p. 250)

73 (72) (a) A short, thick, cylindrical spine on either side of the orifice (Fig. 4.16a); ovicells and avicularia are never present. .... *Aspidelectra melolontha* (p. 184)
(b) Two to six thinner oral spines (or absent); ovicells and avicularia may occur. 74

74 (73) (a) Six to twelve spines form the frontal shield (Fig. 4.16b). .... *Callopora rylandi* (p. 197)
(b) Sixteen to twenty spines form the shield (Fig. 4.16c). .... *Membraniporella nitida* (p. 273)

75(72) (a) Zooids large (0.7 to 1.0 mm long), all costae with a pore. Ovicell with two teardrop membranous fenestrae (Fig. 4.16d). .... *Figularia figularis* (p. 272)
(b) Not as described. 76

76 (75) (a) Autozooids do not grow against each other, uniserial, or they are connected to each other by means of tubes (Fig. 4.16e), interzooidal kenozooids and avicularia; the first pair of costae are slightly directed proximally, making the proximal edge of the orifice slightly V-shaped. .... *Distansescharella* (p. 267)
(b) Not as described. 77

77 (76) (a) Orifice bordered proximally by an arc formed by the first pair of costae (Fig. 4.16g, h). 78
(b) Orifice distinctly D-shaped, proximally with a straight edge (Fig. 4.16f). .... *Puellina/Cribrilaria/Glabrilaria* (p. 252)

78 (77) (a) Ancestrula with opesia. Orifice with a thick border proximally, often with an umbo (Fig. 4.16g). Ovicells without adventitious avicularium. .... *Cribrilina* (p. 246)
(b) Ancestrula as a zooid, ovicells with pseudopores at the tip of conical processes, often a distal avicularium on the ovicell (Fig. 4.16h). .... *Collarina gautieri* (p. 244)

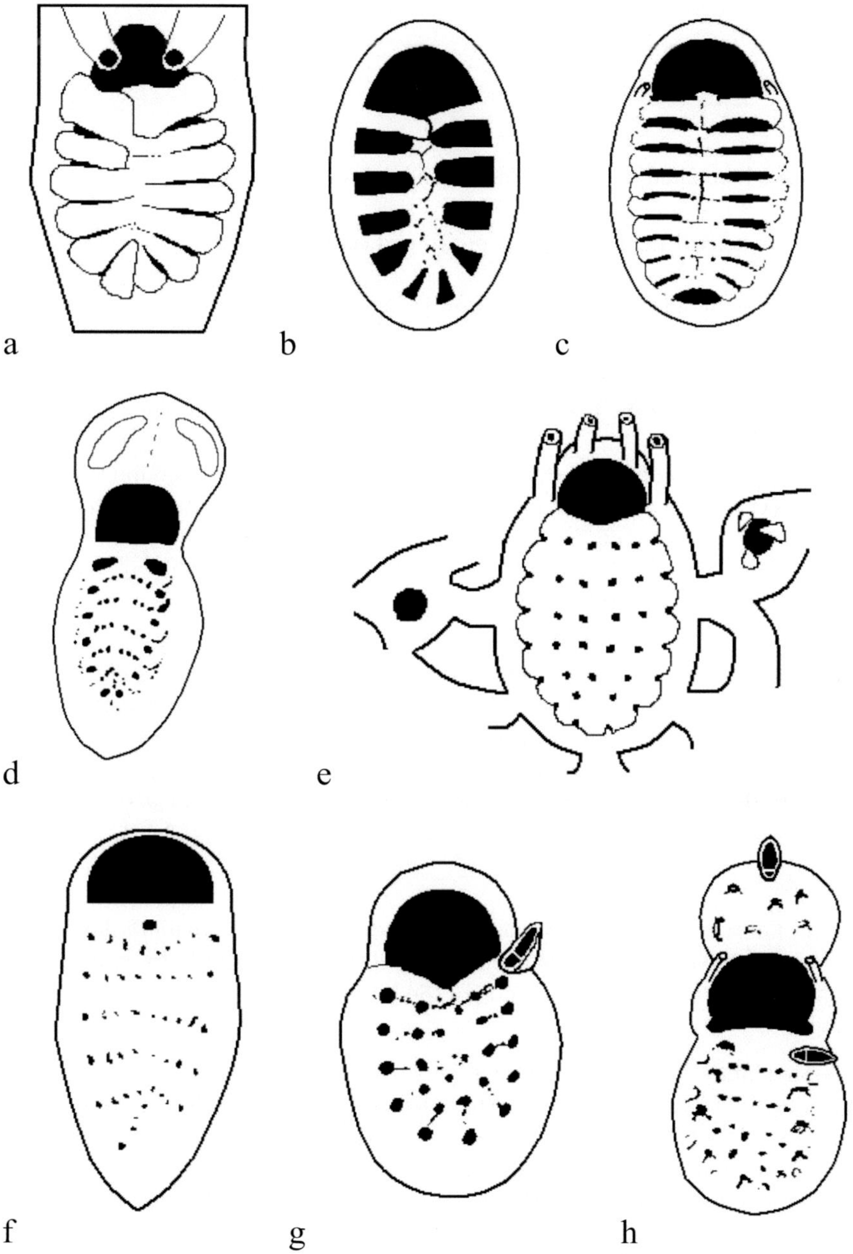

Fig. 4.16 (**a**) *Aspidelectra melolontha*. (**b**) *Callopora rylandi*. (**c**) *Membraniporella nitida*. (**d**) *Figularia figularis*. (**e**) *Distansescharella seguenzai*. (**f**) Orifice D-shaped, *Puellina/Cribrilaria/Glabrilaria*. (**g**) *Cribrilina*. (**h**) *Collarina gautieri*

Systematic Part

Family Aeteidae

Genus *Aetea*

Embryos are incubated in external membranous brood sacs.

1 (a) The stem of the autozooid is not ringed (in backlight). .... *Aetea truncata* (p. 152)
(b) The stem of the autozooid is closely ringed (in backlight). 2

2 (a) Distal portion of the autozooid wider than the stem, spatulate and usually curved so that the opening faces downward. Erect zooid part 0.6–0.8 mm long. .... *Aetea anguina* (p. 152)
(b) Erect zooidal portion straight, the distal part not wider than the stipe. The erect zooid part 0.8–1.8 mm long. .... *Aetea sica* (p. 153)

Aetea anguina Linnaeus, 1758

Description Colony white, the attached parts branching, a zooid stem erecting from any widening. The zooid stem is weak but clearly ringed in transmitted light. The distal part of the zooid bears a frontal membrane occupying one-fourth to one-third of the free length, wider than the stem, spatulate, not ringed. Stem usually curved with the frontal membrane facing down. Embryos are incubated in external membranous, temporary brood sacs, but they are rarely seen (Fig. 4.17).

Occurrence

Belgium On washed-up (beachcast or stranded) objects and *Himanthalia* buttons (De Blauwe 2000b, c, 2005a; Vanhaelen et al. 2006). The earliest observation from Belgium was recorded in Westendorp (1843), on washed-up seaweed in Blankenberge.

The Netherlands Lacourt (1949) mentions washing-up on cork in Scheveningen. Bloklander found this species on material washed up in the Netherlands in the autumn of 1947 (Bloklander and Leenhouts 1948). On washed-up plastic and *Himanthalia* buttons (de Ruijter 2012a, 2017c, 2020b). Found in 2015 for the first time autochthonous in the Netherlands on Duik De Noordzee Schoon (Expedition North Sea) (DDNZS) nets on wreck 2801 and 2803 on the Dogger Bank (Faasse et al. 2016). Again in 2016, numerous worm tubes on nets at the Niponia wreck no 2305 (52°44.520′N–3°58.702′E) (van Leeuwen et al. 2017).

Fig. 4.17 *Aetea anguina.* (**a**) Part of colony washed up on seeweed in Arcachon (France). (**b**) Brooding zooid (from Cook 1977). Abbreviations: *a* attached proximal part of autozooid, *b* brood sac, *e* erect distal part of autozooid, *f* frontal membranous wall, *o* operculum. (**c**) Side view of two autozooids

Aetea sica (Couch, 1844)

Description Colony as in *Aetea anguina*. Stems of the zooids annulate when viewed by transmitted light. The distal part carrying the frontal membrane is slightly wider than the stem and is about one-third of the total length. The erect zooid part is straight and longer than in *Aetea anguina*. Dried zooids may be curved randomly, as in Fig. 4.18b, c where the membranous zone faces upward. In contrast, in *Aetea anguina*, the membranous zone faces always the substrate. Embryos are incubated in external membranous brood sacs (Fig. 4.18).

Occurrence

Belgium On washed-up plastic, but less common than *Aetea anguina*.

The Netherlands Lacourt (1949) mentions washing-up on cork in Scheveningen. On washed-up plastic (de Ruijter 2014b, 2018b, 2019a, 2020b, 2024b).

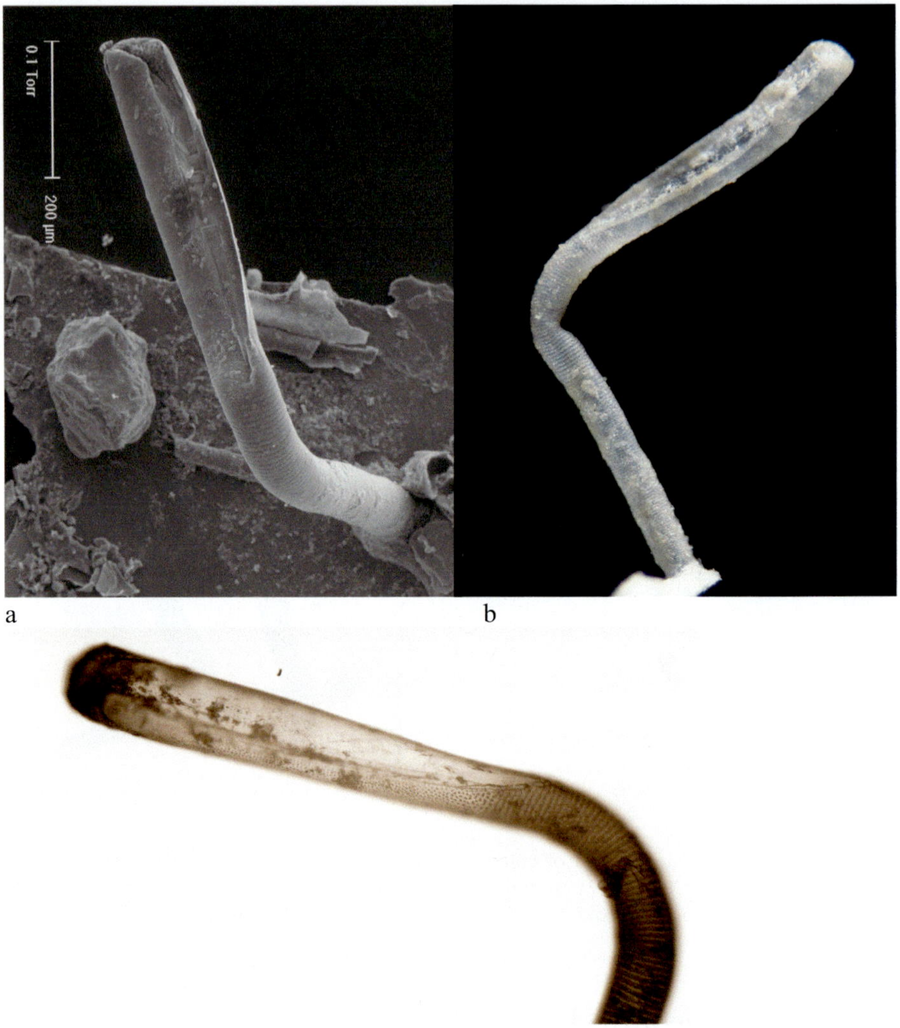

a b

c

Fig. 4.18 *Aetea sica.* (**a**) Zooid from colony growing on beachcast plastic in St-Idesbald (Belgium). (**b**) Erect part of an autozooid. (**c**) Shown in transmitted light

Aetea truncata (Landsborough, 1852)

Description

Colony as *Aetea anguina*. Erect parts of the autozooids straight. Zooid stem punctate, not striated or annulate when viewed by transmitted light. Distal part occupies one-third to one-half of the free length (Fig. 4.19).

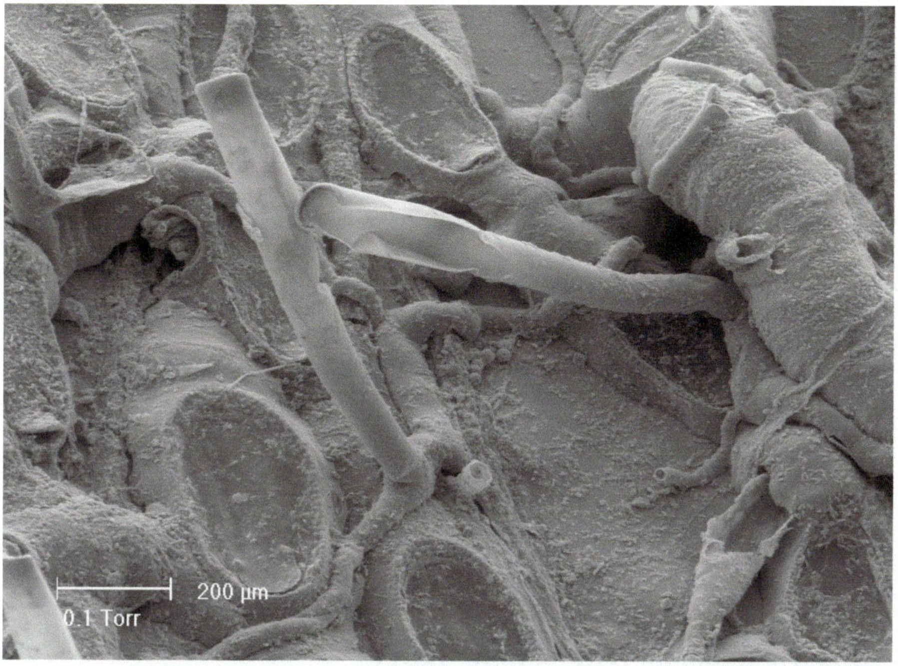

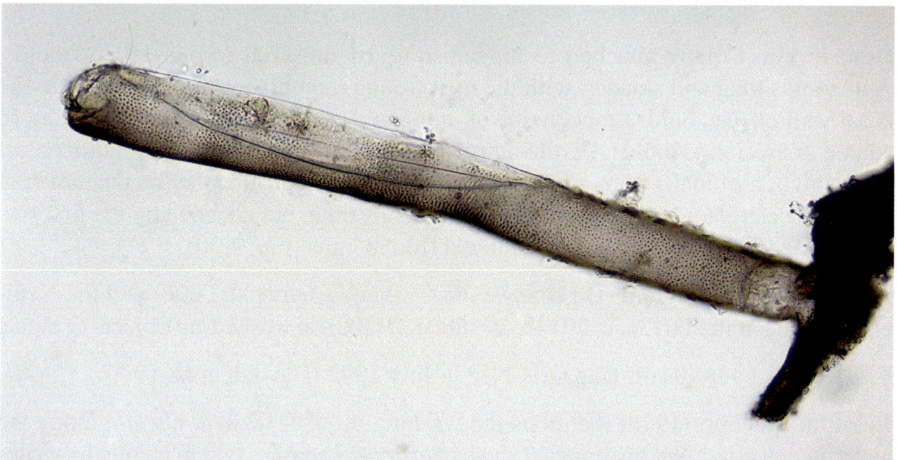

b

Fig. 4.19 *Aetea truncata*. (**a**) Colony growing with *Pyripora catenularia* on beached plastic in St-Idesbald (Belgium). (**b**) Erect part of an autozooid with transmitted light

Occurrence

Belgium Found on a beached *Himanthalia* button in September 2001 (De Blauwe 2003a) and on plastic washed up in St-Idesbald and Oostende in January 2005.

The Netherlands de Ruijter (2013c, 2014b, d, 2018d) reports a colony on washed-up plastic.

Germany Reported from the German North Sea in Zettler et al. (2018).

Family Scrupariidae

Embryos are brooded in a large terminal helmet-like brood chamber.

Genus *Scruparia*

1 (a) Adnate part of the colony is a chain of zooids. Frontal membrane almost parallel to the zooidal axis. .... *Scruparia ambigua* (p. 156)
(b) Colony attached to the substrate by a stolon. Frontal membrane at an angle to the zooidal axis. .... *Scruparia chelata* (p. 158)

Scruparia ambigua (D'Orbigny, 1841)

Description Colony attached to the substrate by uniserial chains of autozooids. Autozooids long and slender, with the oval frontal membrane almost parallel to the basal wall of the zooid. Erect chains of autozooids originate by frontal budding of adnate or erect autozooids. Fertile autozooids bud from adnate or erect autozooids, with a high, terminally pointed, bilobate ovicell-like brood chamber on the maternal zooid. Ancestrula not attached directly to the substrate, but attached by the first two daughter zooids, which arise as distal and frontal buds (Fig. 4.20).

Occurrence In Belgium (De Blauwe 2000c; Vanhaelen et al. 2006) and the Netherlands (de Ruijter 2013a, c, 2014b, 2018b, d, 2020c) on washed-up objects or algae.

France On hydroids off Cap Gris-Nez in July 1992 (Davoult et al. 1999).

Belgium Lacourt (1978) mentions the washing-up of *Scruparia chelata* along the entire Dutch coast, but his image 2 shows *Scruparia ambigua*. In his table, he writes "zooids in linear series, adherent at first, ..." which in turn relates to *S. ambigua*. The Royal Belgian Institute of Natural Sciences (RBINS) owns two colonies from 1900 and one from 1905, which are collected on the Flemish Banks.

The Netherlands Observed in Zeeland near Katsehoek on Noord-Beveland on 1 February 1968 (Heerebout 1970) and Wissenkerke (Faasse and De Blauwe 2004).

Scruparia ambigua was collected at 10 m depth at platform L10-G. Collected at the Princes Amalia Wind Farm (Vanagt et al. 2013) and the Borkum Reef Grounds

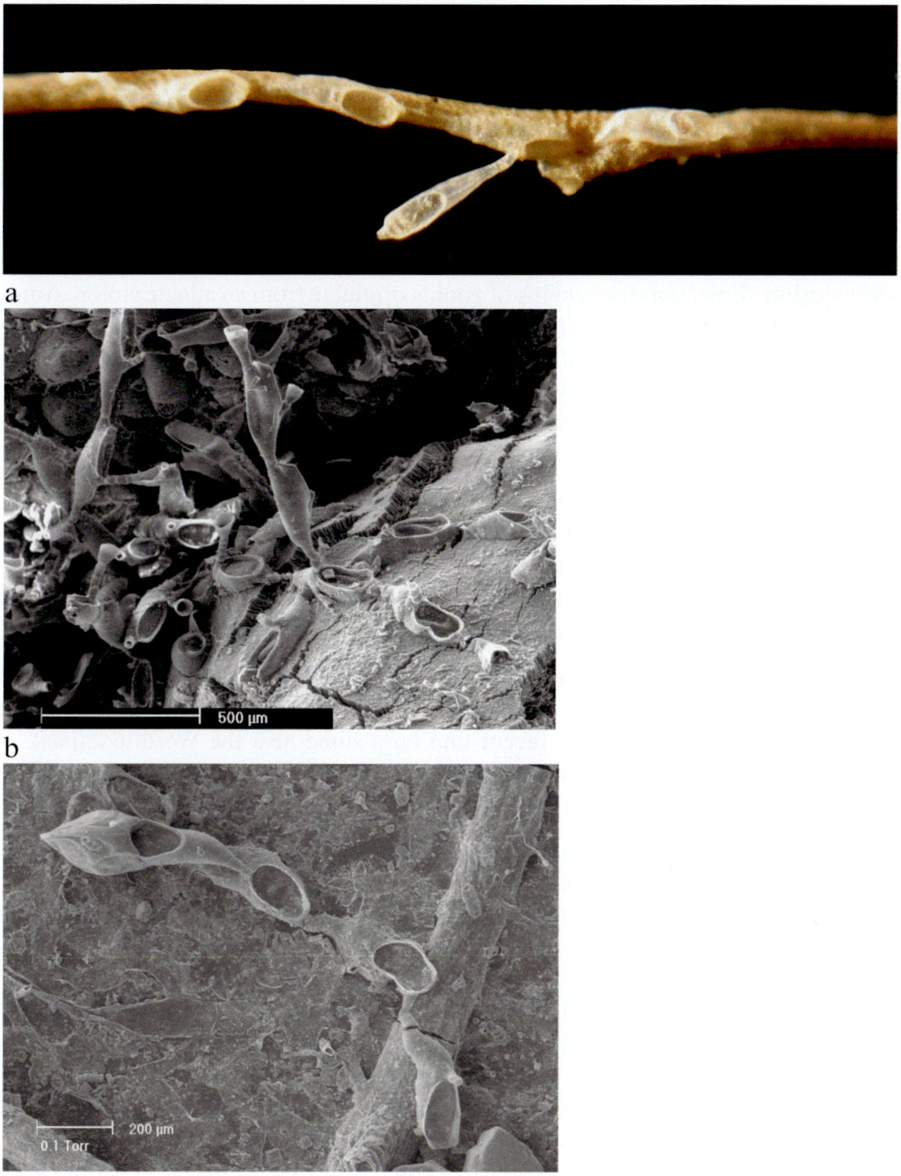

Fig. 4.20 *Scruparia ambigua*. (**a**) Autozooids on hydroid stem. (**b**) Colony growing on plastic fishing gear, St-Malo (France). (**c**) Autozooid with terminal helmet-like brood chamber (left), on beached plastic at Blankenberge (Belgium)

(Coolen et al. 2015). Several samples from platform L10-A taken between 5 and 15 m depth also contained a number of *S. ambigua* colonies (Beukhof et al. 2016).

Germany Collected German North Sea (Zettler et al. 2018) and at Heligoland (pers. comm. Britta Kind).

Scruparia chelata (Linnaeus, 1758)

Description Erect uniserial chains of zooids originate from an adnate stolon. Autozooids horn-shaped, with oval frontal membrane descending to the distal end of the basal wall. Reproductive zooids arise from frontal budding on the zooids. Helmet-like brood chambers nearly spherical, wider than the bearing zooid, and about the same length (not seen in our material) (Fig. 4.21).

Occurrence Beached specimens are common in Belgium and the Netherlands on washed-up plastic, *Himanthalia* buttons (De Blauwe 2005a; Vanhaelen et al. 2006; Verkuil 1998a, b; de Ruijter 2012a, 2014a, 2015d, 2018d, 2020a), *Bugulina flabellata* (de Ruijter 2006a), *Crisularia plumosa* (de Ruijter, unpubl.) and *Sargassum muticum* (de Ruijter 2019b), most likely originating from the English Channel. At Katwijk on *Tubularia* (Maitland 1851).

Belgium Loppens (1906) listed this species as rare on algae, rocks, shells, and Crustacea. Two colonies were collected at the Flemish Banks in 1900 and 1905 (RBINS collection). There is a recent find on a stone near the Westhinderpaal. A colony was found in 2002 on the wreck of the Birkenfels and in 2004 on the Kilmore (Zintzen 2007); in 2021 on the SS. Ptarmigan (van Leeuwen 2023).

The Netherlands The first autochthonous colony of *Scruparia chelata* was discovered on the shipwreck of the SS Nautilus (24 km off the coast of Texel) by Lengkeek et al. (2013) who investigated 10 shipwrecks on the Dutch continental shelf. Found in 2015 on DDNZS nets on wreck 2801, 2803, Rode Poon, and Jeanette Kristina (Faasse et al. 2016).

Collected on wormtubes and *Mytilus edulis* on nets taken from the wreck of Niponia nr 2305 (52°44.520′N–3°58.702′E) (DDNZS 2016—van Leeuwen et al. 2017).

Scruparia chelata was observed at different depths (5–25 m) on platforms L10-G and L10-A, both as detached specimens and attached to *M. edulis* (Beukhof et al. 2016). Collected in 2021 on the wreck of a submarine (51°52.207′ N, 02°52.714′ E) (van Leeuwen 2023).

Germany Collected in the German North Sea (Zettler et al. 2018) and at Heligoland (pers. comm. Britta Kind).

Fig. 4.21 *Scruparia chelata.* (**a**) On coral weed washed up in Oostende (Belgium). (**b**) SEM of the same specimen

Family Eucrateidae

Genus *Eucratea*

Eucratea loricata (Linnaeus, 1758)

Description

The colony forms erect tufts, attached to the substrate with rhizoids. The colony resembles a miniature poplar, about 10–25 cm high, whitish brown. Branches composed of zooid pairs standing back to back, all pairs oriented in the same direction. Zooids elongated, widening distally. Polypide with 10 to 14 tentacles. Spines and avicularia absent. Embryos develop in external membranous sacs, but they are rarely seen (Fig. 4.22).

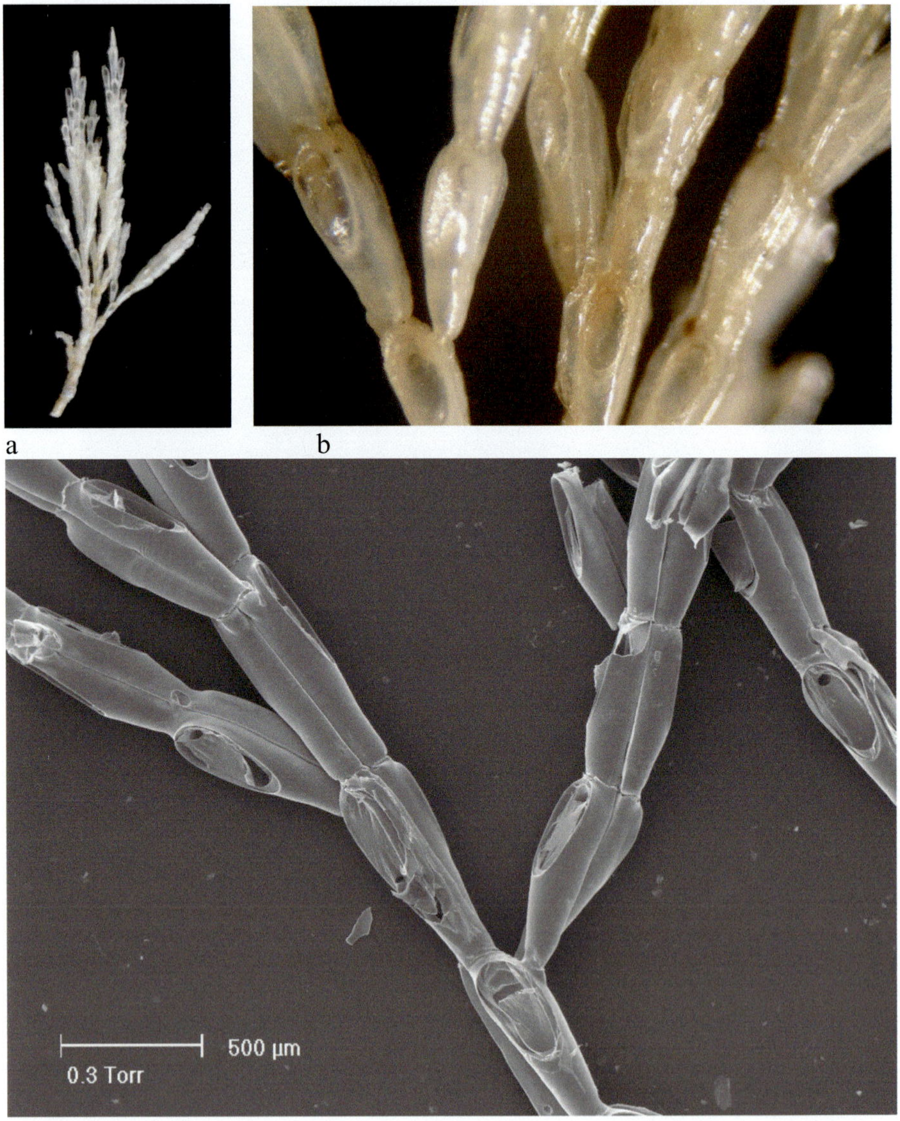

Fig. 4.22 *Eucratea loricata.* (**a**–**c**) Colony from Norway. (Collected in 2004 by M. Faasse)

Occurrence

Belgium Loppens (1906) considered this species as common sublittorally. It was regularly collected in 1905 and 1908 on the Flemish Banks in the area delineated by 51°22′ N, 51°27′ N, 2°25′ E, and 2°31′ E (Gilson collection in the RBINS) (De Blauwe et al. 2006). No observations since 1908, so it is probably locally extinct.

The Netherlands Maitland (1851) reports this species from the beach at Domburg, Lacourt (1949) adds Scheveningen, Den Helder, and on a trawler in IJmuiden. No recent sightings.

France A beached colony was found in Wissant (Northern France) on August 5, 1987 (De Grave and d'Udekem d'Acoz 1988).

Germany Reported by Zettler et al. 2018.

Superfamily Membraniporoidea

Family Membraniporidae

The Membraniporidae have a twinned ancestrula.

Genus *Membranipora*

Membranipora membranacea (Linnaeus, 1767)

Description

The colony forms an extensive, leathery cover on brown algae. Zooids rectangular, with tubercles or short spines (the apex of which is not calcified) at the corners. The membrane covers the entire frontal plane. Colony flexible, so it can cling to seaweeds such as *Laminaria* without breaking when the leaves move in the waves. "Tower zooids," where the frontal membrane faces upward like a tube may be present. Polypide with about 17 tentacles (Fig. 4.23).

Occurrence Lives on brown algae along rocky coasts. On *Laminaria digitata* and *L. hyperborea* the colonies reach large sizes; small colonies can be found on *Fucus serratus*, *Ascophyllum nodosum*, and *Himanthalia elongata* buttons. Occasionally washes up on these seaweeds along the Belgian and Dutch coast. Rarely found on plastic and pebbles covered with green algae.

France Common along the rocky coast of Brittany and Normandy. Detached seaweeds drift to the north and get beached there with their epiphytes as far as Heligoland.

Belgium Not native.

The Netherlands Two records from the Oosterschelde: Canal van Goes on seagrass (Heerebout 1970) and Wemeldinge on sea lettuce (Mulder 1983) are probably misidentifications. Conclusion: not native.

Fig. 4.23 *Membranipora membranacea.* (**a**, **b**) Colony with regularly spaced tower zooids (Damgan, France). (**c**) Colony area with tower zooid (due to the drying of seaweed, zooids are no longer perfectly rectangular) (Brittany, France). (**d**) Colony on *Fucus vesiculosus*, beached in Belgium

Germany Collected in the German North Sea (Zettler et al. 2018) and at Heligoland (pers. comm. Britta Kind). Common on *Laminaria* on Heligoland (pers. obs. in 2023 and 2024).

Genus *Biflustra*

Biflustra tenuis Desor, 1848

Description The colony forms a crust on pebbles or shells offshore. Often several layers grow on top of each other and thus form a thick crust on the substrate. Zooids rectangular to octagonal. Beneath the frontal membrane lies a calcareous cryptocystal shelf that is mainly well developed proximally, but also laterally. This plate is granulated and its edge with the opesia is finely serrated. The size of the opesia is variable. Small spiny nodules may appear at the zooid angles (Fig. 4.24).

Occurrence Prenant and Bobin (1966) report the species from the Pacific, warm and temperate Atlantic coasts of America, West African coasts, the Mediterranean, and Denmark; everywhere in shallow water (0–60 m) and they find it curious that it

Fig. 4.24 *Biflustra tenuis*. (**a**) Colony area. (**b**) Twin ancestrula and initial part of the colony. Both images from a colony growing in a dead shell on the Kwinte Bank (Belgium)

has not yet been encountered on the western coast of Europe. Hayward and Ryland (1998) consider this species to be tropical and are not aware of its occurrence in the North Sea. Colonies can be found on the beach, mainly on discolored cockles in Oostende, Wenduine, Knokke, Nieuwvliet, Domburg, and Westkapelle, washed up or after sand replenishment.

Belgium In the period 1899–1914, Gustave Gilson collected several colonies on the Flemish Banks. Unfortunately, they remained unrecognized until 2005 (De Blauwe et al. 2006). Lives on stones and shells between the Goote and Thornton Banks (Kerckhof and De Blauwe 2006), the Kwinte Bank, the Hinder Banks, and on the erosion stones of the Belwind wind farm after 2012. A large living colony on a stone was collected with a beam trawl within 51°30.511′ N, 2°35.273′ E and 51°29.559′ N, 2°36.378′ E in August 2023.

The Netherlands Reports by Lacourt (1949, 1978) on seaweed must regarded as misidentifications. The species lives on shells off the coast near Zeeland. One colony was collected in the Wadden Sea in March 2022, pers. comm. D. Oude Luttikhuis (Royal Netherlands Institute for Sea Research, project Waddenmozaïek) WM 2022, Grid, PosKey 50,171, water depth 710 cm, 53°4.27′ N, 5°4.25′ E.

Biflustra grandicella (Canu and Bassler, 1929)

Description Colony pale to yellowish brown, unilaminar, encrusting cobbles, later forming erect, folded, and anastomosed bilaminar plates, ca 0.6 mm thick. Forming a coral-like mass as big as a basketball, rigid and brittle (similar to colonies of *Pentapora foliacea*). Zooids rectangular, sometimes wider distally than proximally, 0.64–0.80 mm long and 0.29–0.52 mm wide, distal end rounded and proximal margins concave, with beaded rims along thick and salient lateral walls. Frontal membrane without spinules, covering the entire zooid. Operculum with a thin marginal sclerite. Opesia oval, mostly narrower proximally, 0.46–0.52 mm long, occupying about 70% of the zooid length. Cryptocyst granulose and slightly beaded; proximal cryptocyst without proximomedial plate but sometimes with minute denticles projecting into the opesia, frequently placed proximally rather than distally. Tubercles are rare, one or two in proximal corners, often occurring in small clusters in a colony. Lateral transverse walls with 4 multiparous mural septula, each one with 3 to 12 perforations in the septulum. Distal transverse walls with one to three series of uniparous mural septula at the middle of the vertical wall; sometimes a few multiparous septula. Tentacle number without exceptions is 14 (Figs. 4.25 and 4.26).

Distribution *Biflustra grandicella* was described from the vicinity of Hong Kong (Canu and Bassler 1929). Liu (1992) regarded this species as endemic to Chinese coastal waters. The species has been reported outside China from New Zealand (Grange and Gordon 2005), Brazil (Almeida et al. 2017), Australia (Tilbrook 2012), Singapore (Tilbrook and Gordon 2016), and the Netherlands (De Blauwe 2017).

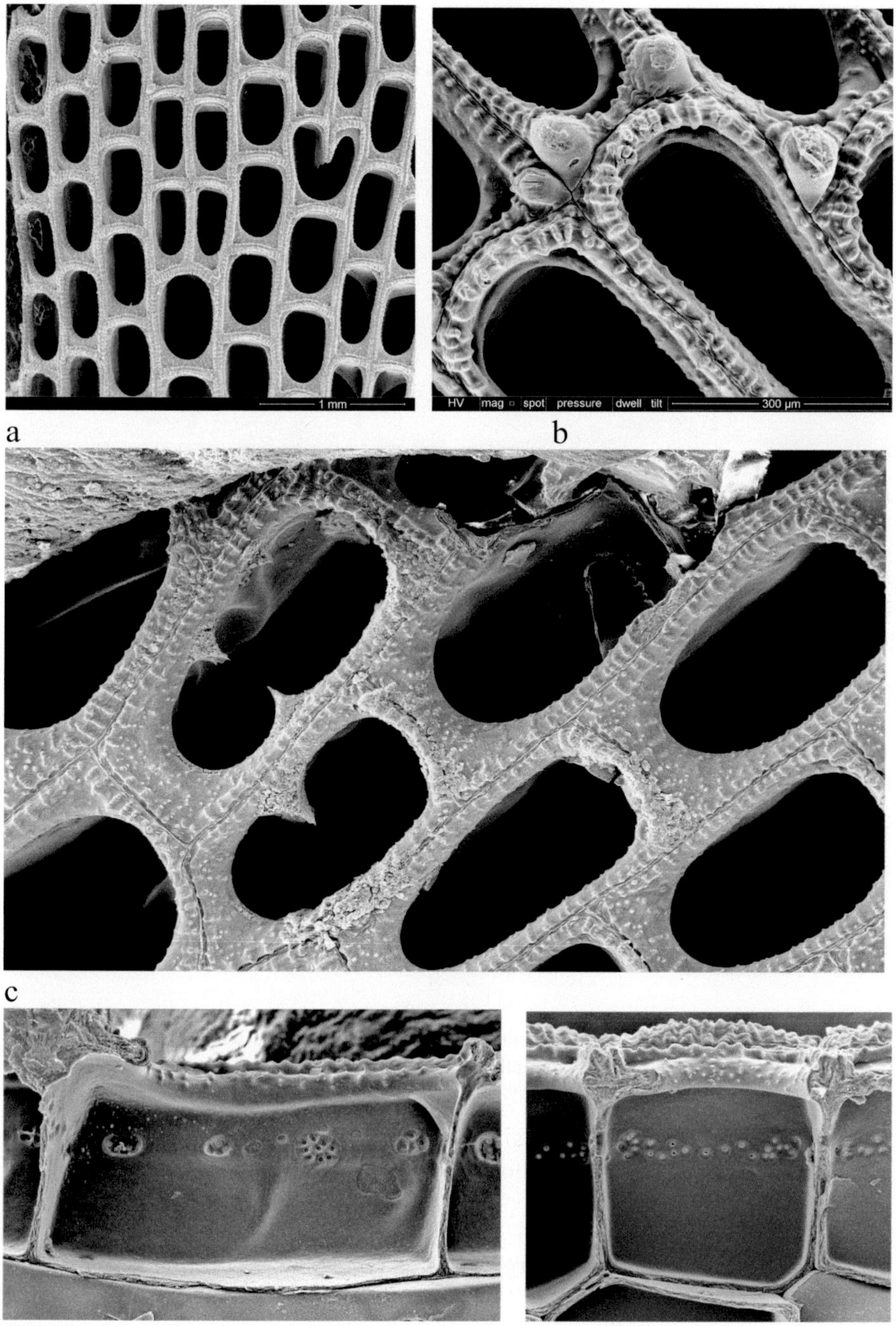

a b c d e

Fig. 4.25 *Biflustra grandicella.* (**a**) Colony fragment. (**b**) Close-up of two zooids, showing rare tubercles, one or two on proximal corners. (**c**) Zooids showing minute denticles projecting into the opesia. (**d**) Lateral wall with multiparous mural septula. (**d**) Distal wall with uniparous mural septula. Nuclear powerplant near Vlissingen (the Netherlands)

Fig. 4.26 *Biflustra grandicella.* (**a–d**) In situ population at the warm-water outlet of the nuclear power plant near Vlissingen (the Netherlands). (**c**) Colony survived freezing temperatures

During three events the species was reported in the Netherlands. This will be discussed in more detail on the following pages.

The Netherlands:

First Event (Not Published Before) Herman Nijhuis worked at the Verolme shipyard in Rozenburg (Rotterdam) in 1968. He collected a sample (20 × 15 cm) scraped off the wall of a mammoth tanker. A piece of the material was sent to the author in 2020 and identified as *Biflustra grandicella*. According to Herman, there was a heap of bryozoans at the wharf, measuring 20 by 20 meters and 3 m high before it was crumbled! And only 10% of the tanker had been cleaned at that point.

Second Event A colony fragment was found on the beach near the nuclear power plant in Vlissingen in December 2016. The beach is very well known for its fossils, and it would be strange that earlier beached material stayed unnoticed. Several colonies in situ were found at extreme low tide in February 2017 next to the warm-water outlet of the power plant (De Blauwe 2017). The population remained stable in the following years and seems not to have expanded to other regions. The temperature at this spot is higher than the surrounding area and the colonies were growing in the small area with pebbles surrounded by sand. A colony was collected at freezing temperatures of −7 °C, kept in seawater, and an hour later the bucket was frozen. The bucket thawed during the trip home and the colony was brought into seawater at room temperature. A week later, the colony was investigated under a light microscope and still alive. This species formerly only known from (sub)tropical areas can thus survive a short period of freezing. The local temperature increase only seems important for the settlement of the larvae, an established colony survives brief freezing temperatures.

Third Event Between the end of August and the beginning of December 2021, 12 colony fragments washed up upon the shores of the Netherlands between Den Haag and Langervelderslag and another fragment beached more to the south at Goeree in February 2021. Where those colony fragments come from remains a mystery. It is not evident that a new population has settled off the coast. It is rather plausible that fragments of a ship's hull have been broken off by, for example, sanding over a sandbank. Another possibility is that the fragments were scraped off in a harbor and then dumped into the sea. However, there is no evidence for such practices. There was an identical mass stranding of *Bugula neritina* in roughly the same area a few years earlier.

Genus *Jellyella*

Jellyella tuberculata (Bosc, 1802)

Description Colony encrusting, unilaminar, white. Zooids rectangular, frontal surface of autozooids mostly membranous. Lateral wall slightly calcified. Cryptocyst irregular, occupying less than one-fifth of zooidal length, with minute tubercles on its surface. Gymnocyst often forming two rounded tubercles proximally. Operculum with heavily sclerotized outer rim, wider than long. Polypides with 15 translucent white tentacles. Avicularia absent. No ovicells or avicularia. Colonies develop from a twinned ancestrula. Resembles *Biflustra tenuis* but the proximal calcareous shelf is smaller and the tubercles can be quite long and obliquely directed proximally (Fig. 4.27).

Occurrence Widespread in all tropical and warm-temperate ocean waters on brown algae. It is the most abundant encruster of drift *Sargassum* in the tropical Western Atlantic, and it is found more occasionally on hard substrata, especially floating plastic debris (Winston 2012).

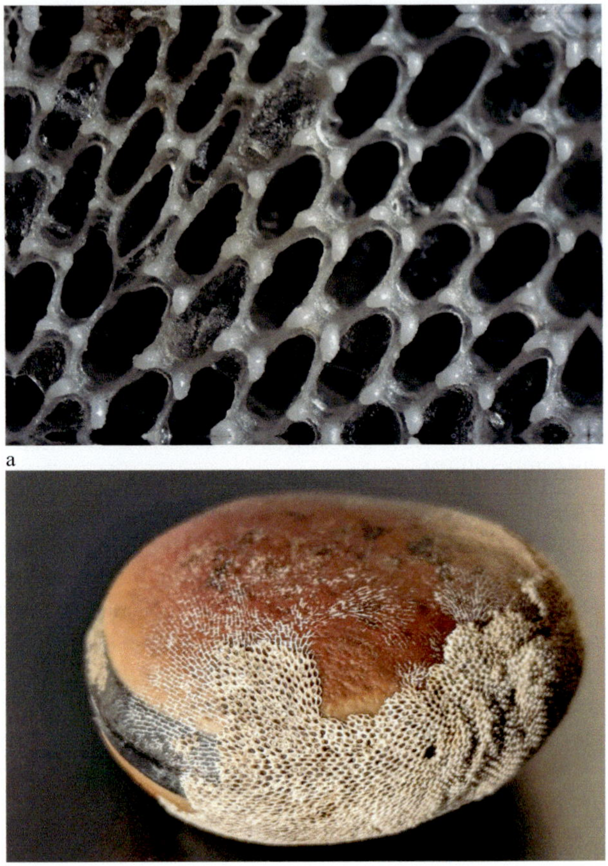

Fig. 4.27 *Jellyella tuberculata*. (**a**) Part of the colony growing on a plastic bottle with *Lepas anatifera,* Bredene (Belgium). (**b**) Colony on a floating seed of *Mucuna*

Belgium This species washed ashore:

November 25, 2008, on plastic between Zeebrugge and Blankenberge
October 29, 2017, Oostende on a washed-up buoy with the crab *Planes minutus*
March 26, 2019, Bredene on a plastic bottle with *Lepas anatifera*
September 13, 2019, Lombardsijde, on *Lepas anatifera*, not verified
January 17, 2024, on a seed of *Mucuna urens* (Red Hamburgerbean), beached at Oostende (pers. comm. Aaron Fabrice).

The Netherlands On September 2, 2014, on barnacles near Egmond (de Ruijter 2015b). In February 2024 on plastic at Texel (pers. comm. Sytske Dijksen).

Jellyella eburnea (Hincks, 1891)

Description

Colonies encrusting shells of *Spirula*. Twinned ancestrular zooids with gymnocyst occupying at least two-thirds of the frontal length. Autozooids indistinctly rhombic in frontal outline, and longer than wide. Basal walls uncalcified apart from a marginal rim surrounding a basal window; substrate may be etched beneath the basal window. Gymnocyst well developed, proximal gymnocyst generally occupying about one-half to one-third of the frontal area, tapering distally and absent at the distal end of zooids; smoothly calcified surface fluted or folded into rucks, tubercles, or spines variable in prominence and number; spines short, hollow, sometimes open-ended but more often rounded and terminally closed, hook-like, those adjacent to the opesia bending over frontal membrane but not reaching the mid-line, occasionally with transverse furrows; one or sometimes more distally directed tubercles present in the proximal part of gymnocyst, often originating along the border with proximal neighboring zooid, very occasionally longer and spine-like. Cryptocyst lacking. Opesia ovoidal, longer than wide. Operculum crescentic, with a marginal sclerite. Vertical walls forming a narrow shelf at the distal end of the zooid, and often with minor buttresses in corners of zooids and at junctions with basal walls. Pore chambers lacking. Simple pores present in lateral vertical walls separating zooidal series, located close to basal walls. Multiparous septula developed in transverse vertical walls adjacent to basal walls; individual pores about 1 μm in diameter, centered within depressions separated by slight buttresses, sometimes two or more pores sharing the same depression. No ovicells or avicularia. Kenozooids present in areas of crowding, particularly where growing edges coalesce (Fig. 4.28).

Occurrence This is the sole bryozoan found on *Spirula* shells. Colonies are known from Florida, South Africa, East Africa, Madagascar, Mauritius, Fiji, Australia, and New Caledonia (Taylor and Monks 1997).

The Netherlands Frank Husslage found a *Spirula* on the beach of Oost-Kapelle in the spring of 2014. Arie Twigt, who has the shell in his collection brought a colony of *J. eburnea* attached to the *Spirula* to my attention.

The species described by Winston (1982, p. 120) as *Membranipora* sp. is probably closely related to *J. eburnea*. This un-named species from the east coast of Florida was found only on wood and plastic objects in beach drift. The proximal gymnocyst is not as well developed as that in typical *J. eburnea*, and Winston mentions that, unlike those in *J. eburnea*, the ends of the gymnocystal spines may be bifid, but other aspects of skeletal morphology are strikingly similar.

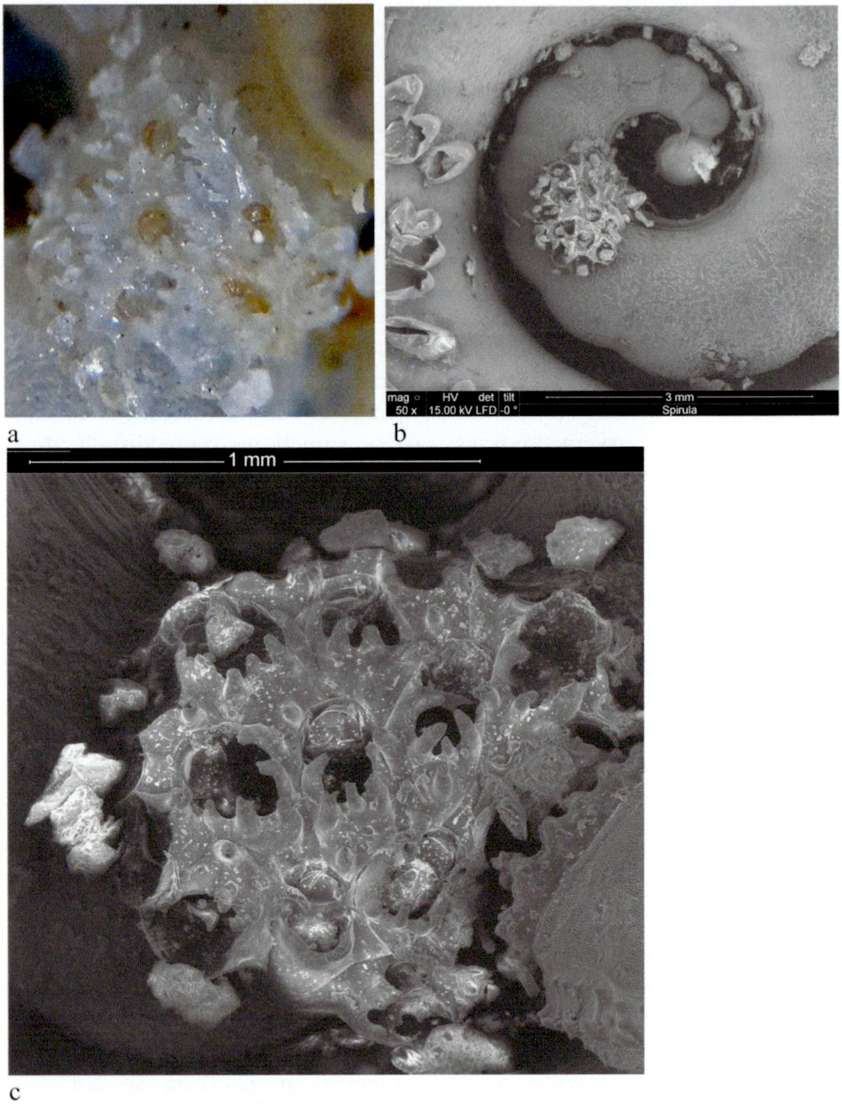

Fig. 4.28 *Jellyella eburnea.* (**a**) General view of the colony (photo by A. Twigt). (**b**, **c**) SEM images showing the same colony

Family Electridae

Genus *Conopeum*

1 (a) Marine or estuarine. Frontal calcification present, but reduced. Spines, if any, placed around the opesia; often a pair of triangular chambers (kenozooids) at the distal end of each zoid; encrusting on hard substrates. .... *Conopeum reticulum* (p. 171)

(b) Only in brackish water. Frontal calcification minimal. Spines limited to a small pair at the distal end of the zoid; kenozooids rare and not in pairs; often on plants in estuaries. Erect colony parts not flexible. .... *Conopeum seurati* (p. 174)

(c) Marine and estuarine. Erect, flexible colony with small basal portion, giving rise to bilaminar fronds with cylindrical proximal port........................ .... *Conopeum chesapeakensis* (p. 171)

Conopeum chesapeakensis (Banta, Perez, and Santagata, 1995)

Description

Colonies tan-colored, erect, and branching. Branches of the colony are cylindrical, comprising four or more rows of zooids arranged in a quincunx and arise from the encrusting base or from flat bilaminar fronds. The cylindrical portion expands along its length, becoming a two-layered ribbon or frond, with 20 or more rows of zooids. Autozooids lightly calcified, quadrangular, oval opesia bordered by a coarsely crenulated cryptocyst. The gymnocyst comprises the smooth lateroproximal areas outside the crenulations. The zooids of the two layers share their basal cuticle and develop in unison, but are also offset, so that the lateral wall between two zooids in a row on one side of the colony lies adjacent to the midpoint of a zooid on the opposite side. The branch is bordered by a row of zooids perpendicular to the zooids of the two layers. There are no spines, ovicells, or avicularia. The polypide has 12 to 14 tentacles. This bryozoan is characteristic of hard substrates and seagrasses in brackish water (description after: Banta et al. 1995; Winston and Hayward 2012) (Fig. 4.29).

Occurrence

Belgium First reported from Antwerp by Gittenberger et al. (2018). Scraped from both jetties in Lillo and Liefkenshoek in 2023 by the present author. The colonies had a hydroid as substrate.

The Netherlands The first colonies were collected in 2022 in the Netherlands within the Marine Alien Species Detection Network, from several floating docks in the North Sea Canal in Ijmuiden (Gittenberger et al. 2022b, 2023a). Within the network in 2023, colonies were also found at a site in Amsterdam, and within the Wadden Sea, in the ports of Den Helder and Eemshaven (Gittenberger et al. 2023b, 2024). All identifications were confirmed by DNA analyses.

Conopeum reticulum (Linnaeus, 1767)

Description The colony forms a mesh-like crust. Zooids flat. Lateral walls strongly calcified. Frontal surface largely membranous, opesia elliptic to ovate. There may be a varying number of small spines around the opesia, typically thin and pointed and sometimes sloping over the frontal membrane.

Fig. 4.29 *Conopeum chesapeakensis*. (**a**) Colony on the hydroid stem. (**b**) Lateral view of two zooids, one with semiretracted lophophore (crenelated cryptocyst visible). (**c**–**d**) Colony area showing zooidal arrangement. (**d**) Zooid showing communication pores in the distal transverse wall

Characteristic is the presence of a pair of triangular kenozooids at the distal end of each zooid. Occasionally the two kenozooids have fused into a bar along the distal end of the zooid. Unfortunately, the kenozooids are not always present, which leads to confusion with other species. Colonies in empty bivalve shells usually have no kenozooids (Fig. 4.30f). Avicularia are absent, although beginners may mistake the kenozooids as avicularia. Polypide with 10 or 11 tentacles. No ovicells. Ancestrula single.

The best character (if spines and kenozooids are lacking) is a large septulum (porous zone where zooids are connected to each other), in the terminal wall typically supported by thickened vertical calcified struts. These vertical struts are absent

Fig. 4.30 *Conopeum reticulum.* (**a**) Colony on *Mytilus edulis* (marina Zeebrugge, Belgium) (photo by M. Decleer). (**b**) Some kenozooids have fused into a bar. (**c**) Terminal wall with septula, surrounded by thickened calcified struts. (**d**) Zooids and kenozooids. (**e**) Zooids with spines surrounding the opesia. (**f**) Part of colony in dead shell, without kenozooids. Kwintebank (Belgium)

in similar species such as *C. seurati*, *Electra monostachys,* or *Einhornia crustulenta*. The struts are best seen holding a cleaned colony at an angle and examining the terminal wall (Fig. 4.30).

Occurrence Native and also often present on washed-up plastic.

Belgium Very common along the coast mainly on mussels, also on crab shells, off the coast mainly on the inside of empty bivalves and on stones. Subtidal on rocks of the erosion protection in Belwind, August 2018 (pers. comm. F. Kerckhof).

The Netherlands Very common along and off the coast and in the estuaries. It was found on only one of the 10 shipwrecks examined in Dutch waters, namely on the "Victoria City" (Lengkeek et al. 2013). Collected at the Dutch Borkum Reef Grounds (Coolen et al. 2015).

Germany Collected in the German North Sea (Zettler et al. 2018) and at Heligoland (pers. comm. Britta Kind).

Conopeum seurati (Canu, 1928)

Description Encrusting colony, the shape is determined by the nature of the substrate; on flat surfaces, it forms leathery crusts, on irregular surfaces (as on plant stems); it often rises in two-layered lobes. Zooids are typically elongate-rectangular, shape and size vary according to the substrate.

Calcification thinner than in *C. reticulum*. Frontal calcification much more reduced, present as small triangular areas in the corners of the zooid and a narrow, granular cryptocyst within the erect lateral walls flanked by a narrow groove on its outer side.

Spines very variable, usually a single small pair at the distal end of the zooid, often absent, rarely long and curved. No ovicells. Kenozooids can occur in disturbed places; they are usually elongated, triangular, irregularly placed, and never in pairs at the distal end of the zooid. Polypide with 12 to 14 tentacles. Often very similar to *Einhornia crustulenta*. The operculum is translucent except for the brown margin, while in *E. crustulenta* the operculum is completely opaque (Fig. 4.31).

Occurrence

Belgium Not common, probably strongly reduced in distribution due to pollution and loss of biotope. In the past 20 years, some colonies were collected on mussels in the marina of Zeebrugge, at Nieuwpoort, and in May 2023 in the Doeldok in Antwerp.

The Netherlands On reed stems and other substrates. Tolerates very low salinity and very large fluctuations in salinity. There are numerous sites in the Delta area (Faasse and De Blauwe 2004). Reported from Texel in 1966 (Jebram 1968), Oostvoornse Meer (de Kluijver 1997), the marina, and the inner sluice canal in IJmuiden (Faasse and De Blauwe 2004).

Germany Collected in the German estuaries/Wadden Sea (Zettler et al. 2018).

Genera *Electra* and *Einhornia*

The genus *Einhornia* is characterized by a single proximedial spine and a calcified operculum. Molecular research confirms that the species in this genus stand out from other *Electra* species (Nikulina 2007).

1 (a) Frontal calcification with many pseudopores. .... *Electra pilosa* (p. 179)
  (b) Frontal calcification without pseudopores, kenozooids absent. 2

Fig. 4.31 *Conopeum seurati*. (**a**) Colony on reed, Nieuwpoort (Belgium). (**b**) Area of the colony growing on oysters in the harbor of Vlissingen (the Netherlands). (**c**) SEM image of an area of the same colony

(c) Frontal calcification with few or no pseudopores, kenozooids frequent. .... *Electra* cf. *asiatica* (p. 175)

2 (a) Proximal spine short and thick, other spines absent. Operculum calcified, like an opaque semicircular valve. Estuarine habitats. ....*Einhornia crustulenta* (p. 181)

(b) Spines present or absent. Operculum uncalcified. .... *Electra monostachys* (p. 176)

Electra cf. *asiatica* Grischenko, Dick and Mawatri, 2007?

Description Colony encrusting, unilaminar, radiating in fan-shaped lobes, sometimes in single or double rows of zooids at the colony edge. Zooids elongated, rectangular, or oval, separated by narrow grooves. Mural rim raised, sharp. Cryptocyst narrower distally, wider and gradually sloping proximally, granulated just inside mural rim, smooth around opesial margin. Basal wall calcified. Opesia oval, elliptical, occupying 60–80% of zooidal length. Operculum with chitinized margin, semicircular, with concave proximal margin. Gymnocyst smooth, narrow distally and laterally, usually elongate and tapering proximally, often with minute transverse striations. A few pseudopores are sometimes visible with a light microscope. In most zooids, the gymnocyst proximal to opesia rises to a conspicuous boss bearing a pointed chitinous spine. Some zooids bear one proximal and two distal spines. Zooids interconnect by circular multiparous septula in each wall. No ovicells. Kenozooids of irregular shape and size occur frequently among autozooids; these are completely ringed by a granular cryptocyst and have roughly circular opesia. Ancestrula tatiform (Fig. 4.32).

Occurrence

Germany Only reported by personal communication by Lisa Schüler (IFAOE) from the Wadden Sea and North Sea harbors.

Electra monostachys (Busk, 1854)

Description Colony forms a star-shaped crust. Zooids elongated, proximally narrower. Frontal calcification one-fourth to one-third of the surface, smooth, without pseudopores. Oval to elliptical opesia. Operculum with a thin marginal induration. The number of spines varies considerably; the most constant is the medial proximal spine, which is thin, pointed, and curved over the frontal membrane. A pair of equally slender but straight spines at the distal end of the opesia. Up to four to six pairs of short, curved spines may surround the frontal membrane, but they are often absent. No ovicells.

The zooids often form an oval scar on the inside of shells. Young colonies of *Conopeum reticulum* resemble *Electra monostachys* in that the first zooids grow in a single row in the opposite direction of the ancestrula and have proximal frontal calcification. A good distinction is the fact that with *Electra monostachys* the zooids make a scar in the substrate, which does not happen with *Conopeum reticulum* (Fig. 4.33).

Occurrence Usually on stones or the inner side of empty bivalve shells, often together with *Aspidelectra melolontha*. Occurs in the sea and in river mouths. Rarely washes up on plastic.

Belgium Collected alive on shells on the Kwinte Bank and the Hinder Banks.

The Netherlands Records from Texel (Mulder 1983), shell banks near Terschelling and Vlieland (Dankers and Van Moorsel 2001), south-west of Terschelling and west of Schouwen (Jebram 1968) and from the Goese Sas and from the Western Scheldt (Faasse and De Blauwe 2004). Collected in March 2008 on a roof tile in the tidal pool in Kattendijke. In 2013 in an old cockle in Westkapelle (de Ruijter 2013b).

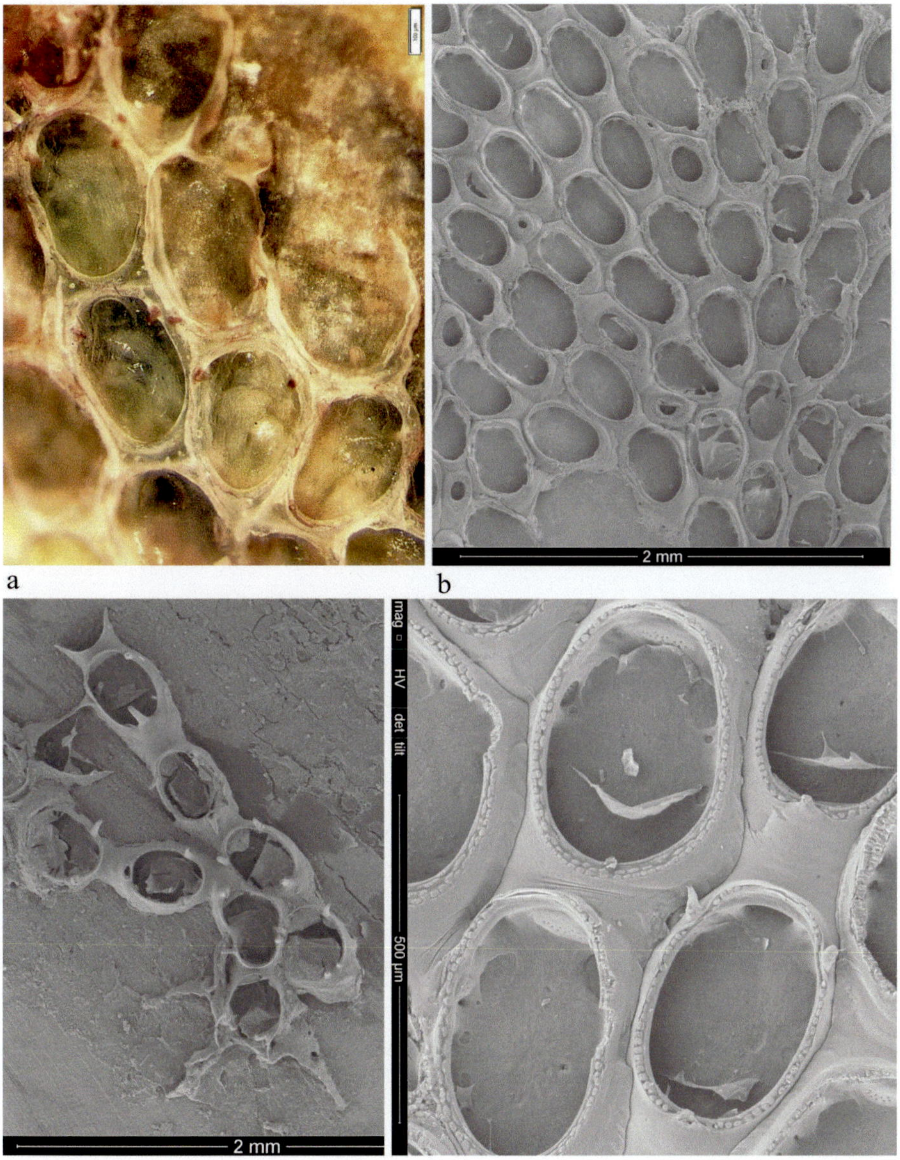

Fig. 4.32 *Electra* cf. *asiatica*. (**a**) Zooids with few pseudopores (photo by J. Leitinger). (**b**) Multiserial colony area with kenozooids. (**c**) Uniserial colony part showing zooids with spines. (**d**) Zooids with multiparous septula

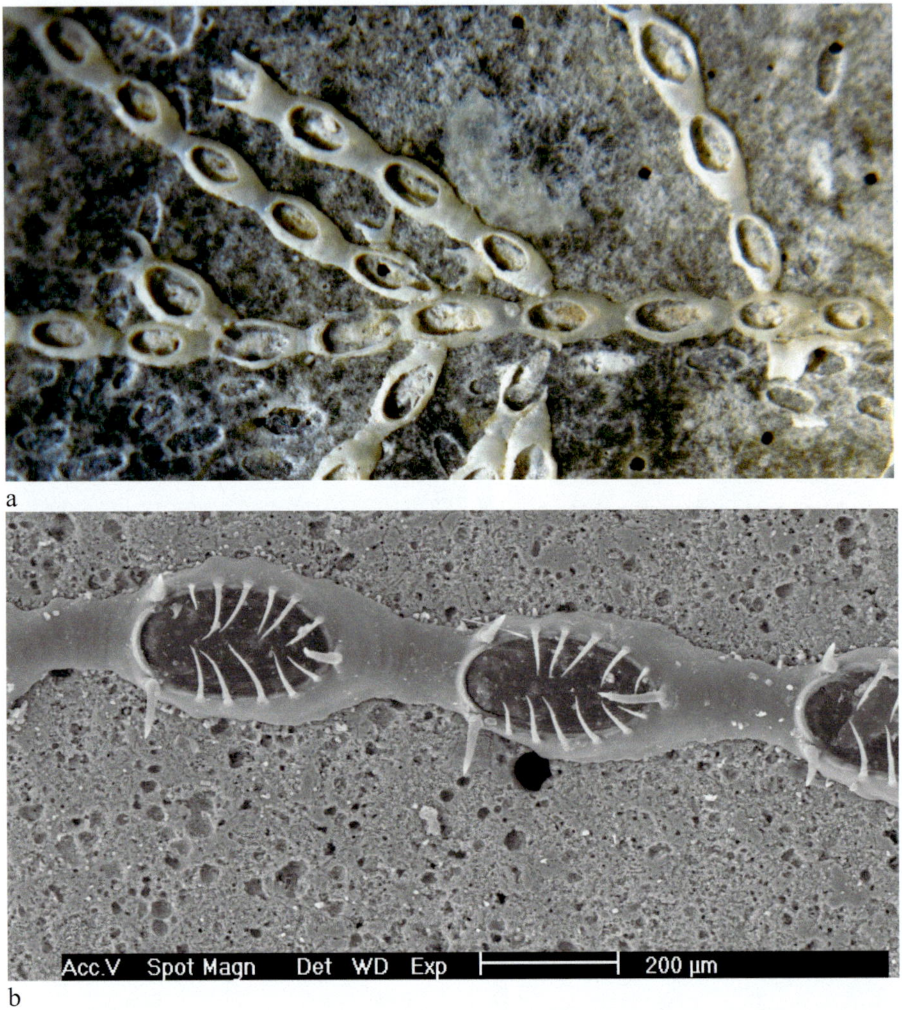

Fig. 4.33 *Electra monostachys*. (**a**) Damaged colony on a shell valve showing the typical growth pattern. (**b**) Zooids with opesia protected by periopesial spines. Kwinte Bank (Belgium)

Collected in 2021 in a valve of *Acanthocardia echinata* on Ameland (pers. comm. Ellen van der Niet). From Ameland to Rottumeroog traces of this species present in old and fossil bivalves Staal (2020). A colony of *Electra monostachys* comprising several dozen zooids was found on one of the scour-protection rocks from platform L10-G (Beukhof et al. 2016).

Germany Collected in 2015 on the inner side of a bivalve on the seaside of Borkum (by de Regt on observado.org with photo). Collected in the German North Sea (Zettler et al. 2018) and at Heligoland (pers. comm. Britta Kind).

Electra pilosa (Linnaeus, 1761)

Description Colony at first encrusting. In the absence of enough substrate (on hydroid stems), the colony forms erect blades of zooids growing back to back. The thin translucent frontal calcification occupies up to half of the frontal surface, often much less. Striking and characteristic are the large round pseudopores in the frontal calcification. 1 to 12 spines surround the oval opesia. The proximal spine is always present, usually longer than other spines and sometimes two to three times as long as the zooid. In this case, the colony will look hairy. Operculum transparent, with a thin cuticular thickening. No ovicells (Fig. 4.34).

Occurrence Grows on just about anything: rocks, shells, hydroids, plastic, algae, and other bryozoans. Definitely the most common species in the area. Encrusting colonies are often present on washed-up material such as plastic and *Himanthalia* buttons.

Belgium Loppens (1906) and the collection in the RBINS testify that this species was also very common in the past. Very common on beachheads and in marinas, as well as off the coast on stones, shells, and shipwrecks (Zintzen 2007), and in inland waters such as the inner harbor of Zeebrugge. Subtidal on rocks of the erosion protection in Belwind, Aug 2018 (pers. comm. F. Kerckhof).

The Netherlands Very common in Zeeland on beachheads and in marinas, as well as in inland waters such as the Sas van Goes. Lengkeek et al. (2013) found it on all of the ten examined shipwrecks in Dutch waters. Collected at the Dutch Borkum Reef Grounds (Coolen et al. 2015).

Germany Collected in the German North Sea (Zettler et al. 2018) and at Heligoland (pers. comm. Britta Kind).

Mass Strandings of *Electra pilosa* in 1965/66 and 2020/21

Prigge (1967) described a remarkable growth form of *Electra pilosa*, in particular, spherical colonies with a diameter of 3–7 cm that can be carried along by the current over the seabed and meanwhile growing in all directions. This growth form washed up on the Dutch coast and on the German Wadden Islands.

He witnessed a massive wash-up on the island of Norderney, the old island residents never saw anything like this. That means 1965 was probably the first event since 1900. A second, more limited wave came in 1966. The phenomenon was so drastic that there was odor nuisance that caused problems for tourism and fishing.

After that, a massive wash-up was not reported, until de Ruijter (2019d) mentioned lots of colonies washed up at the end of September 2019 at Terschelling. The NIOZ (Royal Netherlands Institute for Sea Research) has seen the amount of *Electra pilosa* increasing in samples in recent years, even at a distance from the coast.

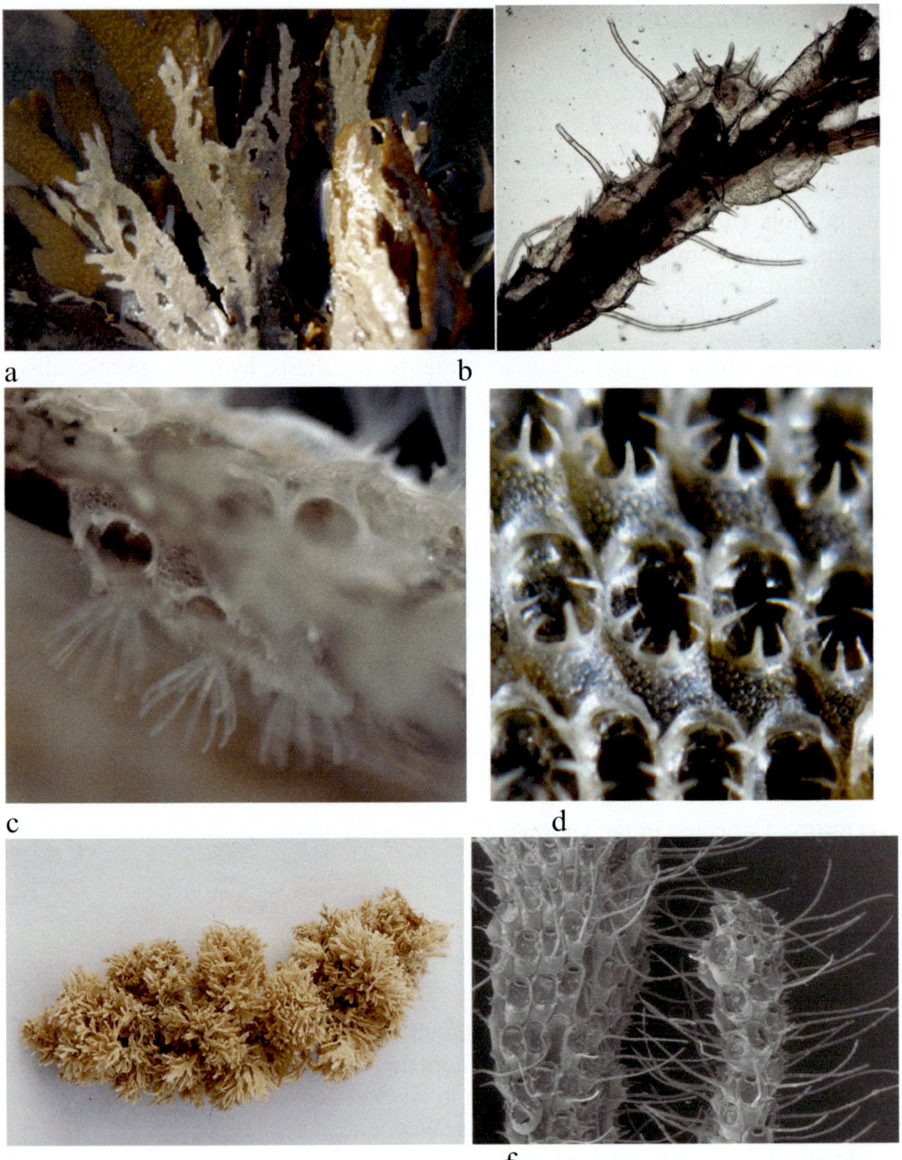

Fig. 4.34 *Electra pilosa.* (**a**) Colony on *Fucus* at Audresselles (France). (**b**) Zooids on hydroid stem, showing both short periopesial and long proximal spine. (**c**) Living colony on hydroid, zooids with expanded lophophores, proximal spine short. (**d**) Zooids showing periopesial spines and typical pseudopores in proximal calcified frontal wall (gymnocyst). (**e**) Colony with many erect bilaminar colony fronds. (**f**) Erect colony parts with very long proximal spine (**b**–**f**) colonies collected on Belgian beaches

A significant mass beaching in North Holland started on November 6, 2020. *Electra* was deposited in carpets hundreds of meters long, tens of meters wide, and from 15 cm to knee height. From mid-December to February 2021, entire beach strips on Texel were covered with *Electra pilosa*. At the end of June and in summer, the beaches of Ameland are particularly covered with long carpets of *Electra pilosa*. Because the event occurred in the summer, it had a severe impact on tourism and was mentioned in the newspapers and on the radio in Friesland (De Blauwe et al. 2021).

Benthos research in December 2020 by NIOZ along a large part of the North Holland coast up to the west point of Terschelling suggests that the massive presence of material on the Texel beach was fed by transport from the south. We probably only see a small part of it washing-up on Dutch beaches.

A mass occurrence of *Electra pilosa* on the German North Sea Coast started in September 2021 (Britta Kind, pers. comm.). *Electra pilosa* was found en masse in dredges in the German North Sea area in late summer 2021, and it could also be seen on underwater photos of the Steingrund, an area northeast of Heligoland.

Nauta et al. (2023) determined by means of molecular analyses that two *Electra pilosa* species are present in Europe (Bleed1172 and Bleed819) based on material collected in 2021 and 2022. The mass occurrence in the Netherlands is due to *Electra pilosa* Bleed819, probably an introduced species, morphologically identical to the *E. pilosa* sensu stricto (Fig. 4.35).

Einhornia crustulenta (Pallas, 1766)

Description Colony encrusting, white. Zooids elongated, rectangular, or oval, separated by narrow grooves. The frontal calcification comprises less than one-fourth of the total length. A narrow fine-grained calcareous margin extends around the proximal half of the elliptical opesia. Operculum very conspicuous, slightly calcified, and visible as an opaque structure. Operculum with transverse proximal border shallowly concave; operculum height distinctly shorter than width (Nikulina 2007). No ovicells. Lateral spines absent; the medial proximal spine (small and rare) is a low chitinous cone on top of a calcified hump; later the top can calcify.

It forms "growing stones" in brackish water in the Netherlands. Often, in these stones, *E. crustulenta* and *C. seurati* grow on top of each other. In the Netherlands, zooids occur with little frontal calcification and without a spine, making it difficult to distinguish between the two species. The calcified operculum is the most obvious feature (Fig. 4.36).

Occurrence On many substrates, limited to estuarine habitats.

Belgium In the brackish part of the Oostende harbor, in the marina of Zeebrugge, in the Nieuw Bedelf in Nieuwpoort, and at the marina of Doel in the Western Scheldt. Colony fragments collected in 2023 from the Edebeek in Damme date from the Middle Ages, when a branch of a salty/brackish creek came inland.

a

b

Fig. 4.35 *Electra pilosa.* (**a**) Spherical colony found at Katwijk, December 3, 2020 (photo by J. van Duijn). (**b**) Mass stranding at Texel on December 7, 2020 (photo by S. Dijksen)

Fig. 4.36 *Einhornia crustulenta.* (**a**) Colony on bark, Den Inkel (the Netherlands). (**b**) SEM image of the colony area showing zooids without proximal spine. (**c**) Zooids with proximal spine (*arrows*) and calcified operculum

The Netherlands Sites are known throughout the coastal region (Faasse and De Blauwe 2004). Bijma and Boekschoten (1985) provide a distribution map of the "growing stones" in the southwest of the Netherlands. Very recent: Wadden Sea on a fragment of *Ensis*, together with *C. reticulum* (pers. comm. L. van Walraven, 2020). Recently found in the Zoutsloot in Harlingen (de Boer et al. 2020). Lauwersmeer? (Lawrentsmeer in Nikulina 2007).

Germany Heligoland, Sylt, Jade Bight.

Genus *Pyripora*

Pyripora catenularia (Fleming, 1828)

Description
The colony forms long chains of zooids, branching across the substrate. Zooids elongated. Frontal gymnocystal calcification about half the zooid length, tapering proximally; smooth, fine-grained, and without pseudopores, some shallow transverse grooves or ridges may occur. Opesia bordered by a very narrow granular cryptocystal margin. Operculum with a distal marginal thickening, placed at the distal end of the frontal membrane. No ovicells. Zooids grow over obstacles but not over each other. A developing autozooid bud that encounters another autozooid, might become a kenozooid. Looks quite similar to spineless *Electra monostachys* but the calcification is thicker (Fig. 4.37).

Occurrence

Belgium Beached in December 2000 on a shoe sole and on plastic in September 2001 (De Blauwe 2003a), autumn 2004 (De Blauwe 2005a), and January 2005.

The Netherlands A colony washed up on rope or plastic (de Ruijter 2013a, 2014b, c, d, 2015d, 2018c).

Genus *Aspidelectra*

Aspidelectra melolontha (Landsborough, 1852)

Description Colony encrusting, unilaminar, multiserial, forming fan-like lobes. Autozooids in regular linear branching series. Gymnocyst well developed. Frontal membrane obscured by a ribcage-like shield comprising 12 to 15 flattened spines. Lateral spines fuse along the midline of the shield, one slit-like lacuna between neighboring spines. Two short, thick cylindrical spines present, each on one side of the orifice. A third may coincidentally be present at the proximal end of the frontal shield. Polypide with nine to ten tentacles. No avicularia or ovicells. Ancestrula tatiform (Fig. 4.38).

Occurrence Often together with *Electra monostachys* on stones and in shells and can occur frequently on the inside of empty oysters, mussels, and cockles.

Belgium Found on the Flemish Banks. Can be found on the beach after sand nourishment (Knokke, Oostende).

The Netherlands Along the entire Dutch coast, in the Western Scheldt and the Oosterschelde (Lacourt 1978). Can be found on the beach after sand nourishments in Nieuwvliet and from Westkapelle to Domburg, mostly in old valves of *Cerasto-*

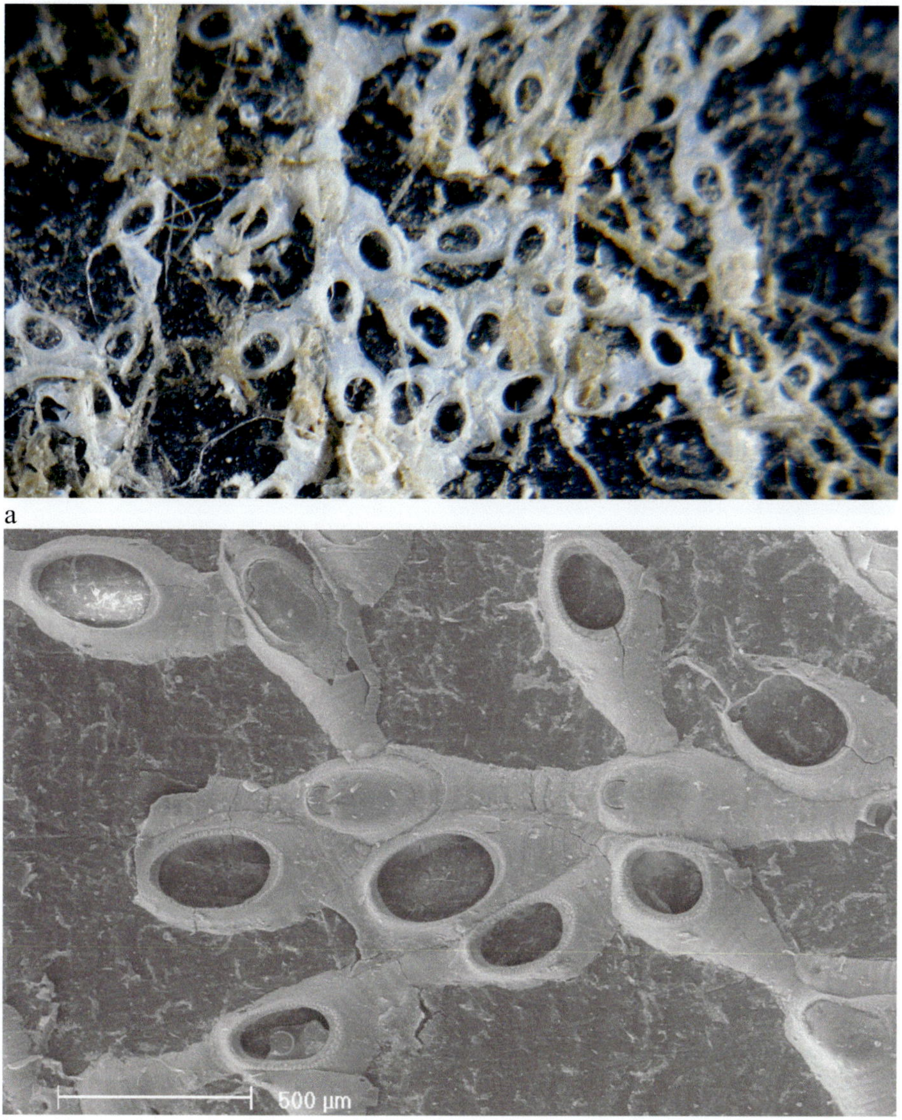

Fig. 4.37 *Pyripora catenularia*. (**a**) Colony on beached plastic (Wales, UK). (**b**) SEM image of zooids from the same colony showing intramural budding (zooid is formed inside older zooid)

derma edule. Sometimes on washed-up plastic (de Ruijter 2011c, 2012b, 2014b, c, 2015a, 2019b, 2020c).

Germany Collected in the German North Sea (Zettler et al. 2018) and at Heligoland (pers. comm. Britta Kind).

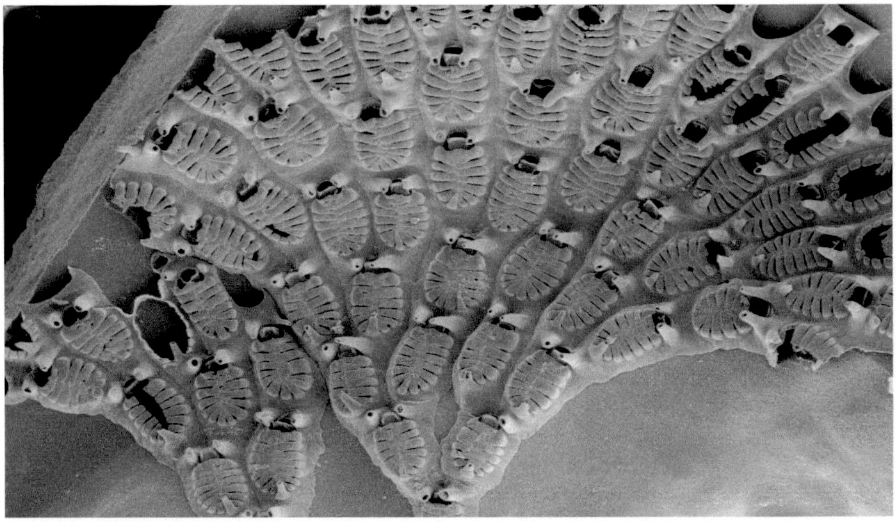

a

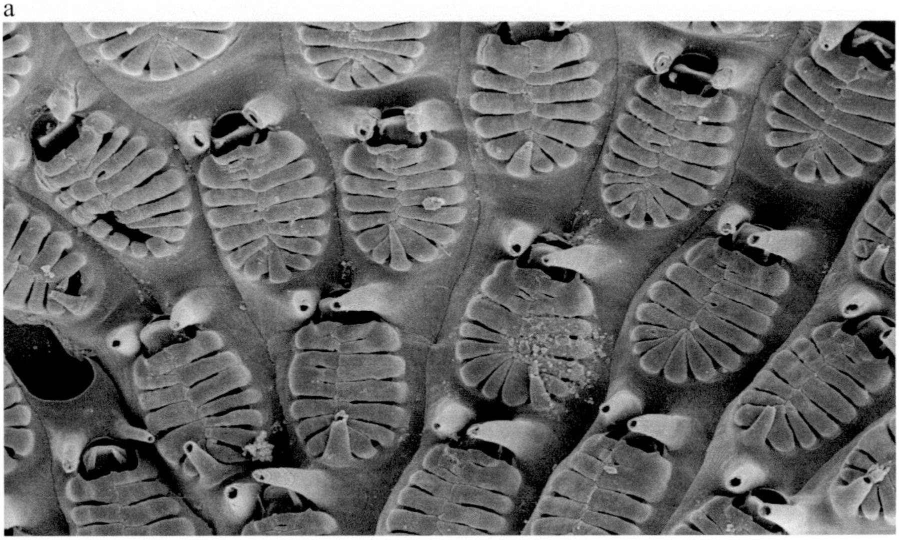

b

Fig. 4.38 *Aspidelectra melolontha*. Colony growing on the inner side of shell valve of *Cerastoderma edule*, Kwinte Bank (Belgium). (**a**) Showing typical growth pattern. (**b**) Zooids with and without proximal spine

Family Flustridae

Genus *Flustra*

Flustra foliacea (Linnaeus, 1758)

Description

The colony begins as a flat crust on the substrate. Erect bilaminar lobes are formed in the second year. Growth from March to September. The winter rest causes growth lines on the leaves. Twelve-year-old colonies have been reported. The colony forms a flexible seaweed-like tuft up to 20 cm high, typically leaf-like and terminating in rounded lobes. Light gray-brown, when fresh with a lemon scent. Zooids of the bilayered fronds are tongue-shaped. four to five short thick distal spines, narrowed at the base and club-shaped.

Avicularia interzooidal, half the size of a zooid; proximal part triangular, with frontal membrane, distal part semicircular and erect with hood-like rostrum and a strongly cuticularized mandible. Ovicells endozooidal, visible as a crescent-shaped cap distal to the parent zooid, its opening separated from the zooid opening, distal to the operculum. Zooids of the first-year incrustation are quite similar to *Conopeum reticulum*, without kenozooids and with two large distal spines that are longer than in *M. membranacea*. Polypide with 13 to 14 tentacles (Fig. 4.39).

Occurrence Sublittoral and mostly on rough bottoms with strong currents. They form a rich microhabitat with *Bugulina flabellata*, *Cradoscrupocellaria* spp. and *Crisia eburnea* as dominant epizoites (Stebbing 1971a). Stebbing (1971b) lists 25 bryozoans found on *Flustra foliacea*. Distribution and abundance significantly reduced, presumably due to bottom disturbance from beam trawling.

Belgium The numerous material preserved in the RBINS shows that this species was very common on the Flemish Banks at the beginning of the twentieth century. During a campaign in June 2005, not a single colony was found on the Hinder Banks (De Blauwe et al. 2006). A young colony on a rock was found off Zeebrugge in July 2001. Collected from the shipwreck of S.S. Ptarmigan (51°41.332′ N, 02°25.87′ E) (van Leeuwen 2023).

The Netherlands Washes ashore less and less often; a strong decline was already reported in Lacourt (1978). Collected on a wreck on the Doggerbank (55°26.090 N, 03°51.170′ E) (van Leeuwen 2023).

Germany Collected in the German North Sea (Zettler et al. 2018) and at Heligoland (pers. comm. Britta Kind).

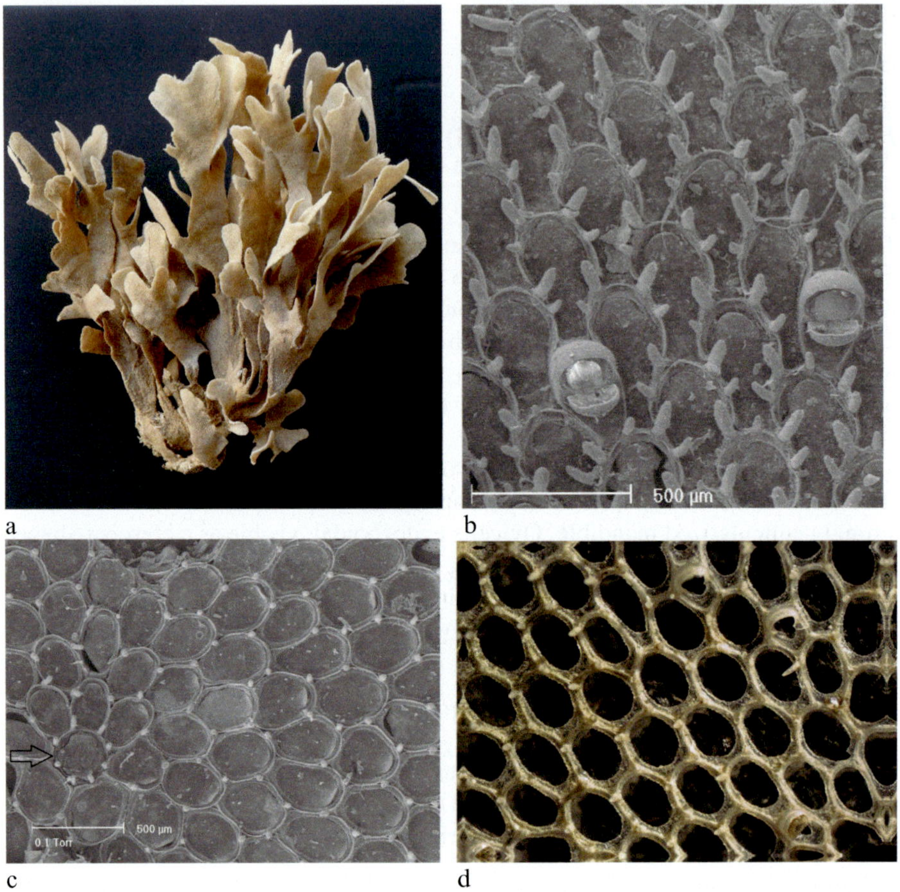

Fig. 4.39 *Flustra foliacea*. (**a**) Erect colony beached at Zeebrugge (Belgium). (**b**) Zooids and two avicularia in an erect colony part. (**c**) Area in young encrusting colony with ancestrula (*arrow*), on beached plastic Northumberland (UK). (**d**) Encrusting colony area with avicularia and a few kenozooids (photo by R. de Ruijter)

Genus *Chartella*

Chartella papyracea (Ellis and Solander, 1786)

Description

The colony forms a whitish, delicate tuft up to 3 cm high. Resembles miniature of *Flustra foliacea*. The narrow fronds are bilaminar. Zooids rectangular 0.5 × 0.2 mm. Walls thin, weakly calcified; a short thick spine on the distal corners of the zooid. Operculum weakly sclerotized, with a thin marginal thickening. Avicularia are absent. Ovicells endozooidal, visible as a small domed hood, the opening distal to the operculum. Embryos pale orange (Fig. 4.40).

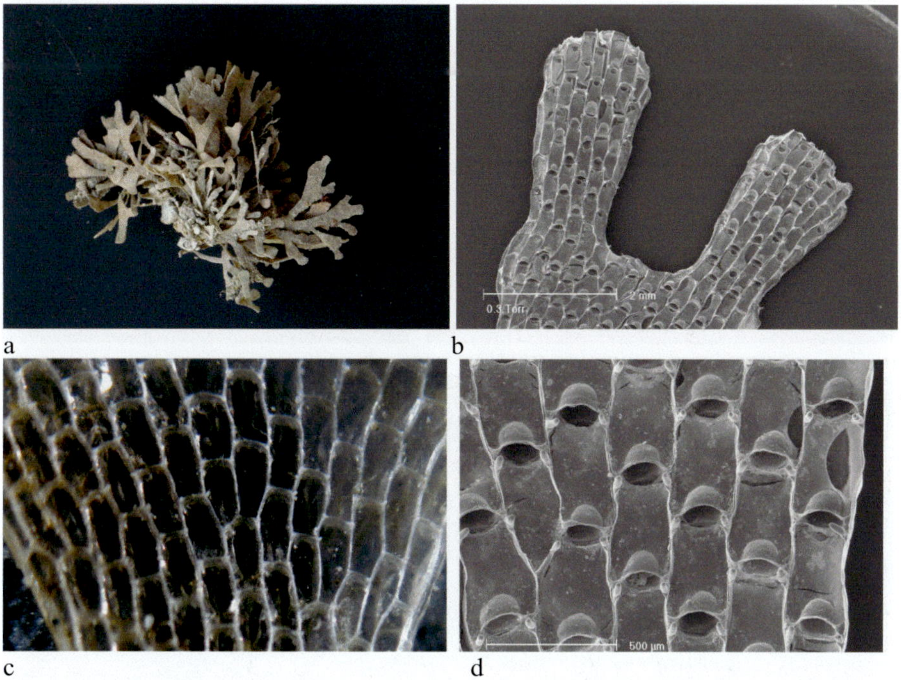

Fig. 4.40 *Chartella papyracea.* (**a**) Colony growing on a stone in St-Jacut (France). (**b**) Typical leaf tip. (**c**) Colony area consisting of zooids without ovicells. (**d**) Colony area with ovicellate zooids

Occurrence

Single colonies were collected off the Belgian coast in 1899, 1900, and 1905 (RBINS collection). In that period, many colonies were also collected from the English Channel. In recent years, a colony washed up on a fishing net in Oostende. Collected on the shipwreck of Rode Poon (SS Trevier, 52°26.201 N, 03°43.970 E) in 2021 (van Leeuwen 2023).

Genus *Securiflustra*

Securiflustra securifrons (Pallas, 1766)

Description The colony forms a whitish tuft up to 10 cm high. Fronds often wedge-shaped with angular apex. Lateral fronds originate on lateral kenozooids. Zooids rectangular, long and narrow, 0.7 to 1.1 mm long, and 0.14 to 0.2 mm wide. The distal angle of the zooid is spineless. The frond margin is bordered by a row of kenozooids. Avicularia sit between the zooids in the zooid series; they are quadrangular, with a raised and thickened rostrum and semicircular, brown mandible. Ovicells endozooidal (Fig. 4.41).

Fig. 4.41 *Securiflustra securifrons*. (**a**) Colony beached at Texel (the Netherlands, photo by R. Dekker). (**b**) Colony beached at Newton by the Sea (UK). (**c**) Colony area showing zooids and a few interzooidal avicularia

Occurrence A well-known species from Northumberland (UK), frequently beached at Seahouses, collected at the wreck of the Somali, 5 miles from the Farne Islands during the DDNZS campaign in June 2018.

Belgium *Securiflustra securifrons* is a northern species not expected here. All samples in the RBINS labeled as *S. securifrons* were in fact *Chartella papyracea*.

France There is one colony in the RBINS, fished up in the Channel on February 7, 1906, at a depth of 28–34 m on the range 50°45′ N–1°3′ E to 50°46′ N–1°9′ E.

The Netherlands In March 2003, a plastic chair washed ashore on Vlieland, on which about 20 colonies were found that were identified as *S. securifrons* (de Ruijter 2003a). This material, which could have been *Chartella papyracea*, was not collected.

In 2021, two colonies were found on the beach of Texel, after sand nourishment with sand collected at 13.4 km off the coast of Texel (May 10 by Sytske Dijksen; May 11 by Rob Dekker (de Ruijter 2021b)). In 2022 on September 25, Rob Dekker found four colonies, up to 13 cm long on the beach near Wassenaarse Slag near Den Haag. Sand nourishment was executed on the beach of Katwijk aan Zee, about 2 km north of the place where he found the *Securiflustra* colonies. The nourishment sand is extracted from the North Sea approximately 14 km west of Katwijk (Rijkswaterstaat information site).

Germany Collected in the German North Sea (Zettler et al. 2018).

Genus *Hincksina*

Revised Diagnosis (Berning et al. 2021) Colony encrusting. Zooecia well calcified, communication via uniparous septula. Narrow gymnocyst present but variably developed, cryptocyst reduced to practically absent. Cylindrical or flattened oral as well as mural spines of variable shape encircling the opesia, spines in late astogenetic zooids may be branched and/or dimorphic to polymorphic in some species, spine bases jointed or fully calcified. Brooding in endozooidal ovicells, not closed by the operculum but by an ooecial vesicle, ooecium a short hood formed by the proximal cystid of the distal zooid or avicularium, ectooecium and endooecium calcified. Avicularia interzooidal, occasionally vicarious; rostrum round or pointed, palatal opesia framed by an immersed calcified shelf, mandible hinged on a pair of lateral teeth. Kenozooids, which replace an autozooid or avicularium, may occasionally be present. Ancestrula tatiform (with membranous frontal wall).

Hincksina flustroides (Hincks, 1877)

Description Colony encrusting, unilaminar, multiserial, forming light brown circular or lobed patches. Autozooids oval to rectangular 0.37–0.52 × 0.2–0.3 mm, in irregular series, separated by shallow grooves, skeleton whitish translucent when dried. Vertical walls with two or more uniparous septula per neighboring zooid. Gymnocyst reduced to practically absent, cryptocyst not developed, opesia extensive. Opercular region laterally framed by a pair of vertically directed, cylindrical or slightly flattened oral spines in both auto- and ovicellate zooids; frontal membrane in mature zooids overarched by 8 to 12 (most often 9 to 10) cylindrical to flattened subclavate mural spines of variable width, the distal ones may be bifid while the proximal spines are generally smaller and shorter as well as more cylindrical, lateral spines of variable length and either leaving a central gap in some zooids or extending beyond the zooidal midline in others; all spines jointed at their base. Ovicells endozooidal, ectoooecium continuous with the proximal gymnocyst, forming a narrow but broad hemispherical cap, proximal margin raised centrally, often producing a central peak (Fig. 4.42).

Occurrence

Belgium On beached plastic in 1999 and twice in 2000. In June 2005, a dead colony was sampled on a shell on the Hinder Banks.

The Netherlands Colonies on beached plastic (de Ruijter 2013c, 2014b, 2017c, 2024b).

Fig. 4.42 *Hincksina flustroides.* (**a**, **b**) Different parts of the same colony on beached plastic Zeebrugge (Belgium)

Superfamily Calloporoidea

Family Calloporidae

Genus *Callopora*

1 (a) Four spines at the distal end of young zooids, only one or two remaining later; two spines in fertile zooids, one or both may be greatly enlarged. .... *Callopora dumerilii* (p. 195)
(b) More than four spines, usually at least eight around the opesia. 2

2 (a) Eight to twelve spines, broad and flattened, spanning the frontal membrane and meeting in the middle; 2 pairs of erect distal spines. .... *Callopora rylandi* (p. 197)
(b) Spines around the opesia sometimes over the membrane, but not flattened and not meeting in the middle. 3

3 (a) Calcified ectooecium of ovicell with a transverse membranous area bounded by a sinuous ridge; 12 to 15 spines around the oval opesia. .... *Callopora craticula* (p. 193)
(b) Ovicell not so. 4

4 (a) Opesia surrounded by 17 to 22 thin spines, curved over the frontal surface. Avicularia absent. .... *Callopora discreta* (p. 194)
(b) Eight to eleven spines; straight, thick, and cylindrical. Avicularia on the proximal part of the zooid. .... *Callopora lineata* (p. 197)

Callopora craticula (Alder, 1856)

Description Colony forms small patches. Autozooids oval, vitreous. Gymnocyst about one-quarter of the frontal surface; 12 to 15 thin, pointed spines, inclined inward over the oval opesia and almost meeting in the midline of the autozooid. A small avicularium present on the gymnocyst: rostrum narrow, pointed, directed distally or proximally. Ovicell prominent, spherical, ectooecium with a transverse membranous area that is bounded by a thickened sinuous ridge. Embryo color red to orange. Ancestrula identical to autozooids but with about nine spines (Fig. 4.43).

Occurrence On stones, shells, and algal holdfasts at extreme low water during spring tides and subtidal. Reported from the German North Sea (Zettler et al. 2018) and from Heligoland (pers. comm. Britta Kind).

Fig. 4.43 *Callopora craticula*, colony on a dead shell at Heligoland. (Photo by B. Kind)

Callopora discreta (Hincks, 1862)

Description Colony encrusting, unilaminar, multiserial, usually less than 50 zooids. Zooids ovate, small, separated by deep grooves. Lateral wall thickened and slightly projecting as a border around the opesia, with 17 to 22 short, thin, pointed spines; two most distal usually erect or curved slightly outward, the rest strongly curved over the frontal membrane. Spine base strongly thickened. Avicularia absent. Ovicells prominent, spherical, thin membranous ectooecium easily removed in dry colonies, endooecium thickly calcified, tuberculate (Fig. 4.44).

Occurrence

Belgium In 2004 and 2005, living colonies were discovered on the Kwinte Bank and deceased colonies on the Hinder Banks; there were no previous reports from the North Sea. A colony was found in the RBINS collection on a shell that Gilson fished up in 1905 on the Hinder Banks.

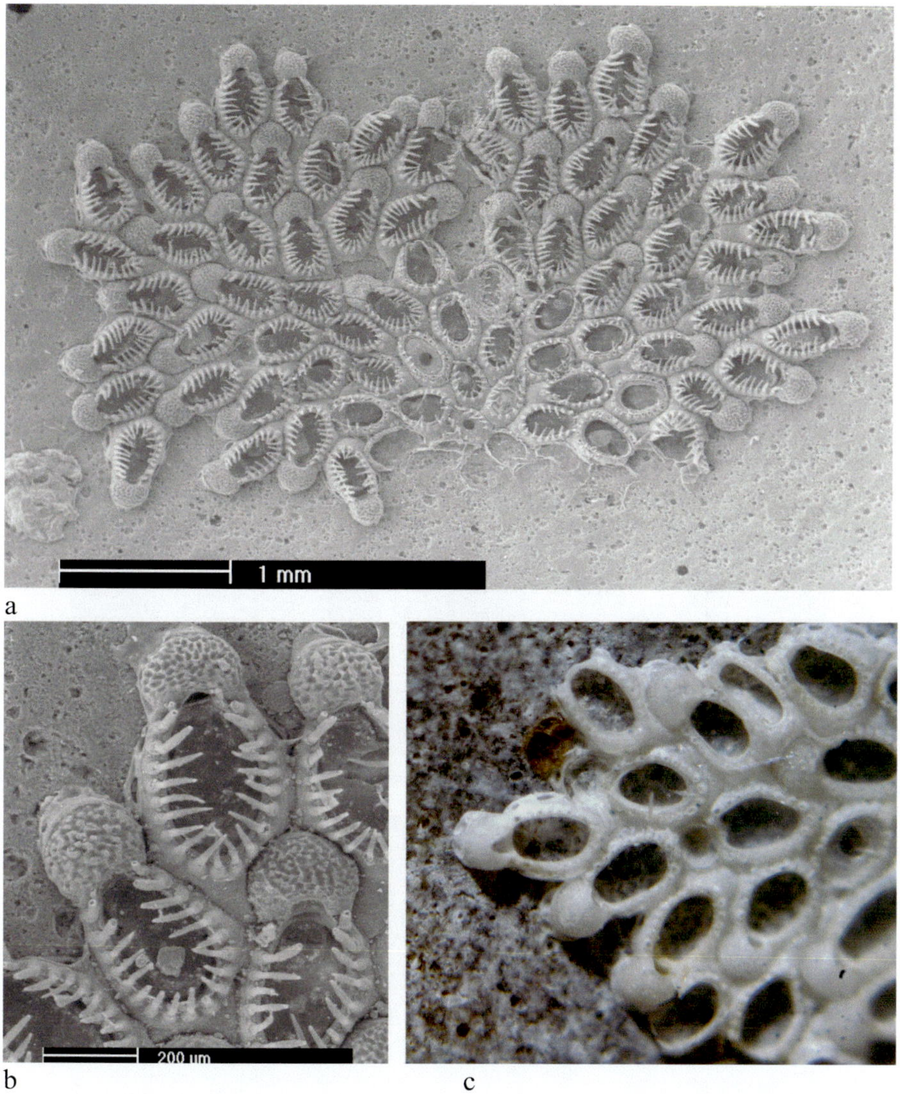

Fig. 4.44 *Callopora discreta*. (**a**) Colony on the shell of *Cerastoderma edule*. (**b**) Ovicellate zooids with tuberculate endooecium and periopesial spines. (**c**) Zooids with ovicells and broken spines. Kwinte Bank (Belgium)

Callopora dumerilii (Audouin, 1826)

Description Colony encrusting, unilaminar, multiserial. Autozooids ovate, sometimes broadened proximally. Separated by deep grooves. Frontal calcification small, triangular, smooth, and vitreous. Four (rarely six) spines at the distal end of the zooid; distal pair small, pointed and erect, usually absent, especially near ovicells, proximal pair always present, one or both may be strongly elongated and thickened, straight or curved, erect or draped over the frontal surface. Avicularia small, placed on the frontal calcification, with a pointed rostrum, often absent. Ovicells prominent, spherical, thin membranous ectooecium easily removed in dry colonies, endooecium thickly calcified, granulated. Ovicell not closed by the autozooidal operculum. Eggs and embryos orange. Colonies without ovicells can be confused with *Amphiblestrum auritum* (Fig. 4.45).

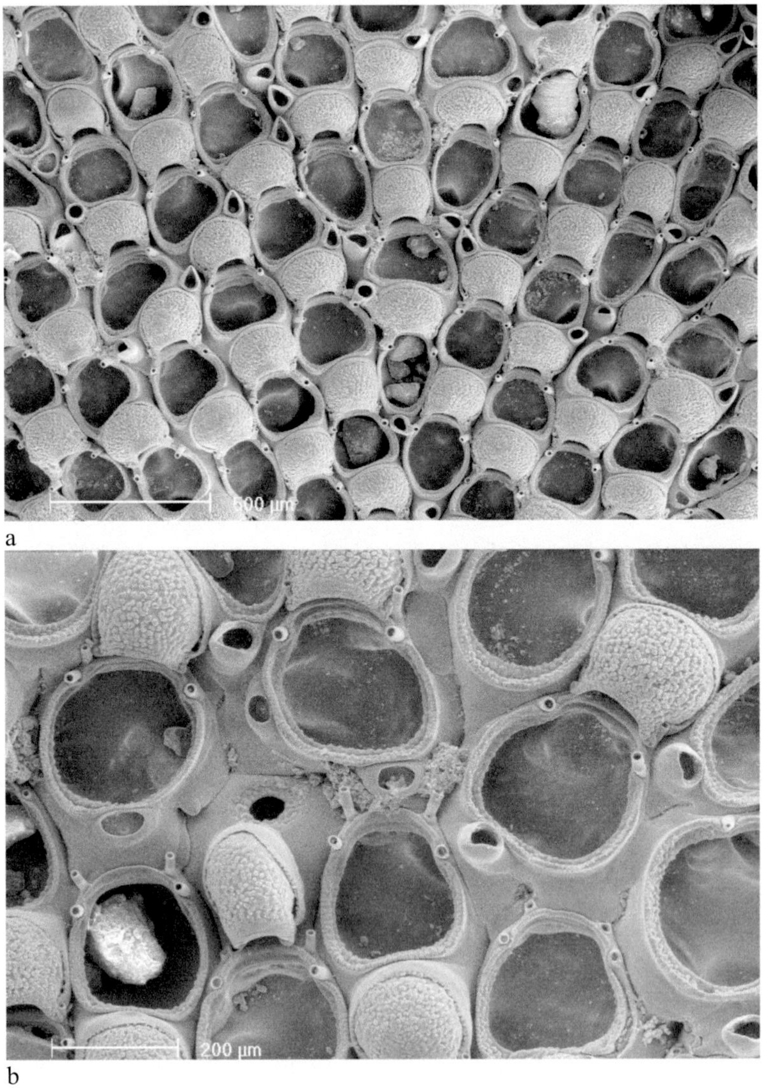

Fig. 4.45 *Callopora dumerilii.* (**a**) SEM image of colony area growing on beached plastic. (**b**) Magnified view showing zooids with ovicells, avicularia, and oral spines

Occurrence In Belgium (De Blauwe 2005a) and the Netherlands (de Ruijter 2011a, 2012a, 2015a) on washed-up objects and *Himanthalia buttons* originating from the Channel coast and Brittany in France.

Belgium Live colonies were found in 2005 on pebbles between the Hinder Banks. Also on calcareous tube worms on buoys at sea (pers. comm. F. Kerckhof). Subtidal on rocks of the erosion protection in Belwind, August 2018 (pers. comm. F. Kerckhof). On the shipwrecks of Birkenfels and Ptarmigan in 2021 (van Leeuwen 2023).

The Netherlands In Zeeland, this species lives near the Goese Sas and on the north side of the Zeeland Bridge (Faasse and De Blauwe 2004), as well as on the Klaver Bank (pers. comm. G. van Moorsel). In 2023 two colonies were collected within the Marine Alien Species Detection Network of the Netherlands, one in the central Oosterschelde just off Zierikzee, and the other more inland in the Krabbenkreek (Gittenberger et al. 2024). A small colony was collected on shipwreck NL 3251 in July 2024 (campaign DDNZS).

Germany Collected in the German North Sea (Zettler et al. 2018) and at Heligoland (pers. comm. Britta Kind).

Callopora lineata (Linnaeus, 1767)

Description Colony encrusting, unilaminar, multiserial. Autozooids rectangular to irregularly polygonal, separated by shallow grooves. Opesia large. Lateral walls erect as a rim around the elongated oval opesia, 8 to 11 (rarely 6 to 14) straight, thick cylindrical spines. Avicularia oriented differently, on the proximal part of the zooid. Ovicells prominent, spherical, closed by zooidal operculum, ectooecium calcified with frontal membranous area. Ovicell not closed by the autozooidal operculum. Embryos pink or orange (Fig. 4.46).

Occurrence

France In 1908 a specimen was dredged just north of Cap Gris-Nez, France (RBINS collection).

In Belgium and the Netherlands (de Ruijter 2007, 2012a, 2020c) on washed-up objects and especially on *Himanthalia* buttons.

Germany Collected in the German North Sea (Zettler et al. 2018) and at Heligoland on beached *Himanthalia*, probably originating from the English Channel.

Callopora rylandi Bobin and Prenant, 1965

Description Colony encrusting, unilaminar, multiserial. Autozooids oval. Frontal membrane obscured by a ribcage-like shield comprising three to six pairs of lateral and one pair of proximal, flattened spines. Lateral spines broadened and fused in the

a

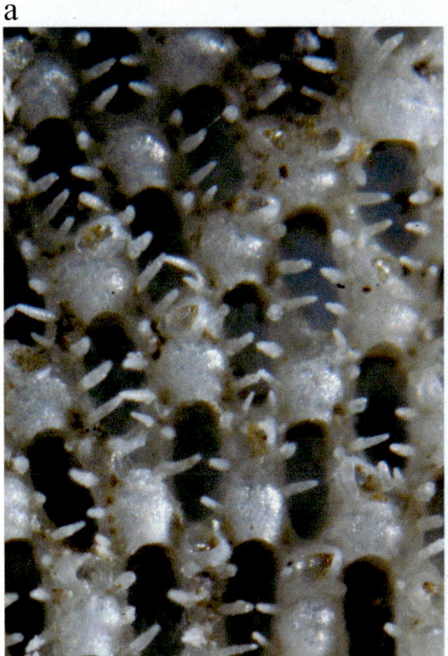

b

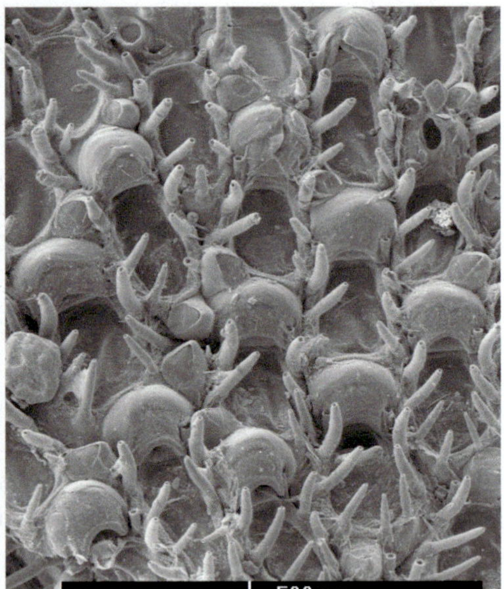

c

Fig. 4.46 *Callopora lineata.* (**a**) Colony area showing ovicellate zooids with embryos, on beached *Himanthalia* button (Zeebrugge, Belgium). (**b**) Colony area, on beached plastic (De Haan, Belgium). (**c**) Zooids with ovicells, and oral and periopesial spines, on beached *Himanthalia* (Oostduinkerke, Belgium) central part of the shield, one wide slit-like lacuna between neighboring spines. Four to six erect cylindrical oral spines, reduced to four in ovicellate zooids. A small avicularium is sometimes present on the proximal gymnocyst of the zooid, rostrum pointed and acute to the frontal plane. Ovicells prominent, globular, thin membranous ectooecium is easily removed in dry colonies, endooecium calcified, slightly granulated, with an inconspicuous rim distally. Ovicell not closed by the autozooidal operculum. Embryos orange. Ancestrula small, tatiform, surrounded by ten erect cylindrical spines.

In the periancestrular zooids, all spines are erect or acute, not meeting in the middle. Young colonies can therefore be confused with other species, for example, *Callopora lineata*. *Callopora rylandi* may also be confused with *Membraniporella nitida* but the frontal shield consists of less-flattened spines with larger intercostal spacing and its ovicell is not closed by the zooidal operculum as it is in *M. nitida* (Fig. 4.47).

Fig. 4.47 *Callopora rylandi.* (**a**) Part of the colony with ovicells and rib-like spines, on beached *Himanthalia*, Oostduinkerke (Belgium). (**b**) SEM image of the same colony

Occurrence Recently found in Belgium and the Netherlands on *Himanthalia* buttons washed ashore (De Blauwe 2000c, 2005a; Vanhaelen et al. 2006; de Ruijter 2011c, 2012a, 2014a, 2017c, 2020c).

Germany Collected at Heligoland (pers. comm. Britta Kind).

Genus *Crassimarginatella*

Crassimarginatella solidula (Hincks, 1860)

Description Colony encrusting, unilaminar, multiserial. Zooids ovate, closely adjacent, separated by shallow grooves. Frontal calcification minor, usually hidden by one or two smooth, calcified nodules; opesia oval. Four distal short spines in young zooids. Only two spines were found in young zooids in Belgian material.

Irregularly formed kenozooids with extensive calcification and reduced opesia may occur. Ovicells with depressed ooecium, wider than long, distally rather flat, smooth calcified ectooecium with a frontal uncalcified area sometimes surrounded by a narrow rim. Avicularia rare, as large as an autozooid, interzooidal, with two sharp pivots medially that grow toward each other from the lateral margin over the avicularium.

Ancestrula is an autozooid, but half the size with 11 to 13 short spines around the opesia (Fig. 4.48).

Occurrence In 2004–2005, five colonies were discovered on the Hinder Banks, of which one on a pebble, the others on the inside of loose shell valves.

Genus *Cauloramphus*

Cauloramphus spinifer (Johnston, 1832)

Description Colony encrusting, unilaminar, multiserial. Autozooids ovate, closely spaced and indistinctly separated by deep grooves. Opesia oval and occupy most of the surface. Lateral wall thick, with 11 to 16 spines, 4 at the distal end, erect, the rest regularly rounded and drooping over the opesia. Spines thick and blunt, hollow, basal with a joint, often short, about half the width of the opesia, but up to twice as long.

Avicularia club-like, with a narrow base, sometimes one or two between the spines, but often absent. Ovicells with internal brood chambers and very small, hardly recognizable ooecium visible like a shallow crescentic furrow at the distal end of the zooid (Ostrovsky et al. 2007). Ancestrula ovate, half the size of an autozooid with about 12 spines around the frontal membrane (Fig. 4.49).

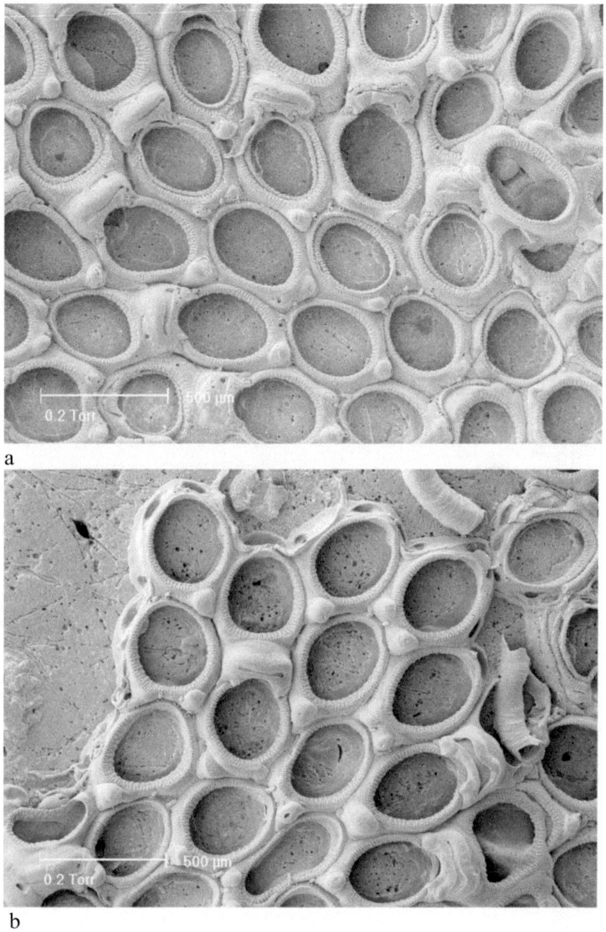

Fig. 4.48 *Crassimarginatella solidula.* (**a**, **b**) Two areas within the same colony, Westhinder Bank (Belgium)

Occurrence In Belgium and the Netherlands on washed-up plastic and *Himanthalia* buttons, originating from the English Channel (De Blauwe 2000c; de Ruijter 2011d, 2015c, 2018a).

Genus *Parellisina*

Parellisina curvirostris (Hincks, 1862)

Description Colony encrusting, unilaminar, multiserial. Autozooids ovate or with an irregular outline, separated by shallow grooves. Lateral walls rise to a granular cryptocystal rim around the opesia. Short spines may surround the opesia, typically one medial and proximal and one on each side of the orifice, frequently absent.

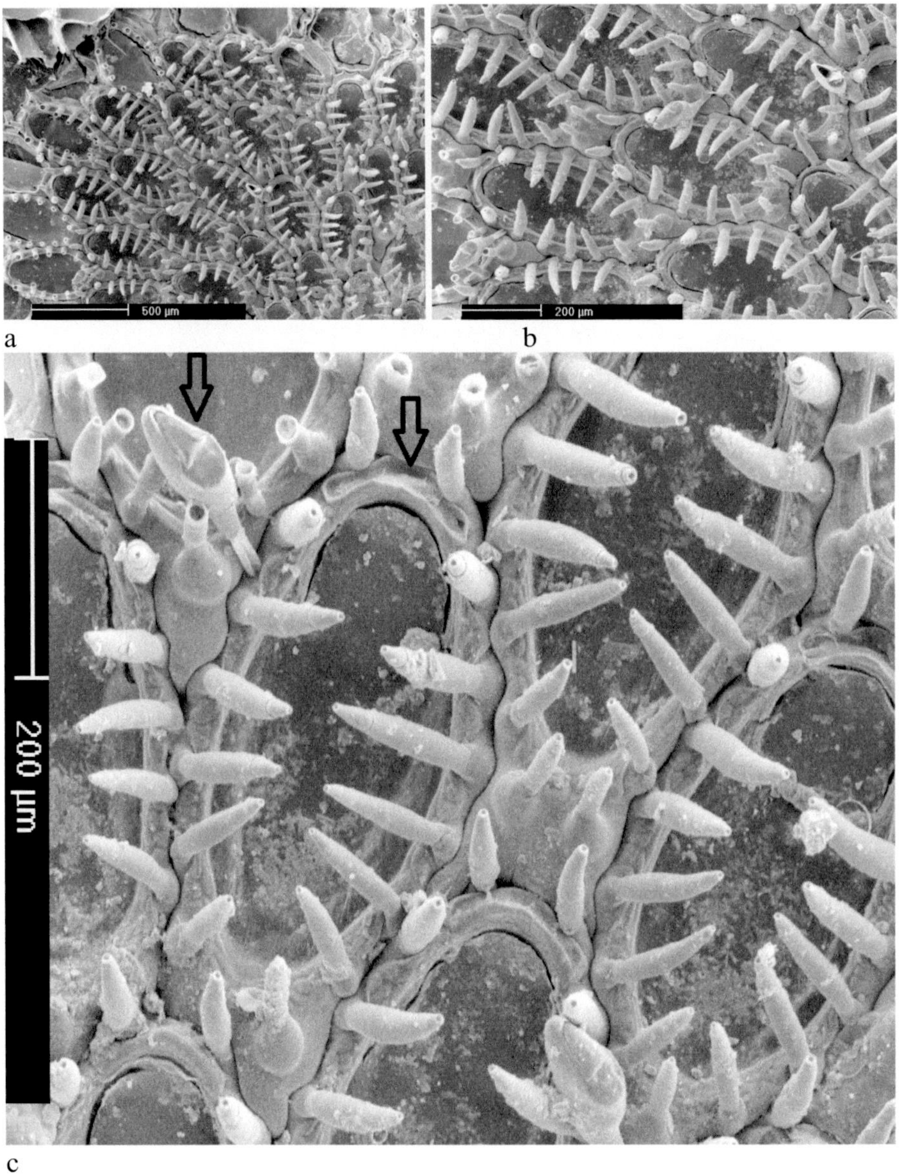

Fig. 4.49 *Cauloramphus spinifer.* (**a**–**c**) Zooids with petiolate avicularia between the spines, on plastic beached in Wales (UK). (**c**) Detail showing petiolate avicularium (*left arrow*) and a hardly visible ooecium (*right arrow*), rudimentary, uncalcified

Rostrum of avicularia erect, directed distally and curved to one side, proximal end with a small rounded or triangular opesia. Each avicularium with a small distal kenozooid. Ovicells prominent, spherical, with thin membranous ectooecium (lost in dry colonies) and calcified endooecium showing granulated surface. Ancestrula

Fig. 4.50 *Parellisina curvirostris.* (**a**) Part of a colony growing on plastic beached at De Haan (Belgium). (**b**) SEM image of the same colony with ovicellate zooids and avicularia

elliptical, half the size of the autozooids, encircled by marginal spines. The first two daughter zooids have very short spines (Fig. 4.50).

Occurrence In Belgium, once on a washed-up plastic object, on September 15, 2001, in De Haan (De Blauwe 2003a).

Genus *Tegella*

Tegella unicornis (Fleming, 1828)

Description Colony encrusting. Autozooids elongate-quadrangular. Calcification thin and delicate. Frontal surface almost completely occupied by an oval frontal membrane. A hollow tubular spine on each side of the orifice. A second pair of spines might be present in young zooids but is soon obscured by the developing ovicell. An avicularium on the gymnocyst and more might be present on the lateral

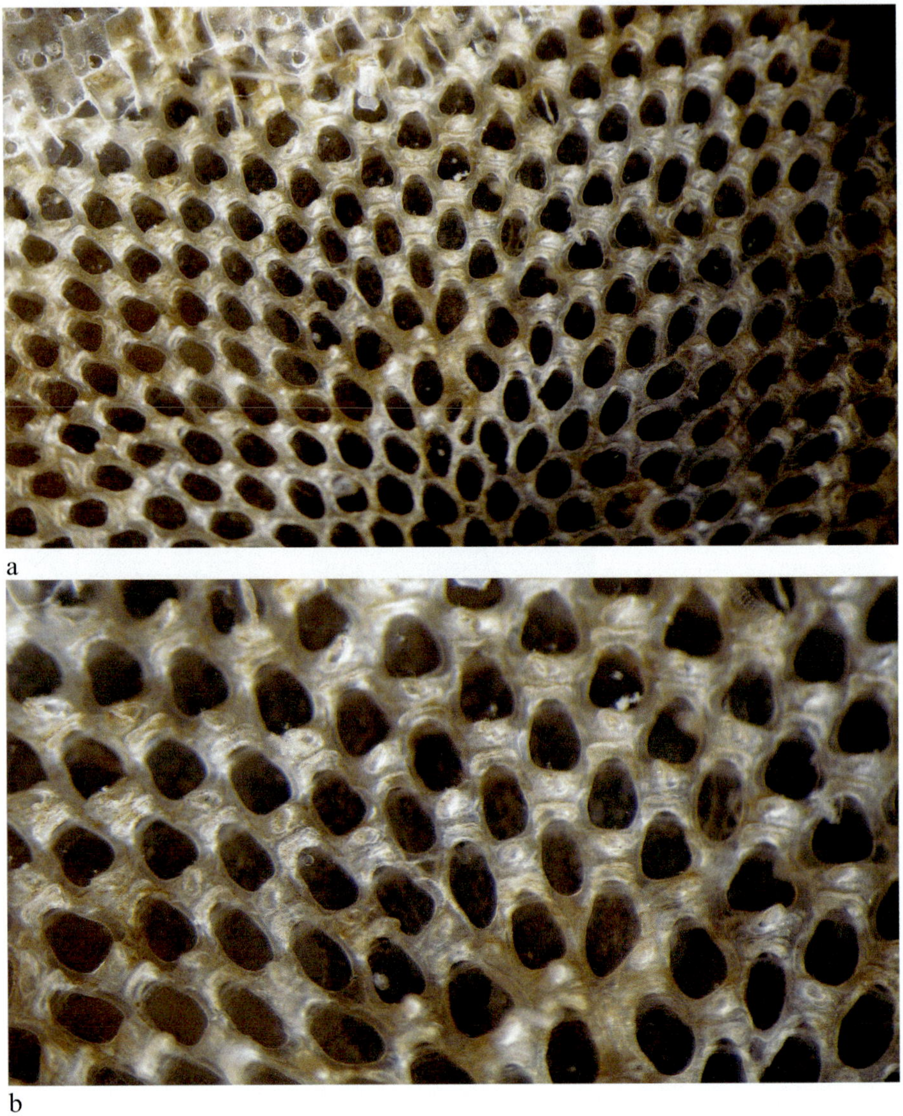

Fig. 4.51 *Tegella unicornis*. (**a**, **b**) Colony area shown by different magnification, Heligoland. (Photos by B. Kind)

walls. Rostrum acute to the frontal plane and directed distally or proximally. If an ovicell develops, the avicularium above becomes intimately associated with it and appears to surmount it. Ovicell prominent, flattened in front with a broad transverse membranous area of ectooecium, bounded by a ridge. Embryos orange (Fig. 4.51).

Occurrence Boreal and Arctic species, widespread in the northern North Sea.

Germany On German coasts (Hayward and Ryland 1998; Zettler et al. 2018) and in Heligoland (pers. comm. Britta Kind).

Genus *Amphiblestrum*

1 (a) Up to six distal spines, one strongly elongated, curved and thickened, draped over the frontal surface. Opesia remarkably three-lobed, formed by the cryptocyst below the membrane. .... *Amphiblestrum flemingii* (p. 205)

(b) Three or four early-stage spines, generally one permanent. Cryptocyst minimal, opesia broadly oval; usually at least two avicularia per zooid. .... *Amphiblestrum auritum* (p. 205)

Amphiblestrum auritum (Hincks, 1877)

Description Colony encrusting, unilaminar, multiserial. Autozooids hexagonal to oval. Frontal calcification minimal, hidden by one, sometimes two avicularia. Opesia ovate, occupying about ¾ of the frontal skeletal surface. Four spines at the distal end of newly formed autozooids, one or two often enlarged. Usually two avicularia on the proximal margin of the zooid, or if the succeeding zooid has an ovicell, they are just distal to it. Ovicells prominent, ectooecium with a ridge marking edges of triangular membranous area. Embryos white. Colonies without ovicells are difficult to distinguish from *Callopora dumerilii* (Fig. 4.52).

Occurrence On the lower beach and shallow sublittoral. In Belgium and the Netherlands sometimes on washed-up objects.

Belgium In the Bay of Heist a colony was found on March 11, 2004, on a rock along the eastern harbor dam of Zeebrugge.

The Netherlands Only recently discovered as autochthonous in Zeeland (Faasse and De Blauwe 2004). Collected on the Dogger Bank in 2016.

Germany Collected in the German North Sea (Zettler et al. 2018) and at Heligoland (pers. comm. Britta Kind).

Colonies collected in the German Baltic Sea show an ancestrula with the opesia surrounded by nine spines, periancestrular zooids with seven or five spines, and zooids at the colony margin with four spines, the proximal pair permanent.

Amphiblestrum flemingii (Busk, 1854)

Description Colony encrusting, unilaminar, multiserial. Autozooids oval, close together, 0.5 × 0.2 mm. Lateral walls rise to a prominent rim around the frontal membrane. Frontal exposure of the gymnocyst minimal, usually hidden by one or

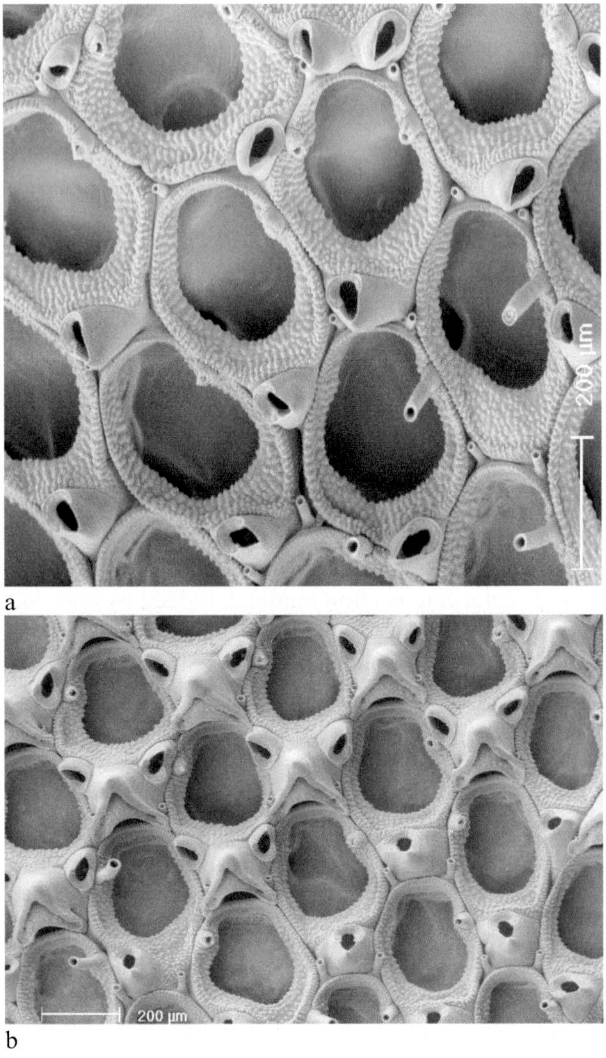

Fig. 4.52 *Amphiblestrum auritum*. (**a**) Colony area showing non-ovicellate autozooids with avicularia and spines. (**b**) Ovicellate autozooids. Damgan (France)

two avicularia. A proximally broad, sunken cryptocystal shelf lies below the frontal membrane; the opesia generally constricted midlength and more or less three-lobed. Up to six oral spines distally, only four in the presence of an ovicell. The distal pair of spines short and thin, one of the proximal pairs very long and thickened, often flattened and curved and draped over the frontal surface of the zooid. One or two avicularia associated with each zooid; the orientation of the pointed mandible is variable.

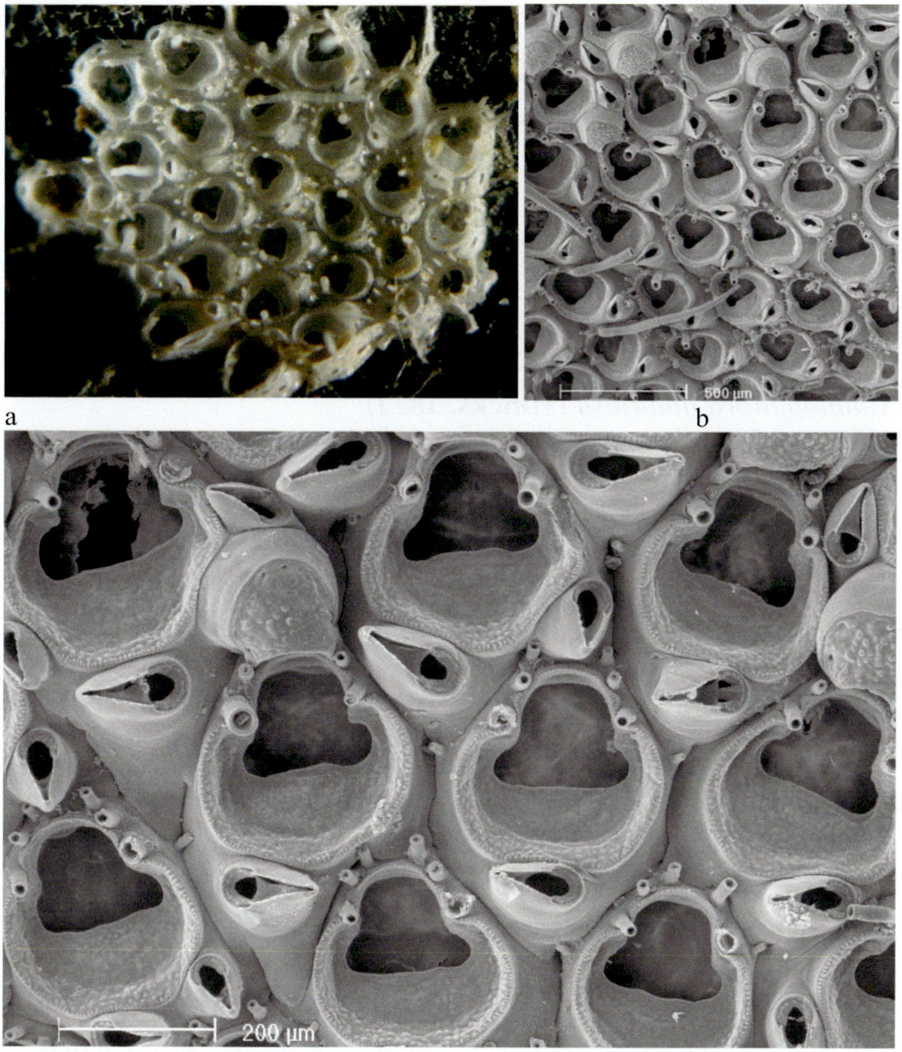

a b c

Fig. 4.53 *Amphiblestrum flemingii*. (**a**) Colony on beached plastic beached in Wales (UK). (**b**, **c**) SEM images of colony areas showing zooids with long spines, avicularia, and ovicells, harbor Le Conquet (France)

Ovicells spherical, ectooecium mostly membranous (lost in dry colonies) and calcified basally, smooth, endooecium with granular surface. Ancestrula smaller than autozooids, round, with eight or nine erect spines around the opesia (Fig. 4.53).

Occurrence

France On boulders near Cap Gris-Nez in June 1992 (Davoult et al. 1999).

Belgium Known from washed-up plastic objects (De Blauwe 2009a).

The Netherlands Washed up on the leg of an Augustine crab *Lithodes maja* (Lacourt 1949) that was probably landed and thrown overboard by fishermen from the northern North Sea. According to Maitland (1897) rare in the Netherlands (washed up?). Washed up at Bergen aan Zee on nylon rope in 2009 (de Ruijter 2011a) and on plastic (de Ruijter 2013a, 2014b, d, 2015a, 2017b).

Germany Collected at Heligoland (pers. comm. Britta Kind).

Genus *Ammatophora*

Ammatophora nodulosa (Hincks, 1877)

Description Colony encrusting, unilaminar, multiserial. Zooidal rim somewhat elevated in the distal half. Under the proximal half of the frontal membrane lies a textured cryptocystal shelf. The opesia is semielliptic to three-lobed. Ovicells subimmersed, calcified ooecium depressed, wider than long. Large, rounded, nodule-like interzooidal kenozooids are typical (Fig. 4.54).

Occurrence

Belgium In 1906, a colony was fished up on the Flemish Banks 51°29′ N, 2°36′ E (RBINS collection). Discovered in 2004 on the inside of shell valves of the Kwinte Bank and a large colony (diameter of 2 cm) on a flint boulder from between the Hinder Banks. In 2005, eight shell valves were fished on the Hinder Banks with one or more colonies of this species on the inside.

Superfamily Buguloidea

Family Bugulidae

Genera *Bugula*, *Bugulina* and *Crisularia*

Bugula senso lato was divided into four clades by Fehlauer-Ale et al. (2015).

Bugula Oken senso stricto comprises one species from our area, e.g., *Bugula neritina*. All species in this clade lack articulated spines and exclusively share the presence of long zooids, bifurcating biserial branches with bifurcation type 4 or a modified type 3 with uniserial appearance, and spherical or hemispherical ooecia attached to the inner distal angle of the autozooid.

Crisularia Gray was resurrected for some species, including the only one from our area, *Crisularia plumosa*. These species share the combined presence of bifurcation type 3 (Harmer 1923) and spherical ooecia with a membranous ectooecium with calcification restricted to its distal edge.

Bugulina Gray was resurrected for, in our area, *B. flabellata*, *B. fulva*, *B. simplex*, *B. stolonifera*, and *B. turbinata*. They are characterized by dimorphic avicularia and

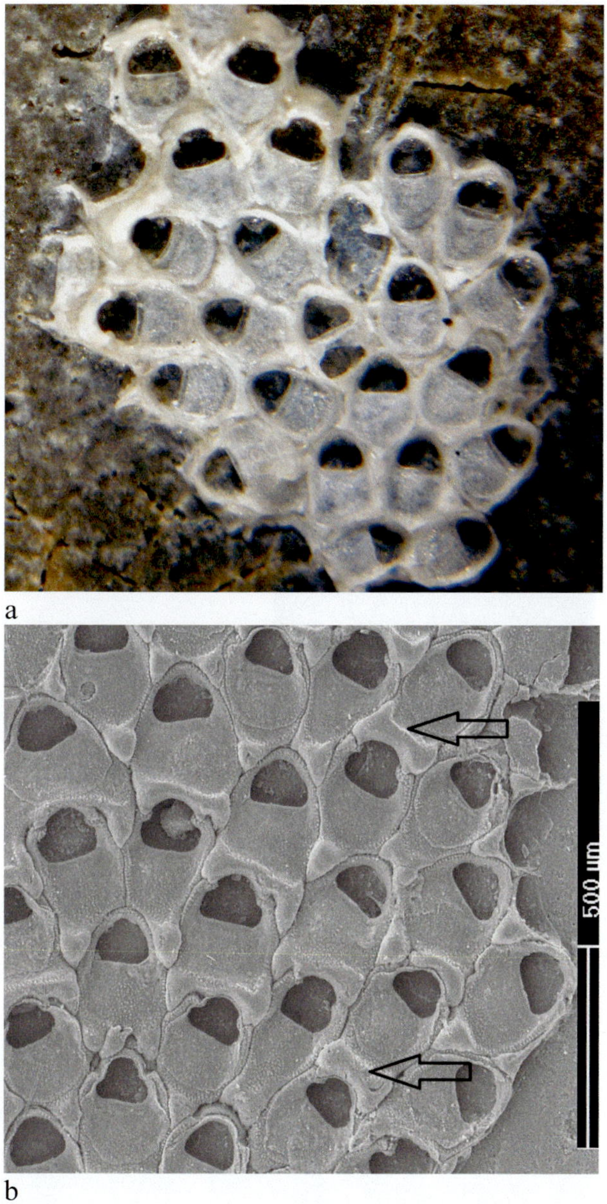

Fig. 4.54 *Ammatophora nodulosa.* (**a**) Young colony on a dead shell from the Kwinte Bank (Belgium). (**b**) SEM image of a colony area showing ovicellate zooids (*arrows* pointing at two ovicells)

globular ooecia centered in the zooid midline and attached to the distal region of the autozooid, with the proximal and lateral ectooecial area calcified, but the rest of the ectooecium membranous, shrunken and folded in dry specimens. Except for *B. avicularia* and *B. stolonifera*, whose colonies have biserial branches (in *B. stolonifera* branching patterns types 3, 4, and 5 may occur in the same colony), species in this clade are characterized by colonies with multiserial branches.

The fourth clade, *Virididentula* does not occur in our area (Fig. 4.55).

1 (a) Avicularia and spines absent. .... *Bugula neritina* (p. 213)
(b) Avicularia present, spines present. 2

Fig. 4.55 Colony forms of *Bugula*, *Bugulina*, and *Crisularia*. (**a**) *Bugulina avicularia*, on a colony fragment of an erect *Porella* species in the fishing harbor of Le Conquet (France). (**b**) *Bugulina flabellata* on beached plastic, Oostende (Belgium). (**c**) *Bugulina fulva*, marina Trébeurden (France). (**d**) *Bugula neritina* (dried colony), marina Hendaye (France). (**e**) *Crisularia plumosa*, Goese Sas (the Netherlands). (**f**) *Bugulina simplex*, marina Trébeurden (France). (**g**) *Bugulina stolonifera*, marina Zeebrugge (Belgium). (**h**) *Bugulina turbinata*, Marloes (UK)

2 (a) Autozooids in two series. 3
(b) Branches with autozooids in four or more series. 7

3 (a) Inner distal angle spineless. Outer distal angle extended to a pointed protuberance. .... *Crisularia plumosa* (p. 221)
(b) Both distal angles bear at least one spine. 4

4 (a) Outer distal angle bears one spine; avicularia short and plump, the beak rectangular hook-shaped. .... *Bugulina turbinata* (p. 221)
(b) Outer distal angle with two to three spines; avicularia not short and plump, beak bent down. 5

5 (a) Outer distal angle bears three spines, inner angle with at least two spines. ... . *Bugulina fulva* (p. 217)
(b) Outer distal angle bears two spines, inner angle with one spine. 6

6 (a) The frontal membrane takes up two-third to three-fourth of the length of the zooid; avicularia small (<200 μm), their length not exceeding the width of a zooid. .... *Bugulina stolonifera* (p. 218)
(b) The frontal membrane occupies almost the entire length of the zooid; avicularia longer (approx. 300 μm), their length exceeds the width of the zooid.. *Bugulina avicularia* (p. 214)

7 (a) Outer distal angle of marginal zooids bears one spine (rarely two), inner angle with one spine. 8
(b) Outer distal angle of marginal zooids bears three spines (rarely two), inner angle with at least two spines... 9

8 (a) avicularia confined to the marginal autozooids, beak bent down. .... *Bugulina simplex* (p. 217)
(b) Large avicularia present on marginal zooids, small ones on the inner zooids, the beak rectangular hooked............................. *Bugulina turbinata* (p. 221)

9 (a) Beak of avicularia rectangular hooked. .... *Bugulina flabellata* (p. 215)
(b) Beak of avicularia bent down. .... *Bugulina fulva* (p. 217) (Fig. 4.56)

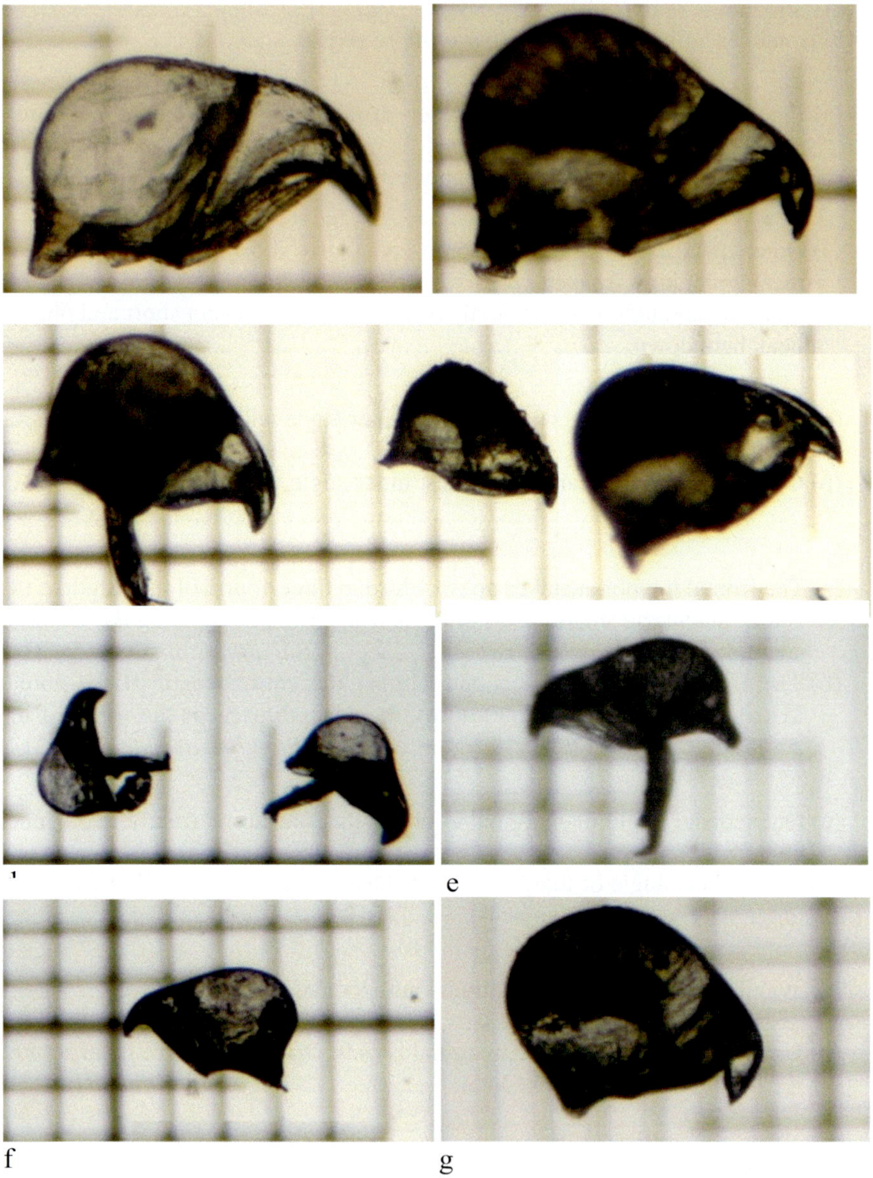

Fig. 4.56 Avicularia of various species on approximately the same scale. (**a**) *Bugulina avicularia*, upper beak bent down. (**b**) *Bugulina flabellata*, upper beak hook-shaped. (**c**) *Bugulina fulva*. (**d**) *Crisulina plumosa*. (**e**) *Bugulina simplex*. (**f**) *Bugulina stolonifera*. (**g**) *Bugulina turbinata*

Bugula neritina (Linnaeus, 1758)

The *Bugula neritina* complex consists of three sibling species (Fehlauer-Ale et al. 2014).

Description

A beginning naturalist might mistake *Bugula neritina* for a small red seaweed. Colonies upright, bushy, branching tufts, up to 15 cm or so in height. They are usually burgundy-colored. Branches with zooids in two series. Zooids large, narrowing proximally; the entire frontal surface is membranous. Spines are absent, but the free outer distal corner protrudes slightly. Polypide with 20 to 24 tentacles. Avicularia are absent. Spherical ooecia, attached to the inner distal angle of the autozooids and oriented obliquely to the axis of the branch. Ancestrula symmetrical, with a U-shaped membranous space; spineless (Fig. 4.57).

Fig. 4.57 *Bugula neritina*. (**a**) Living colony in the Grevelingen (the Netherlands, photo by A. Aleman). (**b**) SEM image of several branches showing ovicellate and sterile zooids. (**c**) Branch bifurcation, Hendaye (France)

Occurrence *Bugula neritina* has a broad global distribution in temperate, subtropical, and tropical waters. The first observation in our neighborhood was in southern England, presumably introduced on ships and first discovered at Plymouth between 1904 and 1912 (Ryland and Hayward 1977).

A species of harbors, ship hulls, buoys, and submerged objects. To be expected on yachts from Southern Europe and as a temporary or permanent introduction to ports.

Belgium Found once on a yacht in the Mercator marina in Oostende in 1999 (Kerckhof 2000) and the next time on that location was on test panels in August 2021. In 2010, as fouling species on the research vessel "Belgica" when it was visiting the inner harbor of Brugge/Zeebrugge. In November 2019, a specimen was found attached to a pontoon in the marina of Zeebrugge. Since then, a population has been established at this location. Extremely numerous in the Vuurtorendok and spuikom at Oostende in September 2024.

The Netherlands *Bugula neritina* was first found in April 2007, in the marina of Burghsluis (Faasse 2007). This observation concerns fresh outgrowths on old stolons. Therefore, the species was probably already present in 2006. Since 2008, the species has been found in "Roompot Marina" near Wissenkerke as well (Faasse et al. 2013).

In November 2015, two colonies were found in the marina of St. Annaland, in the northeast of the Oosterschelde area, and in the marina near the Goese Sas (Faasse and Ligthart 2016). In 2017, one colony in Sophiahaven (Kamperland, pers. obs.), the marina of Wemeldinge (Kuijper and Cadée 2018), and common in Burghsluis (de Ruijter 2019d). In 2020 observed in the marinas of Burghsluis and Scharendijke, and in 2021 at Wemeldinge and in the Grevelingen. In 2020, *Bugula neritina* colonies were collected during an alien species-focused survey of the Veerse Meer. It was found on floating docks off Kamperland, Arnemuiden, Wolfhaartsdijk, and Kortgene (Gittenberger et al. 2021). In 2021, within the Marine Alien Species Detection Network of the Netherlands, colonies were again collected in the Veerse Meer, and additionally in Colijnsplaat marina in the Oosterschelde and in the Westerschelde in the marina of Breskens (Gittenberger et al. 2022a). During subsequent monitoring within the same network in 2022, colonies were collected from floating docks in Bommenede and Herkingen in the Grevelingen (Gittenberger et al. 2023b).

Beached on plastic and in mass on the barnacle *Amphibalanus amphitrite* in 2017–18 (de Ruijter 2017b, 2020c; Kuijper and Cadée 2018). This mystery washing-up is similar to that of *Biflustra grandicella*. It concerns more or less the same coastal area.

Bugulina avicularia (Linnaeus, 1758)

Description Bushy, 2–3 cm high, branches spiral around the main axis. Orange-brown. Branches biserial. Zooids rectangular or narrowing proximally, almost all of the frontal surface membranous. Outer distal angle with two spines, inner distal

a b

Fig. 4.58 *Bugulina avicularia.* (**a**) Part of a colony collected from lobster pot, the harbor of Le Conquet (France). (**b**) SEM image of ovicellate zooids with oral spines and avicularia

angle with one spine. Polypide with 13 to 15 tentacles. Avicularia large, their length exceeds the width of the bearing zoid, attached at one-third or halfway up the lateral wall. Ooecia spherical. Yellow embryos from June to November. Mainly confused with *B. stolonifera*, which is more of a port species (Fig. 4.58).

Occurrence On hydroids and shells where, as with *B. flabellata*, the immediate carrier is usually another bryozoan colony such as *Flustra foliacea*.

Belgium In the RBINS, there are two colonies on *Flustra foliacea*, both dredged in 1905 (51°29′15″ N, 2°32′45″ E and 51°31′30″ N, 2°34′30″ E). No recent observations from the southern North Sea.

The Netherlands Faasse (1998) rightly suggests that reports from the Netherlands of *B. avicularia* in Lacourt (1949), Heerebout (1970), Otten (1992), Platvoet et al. (1995), Faasse (1996) in harbors concern *Bugulina stolonifera*. Figure 29 of "*Bugulina avicularia*" in Lacourt (1978) clearly represents *Bugulina stolonifera* because the avicularia are no longer than the width of a zooid.

Bugulina flabellata (Thompson in Gray, 1848)

Description Forms a tuft 2–3 cm high, of yellowish color when alive, but turning gray or brown when dried. Branches broad, widest distally, somewhat circularly arranged. Zooids in three to eight series, elongated, remaining equal width; the membrane occupies all or most of the frontal surface. Spines three on the distal outer angle, two on the distal inner angle (2:2 on inner zooids), of variable size; these on the marginal zooids are sometimes very large. Polypide with 14 tentacles. Avicularia present on zooids of inner and marginal series, the latter considerably larger; attached one-third to halfway down the side; beak rectangular hooked.

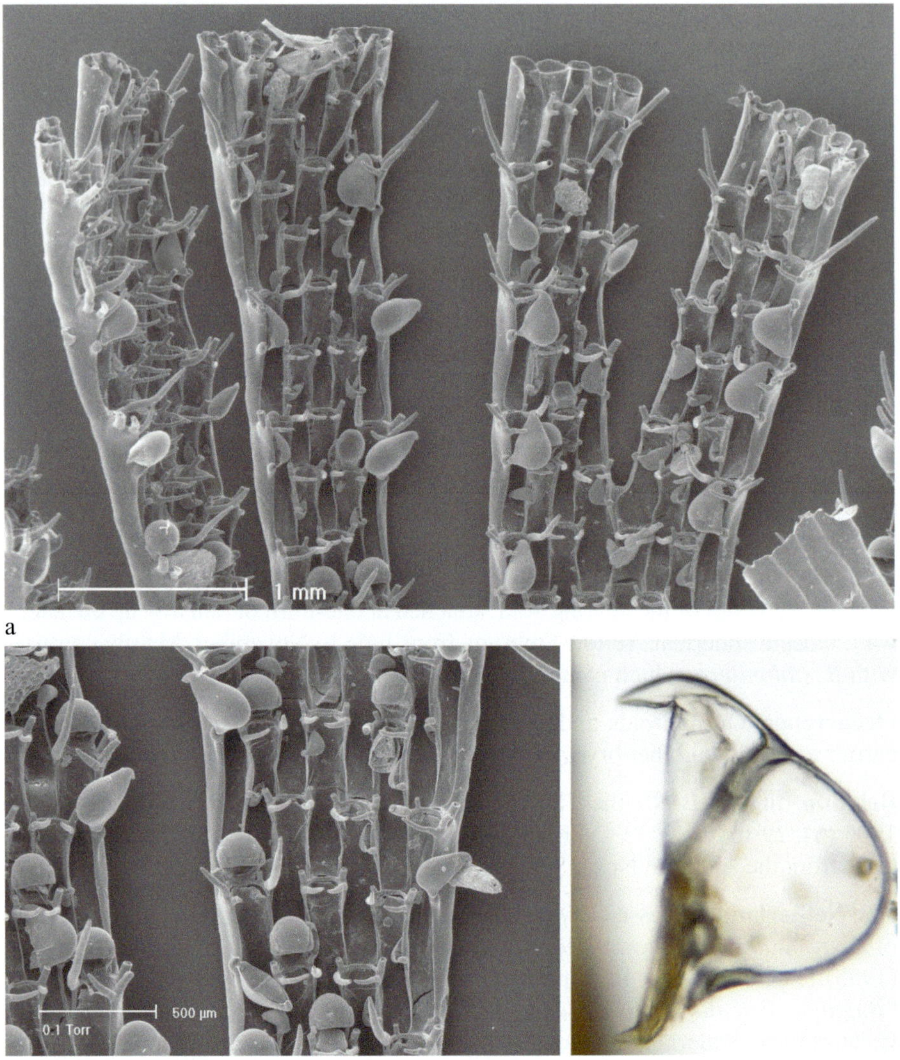

Fig. 4.59 *Bugula flabellata*. (**a**, **b**) Frontal view of branches of a colony collected in the harbor of Le Conquet (France). (**c**) Avicularium in transmitted light

Ooecia almost spherical; embryos yellow, from May to October. Ancestrula with three proximal spines and three at each distal corner (Fig. 4.59).

Occurrence On other Bryozoa, even if they seem to grow on rocks or shells.

Belgium Loppens (1906) mentions this species as common on *Flustra foliacea*, stones, and shells, as well as in the harbor of Nieuwpoort, sublittorally on harbor structures. The large amount of material on *Flustra foliacea* in the RBINS suggests that this species was much more common in the early twentieth century than it is today.

Recently washed up in Belgium on *Flustra foliacea* and plastic objects. During a campaign on the Hinder Banks in 2005, *Bugulina flabellata* was found at only one location. The decline of *Flustra* may be at the base of a decline of *Bugulina flabellata*.

The Netherlands Very few observations from the Netherlands: washed up in Scheveningen on cork (Lacourt 1949), in September 2003 at Callantsoog (de Ruijter 2003b), in August 2005 at Egmond aan Zee (de Ruijter 2006a), on *Flustra* (de Ruijter 2013b, 2014a), on plastic (de Ruijter 2014c), near Camperduin on washed-up plastic lobster trap entrance on May 4, 2020 (de Ruijter 2020b), and at Texel on January 12, 2024 (pers. comm. Rob Dekker).

Collected in 2015 on DDNZS nets on wreck 2801 and 2803 on the Doggers Bank (Faasse et al. 2016).

Germany Collected in the German North Sea (Zettler et al. 2018) and at Heligoland (pers. comm. Britta Kind).

Bugulina fulva (Ryland, 1960)

Description The colony forms a dense yellow-brown tuft, 2–3 cm high. Branches with autozooids in two series, but distally in a colony branches can consist of four series of autozooids and then split into two branches of two zooid series. Autozooids rectangular or slightly narrowed proximally. The opesia occupies the entire frontal surface.

Spine formula on marginal zooids 3:2 or 3:3; formula of inner zooids 3:3 (2:2 in the presence of oecia). Polypide with 13 tentacles. Avicularia attached just below the spines, their length somewhat more than the width of a zooid, or smaller on the inner zooids; beak bent down. Ooecia spherical. Embryos yellow, present during summer and autumn. Ancestrula with three proximal spines, three (rarely two) on each distal corner, and often with one on each side (Fig. 4.60).

Occurrence Rare on washed-up plastic: Zeebrugge (De Blauwe 2005a), Bergen aan Zee (de Ruijter 2017b).

Bugulina simplex (Hincks, 1886)

Description Forms a funnel-shaped colony, which later becomes bushy. Height up to 3 cm. Straw-colored to orange-brown. Branches are narrow proximally and widen distally. They are usually made up of three to six series of zooids. Zooids frontally entirely membranous. Spine formula usually 1:1. Sometimes a second spine on the distal outer corner of marginal zooids. Polypide with 12 to 14 tentacles. Avicularia only present on the marginal zooids. Their length is equal to or slightly more than

a b

Fig. 4.60 *Bugulina fulva*. (**a**) Colony branches showing ovicellate zooids with spines and avicularia. (**b**) Side view of the branch showing spines, avicularia, and ooecia. Colony collected on beached plastic, Zeebrugge (Belgium)

the zooid width. The beak is bent downward. Ooecia hemispherical. Embryos yellow, from July to November. Ancestrula with three distal spines and one proximal spine. Settlement mainly in August and September (Fig. 4.61).

Occurrence

Belgium Discovered in the Mercator marina in Oostende in September 2000 (De Blauwe and Faasse 2001). Recently washed ashore in Zeebrugge on a plastic bucket (De Blauwe 2005a) and found in the Zeebrugge marina since June 2007.

The Netherlands Also discovered in Sas van Goes in September 2000 (De Blauwe and Faasse 2001). In Zeeland expanding to other marinas: Burghsluis, Wemeldinge, Yerseke, Scharendijke, and Ouddorp. Only present for a short time of the year and often in small concentrations or very locally. In 2020 and again in 2021, *Bugulina simplex* was collected in an alien-species-focused survey of the Veerse Meer, from floating docks in the marina of Kortgene (Gittenberger et al. 2021).

Bugulina stolonifera (Ryland, 1960)

Description Forms compact tufts up to 4 cm high. A creeping stolon may develop on which secondary clumps sprout. Color light tan to gray. The branches consist of two series of zooids. Zooids tall and slender. The frontal surface is half to three quarters membranous. Spine formula 2:1. In some colonies the spines are barely developed. Colonies with more spines were found in the fall, causing confusion with other species. Polypide with 13 or 14 tentacles. The length of the avicularium

Fig. 4.61 *Bugulina simplex*. (**a**) Living colony on seaweed in the marina of Yerseke (the Netherlands). (**b**) Part of a colony collected in Goese Sas (the Netherlands). (**c**) Ovicellate zooids with spines and avicularia, Trébeurden (France)

is less than zooid width, attached somewhat lower than the distal outer spine, the beak is curved downward. Ooecia hemispherical. Embryos yellow. Ancestrula with two or three spines on either side distally, one spine proximally. Until 1998 incorrectly identified in our region as *Bugula avicularia* (Faasse 1998; Kerckhof 2000), except by d'Hondt and Cadée (1994) (Fig. 4.62).

Occurrence Tolerates reduced salinity and pollution.

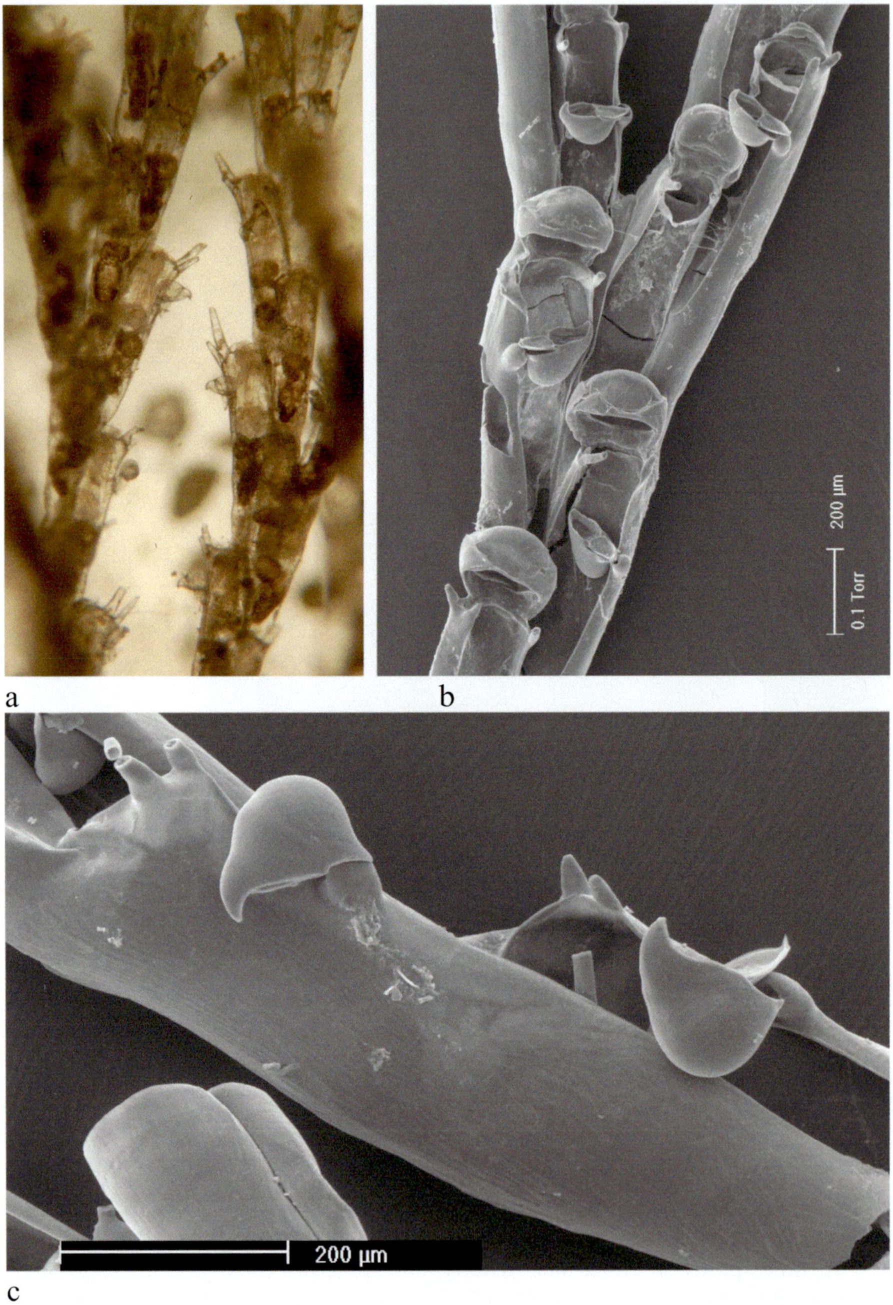

Fig. 4.62 *Bugulina stolonifera*. (**a**) Branches of a living colony on a groyne at Koksijde (Belgium). (**b**, **c**) SEM images of a colony branch showing zooids with ooecia, spines, and avicularia, Goese Sas (the Netherlands)

Belgium Loppens (1906) does not yet mention this species (by its synonyms) from Nieuwpoort. Polk (1976) found colonies in the "Spuikom" of Oostende and mentions them under the name *B. avicularia.* Author found colonies in the marina of Zeebrugge, Blankenberge, Nieuwpoort, in the "Spuikom" of Oostende, Vuurtorendok Oostende and on a groyne in Koksijde.

The Netherlands This species is considered invasive and therefore the first possible report in the Netherlands is interesting: Horst (1885) in the former Zuiderzee (Faasse 1998). Common in marinas, such as Burghsluis (de Ruijter 2019d). Present on the North Sea side of the Brouwersdam (de Kluijver 1989) and in canals and closed water bodies in Zeeland (Faasse and De Blauwe 2004). Also reported from the NIOZ harbor on Texel in d'Hondt and Cadée (1994).

Germany Collected in the German North Sea (Zettler et al. 2018).

Bugulina turbinata (Alder, 1857)

Description Forms tufts up to 6 cm high. The branches are spirally implanted on the main axis. The branches are proximally narrow and consist of two series of zooids. Distally they become wider and multiserial. Most of the frontal zooid surface is membranous. Spine formula 1:1. Polypide with 13 tentacles. Avicularia attached just below the spine. Those on the inner zooids are much smaller than those on the marginal zooids. The beak is short and curved downward like a hook. Ooecia nearly spherical. Embryos yellow, present from August to November (Fig. 4.63).

Occurrence

Belgium In 1904 a specimen was dredged off the Belgian coast at 51°21′ N, 2°32′ E and in 1911 at the Schoonveld Drought (RBINS collection). There are no recent finds from Belgium.

The Netherlands Found on washed-up *Flustra foliacea* on Texel (de Ruijter 2004) and on washed-up plastic (de Ruijter 2006a, 2012b). Recorded by Van Moorsel and Waardenburg (1990) from wreck "K" off the west coast of the Netherlands on August 2, 1991, at a depth of 25 m (52°26′15″N, 03°44′04″E) (Faasse et al. 2013).

Germany Collected in the German North Sea (Zettler et al. 2018).

Crisularia plumosa (Pallas, 1766)

Description Elongated, feathery tufts are spirally implanted on a main stem. Length up to 8 cm. Color light brown-yellow-orange. The twigs consist of two series of autozooids. Autozooids slender, proximally narrowest, three quarters to almost all of the frontal surface membranous. Inside angle without spine. Distal outer angle with one prominent spine.

Fig. 4.63 *Bugulina turbinata.* (**a**) Part of a colony from Marloes Sands (UK). (**b**) SEM image of a colony from Trébeurden (France)

Polypide with 14 tentacles. Avicularia very small, their length less than the width of a zooid. The beak is slightly bent downward. Ooecia spherical. Embryos yellow, from July to September. Ancestrula vase-shaped with a rounded terminal frontal membrane, without spines (Fig. 4.64).

Occurrence

Belgium Loppens (1906) listed this species as rather rare on shells. Three colonies were collected on the Flemish Banks in 1905 and one colony in 1911 (RBINS col-

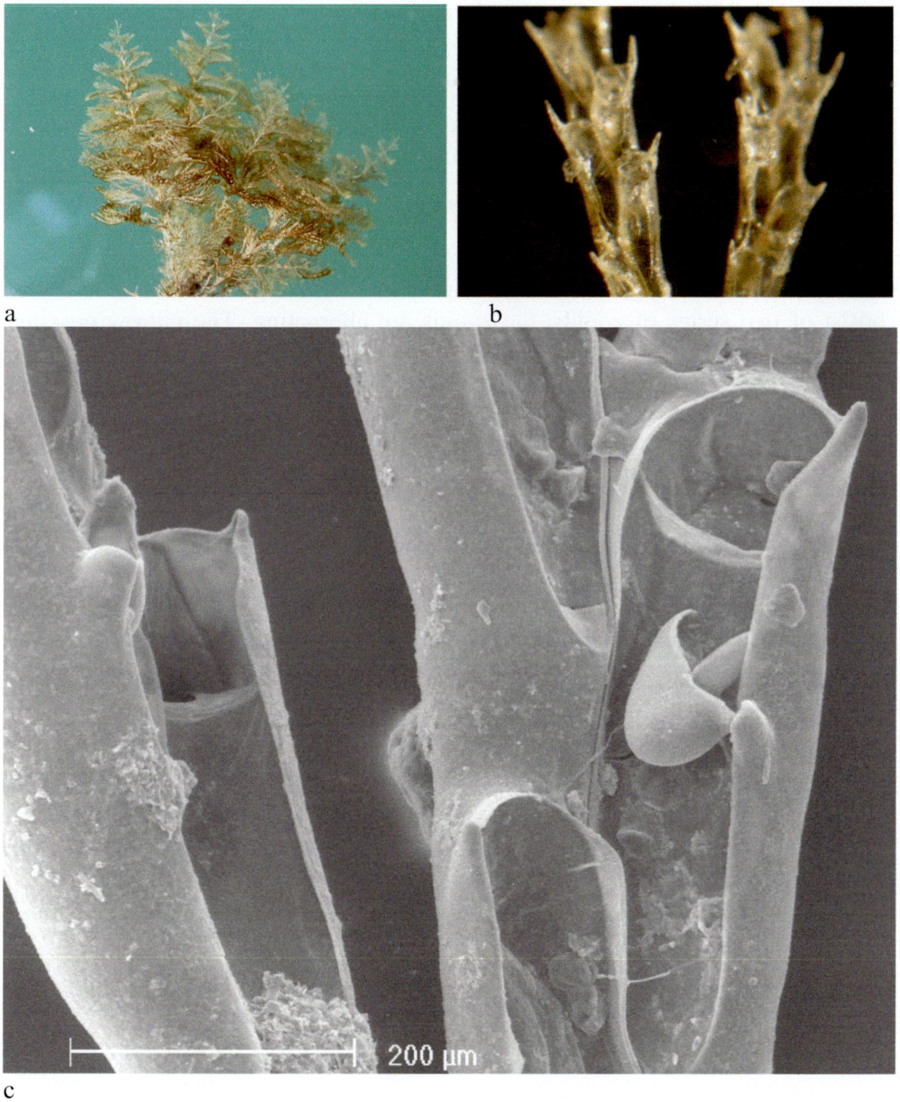

Fig. 4.64 *Crisularia plumosa*. (**a**) Living colony, marina Breskens (The Netherlands). (**b**) Part of a beached colony in Bredene (Belgium). (**c**) SEM image of non-ovicellate zooids with avicularium from the same colony

lection). Found in Belgium in 1999 in the Spuikom of Oostende (Polk 1976; Kerckhof 2000) and on the boulders along the eastern breakwater of the Zeebrugge harbor (Baai van Heist) just below the low water mark.

The Netherlands Common in Zeeland on stones and in harbors. Reported more northerly in the Netherlands from Den Helder and the former Zuiderzee (Lacourt 1949). On washed-up plastic (de Ruijter 2008, 2015b, 2020b).

Germany Collected in the German North Sea (Zettler et al. 2018) and at Heligoland (pers. comm. Britta Kind).

Genus *Bicellariella*

Bicellariella ciliata (Linnaeus, 1758)

Description Forms a white, feathery tuft, up to 25 mm high. The curved branches consist of two series of zooids, the zooids are alternating. The narrow proximal zooid parts grow against each other, the distal parts are bent outward. The frontal membrane is oval and occupies about one-third of the zooid length.

One proximal spine and a row of four to nine inwardly curved spines on the distal margin. The polypide has 12 to 16 tentacles. Avicularia attached to the outside of the zooid, proximal to the opesia. Inner margin of the beak serrated. Ovicells attached to autozooid by means of a short stalk. Ooecium spherical, with an opening on the side of the frontal wall. Embryos pure white. First generation broods embryos in April–May and then dies. Second generation carries embryos in August. Colonies formed from second-generation larvae overwinter. Ancestrula funnel-shaped with a terminal round membrane, around which are 10 to 12 spines (Fig. 4.65).

Occurrence Grows on groynes, stones, shells, hydroids, bryozoans, etc. from the lower beach and in shallow sublittoral areas.

Belgium Loppens (1906) called this species fairly common. Native in the Bay of Heist and on shipwrecks (Zintzen 2007).

The Netherlands In the eastern part of the Oosterschelde, in the Western Scheldt estuary, and one record from the Veerse Meer. De Ruijter (2005a, 2007, 2008, 2014b, c) reports colonies on washed-up plastic at Egmond aan Zee. Previously washed ashore in between Zeeland and Den Helder (Lacourt 1949).

Germany Collected in the German North Sea (Zettler et al. 2018) and at Heligoland (pers. comm. Britta Kind).

Family Beaniidae

Genus *Beania*

Beania mirabilis Johnston, 1840

Description The colony is thin, branching, and attached with rhizoids. The rhizoids originate basally and have an adhesive plate against the substrate. Zooids proximally tubular and abutting the substrate. Erect distally, in the shape of a coal shovel. Frontal membrane elongated, with 5 to11 spines on each side. One or two spines on the distal margin. Avicularia and ovicells are absent in this species (Fig. 4.66).

Fig. 4.65 *Bicellariella ciliata*. (**a**) Colony from Bay of Heist (Belgium). (**b**, **c**) Branches showing zooids with long spines, ooecia, and small avicularia. (**d**) Avicularium with rised mandible and serrated rostrum

Occurrence

Belgium A colony washed ashore in September 2001 on a *Himanthalia* button in Zeebrugge (De Blauwe 2003a) and in January 2005 in Oostende on a plastic object.

The Netherlands Beached in the autumn of 1947 (Bloklander and Leenhouts 1948). On washed-up plastic (de Ruijter 2011d, 2013a, 2014a, d, 2018b, 2020b, 2024b).

Family Candidae

Genus *Caberea*

Caberea boryi (Audouin, 1826)

Description The colony forms a small, fan-shaped tuft, up to 1 cm high, attached with rhizoids. The branches appear hairy. The zooids are in two, rarely three series. The zooid is short, with an oval membrane occupying most of the frontal surface,

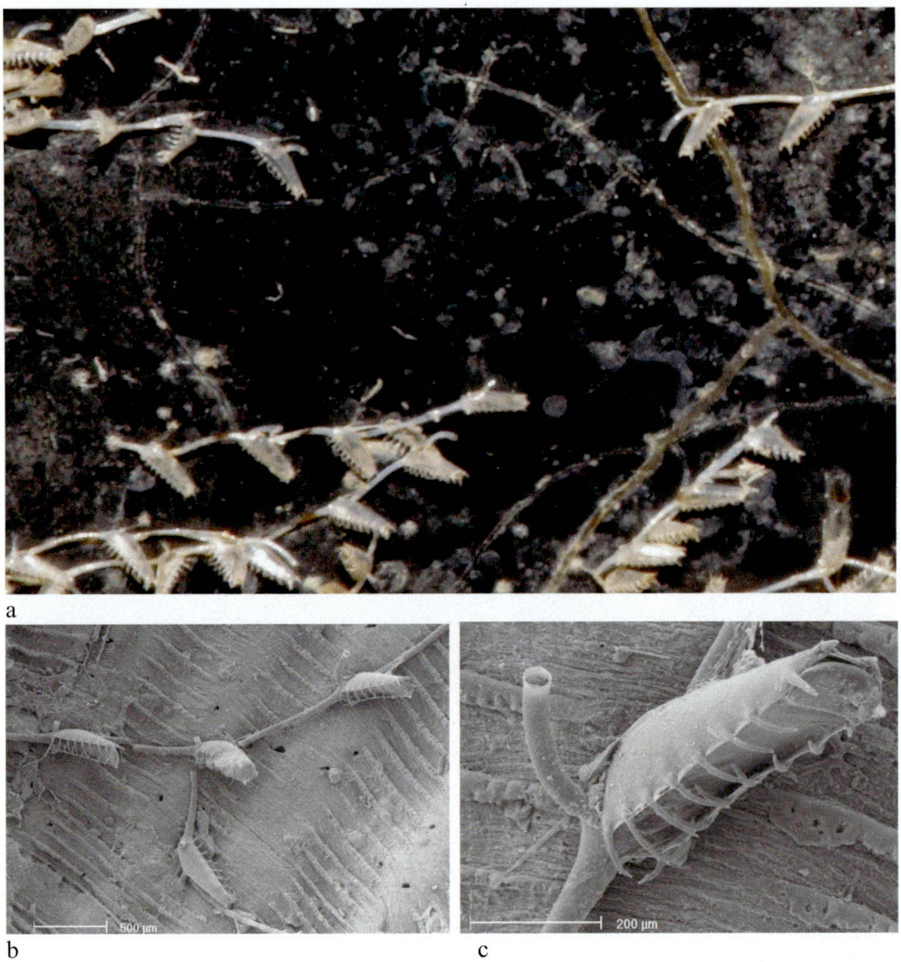

Fig. 4.66 *Beania mirabilis*. (**a**) Colonies on beached plastic, Texel (The Netherlands, photo by S. Dijksen). (**b**) Colony fragment. (**c**) Autozooid, Le Conquet (France)

around which is a broad calcified border. Distally, one spine is on the inner angle and two or three on the outer angle. The scutum covers almost all of the frontal membrane except for a crescentic proximal portion. Polypide with 12 tentacles. Lateral avicularia are small with a rounded mandible. A frontal avicularium on the inner lateral side of the zooid; its rounded mandible ends in a hook. A vibraculum covers the basal surface of the zooid. In the center is a groove, at an angle to the axis of the branch. The seta is finely serrated. Ovicells prominent, globular, slightly bent to the axis of the branch; ectooecium with a small membranous frontal area. A small membranous area lies just distal to its opening (Fig. 4.67).

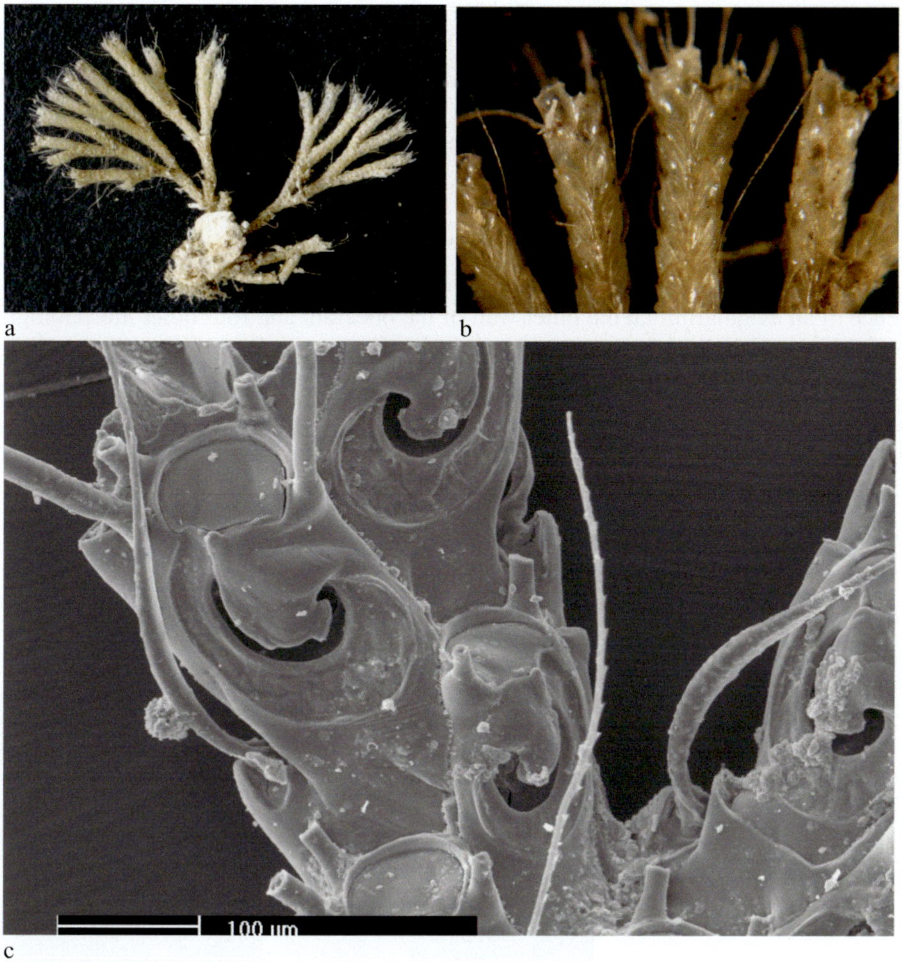

Fig. 4.67 *Caberea boryi.* (**a**) Colony from Le Conquet (France). (**b**) Basal view on distal tips of the branches showing long setae of vibracula. (**c**) Autozooids with operculum, scutum, spines, and vibracular setae, the colony was collected on a lobster pot in St-Malo (France)

Occurrence

Belgium This southern species washed up on plastic in Oostende in January 2005.

The Netherlands Beached in the autumn of 1947 (Bloklander and Leenhouts 1948) and washed up on plastic near Ouddorp in 2014 (de Ruijter 2014c).

Genera *Cradoscrupocellaria* and *Scrupocellaria*

The genus ***Cradoscrupocellaria*** is characterized by the presence of:

- Articulate distal unbranched spines
- A branched scutum arising from the midline of the inner edge of the opesia
- A trapezoidal vibracular chamber
- A single axial vibraculum
- An ooecium with some ectooecial pseudopores

The genus ***Scrupocellaria s. str.*** is redefined for a clade characterized by the presence of an avicularium at the outer wall of the ooecium and four other morphological features:

- Vibracular chamber with curved setal groove
- Ooecium with a single and small ectooecial membranous area (fenestra)
- Two axillary vibracula
- A membranous operculum with a distinct distal rim

1 (a) One axial vibraculum (Fig. 4.68a). Scutum branched. 2
(b) Two axial vibracula (Fig. 4.71c), scutum spatulate or absent. 3

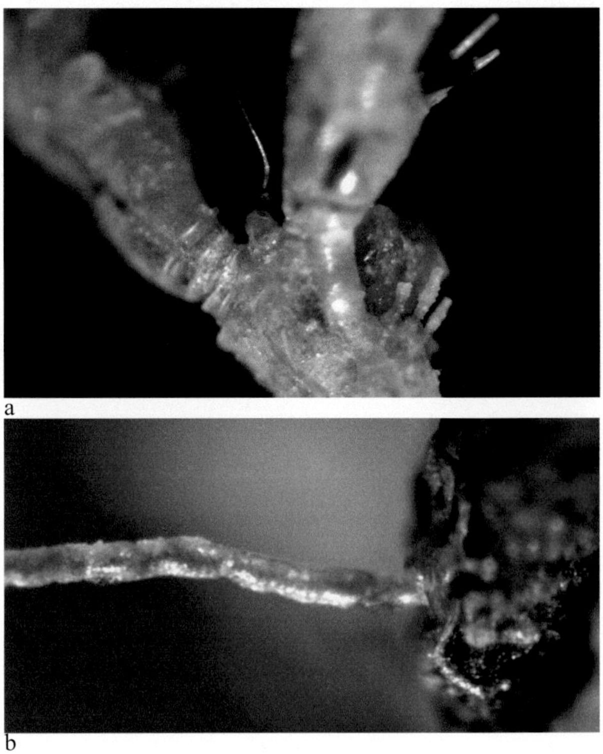

Fig. 4.68 *Cradoscrupocellaria ellisi.* (**a**) Branch bifurcation showing single axial vibraculum with a seta. (**b**) Rhizoid without reverse hooks, with disc-like attachment zone forming thin projections

2 (a) Scutum with 6–9 distal stout projections; angled at 100–120°, with first branches 0.035–0.045 mm wide, and secondary branches about 0.025–0.035 mm. Rhizoids with closely spaced reverse hooks for most of their length (Fig. 4.70a). .... *Cradoscrupocellaria reptans* (p. 231)
(b) Scutum with 8 to 13 stout projections at distal tips; scutum angled at 115–155°, with first branches 0.045–0.060 mm wide, secondary branches about 0.035–0.045 mm. Rhizoids smooth (Fig. 4.68b). .... *Cradoscrupocellaria ellisi* (p. 229)

3 (a) Scutum spatulate. .... *Scrupocellaria scrupea* (p. 232)
(b) Scutum absent. .... *Scrupocellaria scruposa* (p. 234)

Cradoscrupocellaria ellisi (Vieira and Spencer Jones, 2012)

Description Colony erect, branched, internodes comprising 5 to 12 zooids. Opesia oval, occupying the distal half (approximately) of the zooid. Scutum inserted at the midline of the inner edge of opesia, branched, occupying almost the entire opesial area, slender, flattened, branched two to three times, with 8 to 13 stout projections at distal tips; scutum angled at 115–155°, with first branches about 0.045–0.060 mm wide, secondary branches about 0.035–0.045 mm. Zooid spines as follows: one long distal spine, three outer spines, one to two inner spines; most proximal outer and inner spines directed frontally; axial zooid with five distal spines. One distolateral avicularium sometimes present on each zooid, distolaterally directed and obscured by outer distal spines; rostrum triangular, with a slightly serrated lateral edge, mandible triangular. A large avicularium present on the gymnocyst of some zooids. A basal vibracular chamber often present proximally on the basal surface of each zooid, rarely conspicuous in frontal view; the setal groove directed transversely, straight, with smooth seta longer than one zooid length. A single axial vibraculum. Rhizoids smooth, attachment disc-like, with some projections at the distal end. Ovicell prominent, hemispherical, calcified ectooecium with 12 to 18 small rounded pseudopores; two outer and two inner distal spines in ovicellate zooids (Fig. 4.69).

Remarks This species was previously recorded by numerous authors as *Scrupocellaria reptans*. Records from literature before 2012 and even later are unreliable and the material in collections should be re-examined. *Cradoscrupocellaria ellisi*, named after the British naturalist John Ellis (1714–1776), is distinguished by smooth (not hooked) rhizoids, robust more-branched scuta and ovicells with small rounded pseudopores.

Distribution Widespread in the North Sea, British Channel, Irish Sea, Celtic Sea; Adriatic and Tasmania.

a

b

Fig. 4.69 *Cradoscrupocellaria ellisi.* (**a**, **b**) Two colony fragments showing non-ovicellate zooids with scutum and spines, avicularia and vibracularia with their setae. On beached *Himanthalia* button at Oostduinkerke (Belgium)

Occurrence Probably the most common *Cradoscrupocellaria* species beached in our area.

Belgium Five colonies re-examined from their own collection, beached on plastic and *Himanthalia* buttons.

The Netherlands Probably also common on beached material but colonies have not been re-investigated. Reported from beached *Himanthalia* button (de Ruijter 2012a, 2014a) and on 20% of the more than 100 colonies of *Flustra foliacea* washed ashore near Ouddorp (de Ruijter 2015d).

Germany *C. ellisi* was collected at Heligoland (pers. comm. Britta Kind) (on *Himanthalia*?).

Cradoscrupocellaria reptans (Linnaeus, 1767)

Description Colony erect, branched, internodes comprising four to ten zooids. Lateral edge of internode almost straight to slightly curved; chitinous joint passing across gymnocyst and below opesia in both outer zooids (C and D) of bifurcation, and across proximal gymnocyst of inner zooids (F and G). Autozooid subelongate, narrowed proximally, with smooth proximal gymnocyst. Opesia oval, occupying distal half (almost) of zooid, cryptocyst very narrow, inconspicuous. Scutum inserted at the midline of the inner edge of opesia, branched, occupying most of opesia; slender, flattened, branched two to three times, with six to nine distal stout projections; angled at 100–120°, with first branches about 0.035–0.045 mm wide, and secondary branches about 0.025–0.035 mm. Zooid spines as follows: one long distal spine, three outer spines, one to two inner spines; most proximal outer and inner spines directed frontally; axial zooid with five distal spines. One distolateral avicularium sometimes present on each zooid, distolaterally directed and obscured by outer distal spines; rostrum triangular, with a slightly serrated lateral edge, mandible triangular. A very large avicularium present on gymnocyst of some zooids, almost aquiline, with a raised tubular base, rostrum serrated laterally, slightly curved, and directed forward; mandible triangular with hooked tip. A basal vibracular chamber often present proximally on the basal surface of each zooid, inconspicuous in frontal view; setal groove directed transversely, straight, with smooth seta longer than one zooid length. Single axial vibraculum. Rhizoids with several close-spaced reverse hooks for most of their length. Ovicell prominent, hemispherical, with 8 to 13 medium-sized rounded pseudopores; two outer and one to two inner distal spines in ovicellate zooid (Fig. 4.70).

Distribution Western Britain. Rare in the North Sea.

Occurrence

Belgium A colony on a beached *Himanthalia* button in Zeebrugge in November 2023.

Rhizoids with many hooks and some without hooks, all with attachment disc (!) too; first-branch scutum 42 µm, second-branch 24 µm, very large frontal avicularia.

Germany One sample on *Laminaria* holdfast is from July 2017 from the south harbor of Heligoland. Another sample is from 2024 on a barnacle *Verruca stroemia*, from the tidal area on the west coast of Heligoland, has one rhizoid with a few hooks, while all the others are smooth. (pers. comm. Britta Kind).

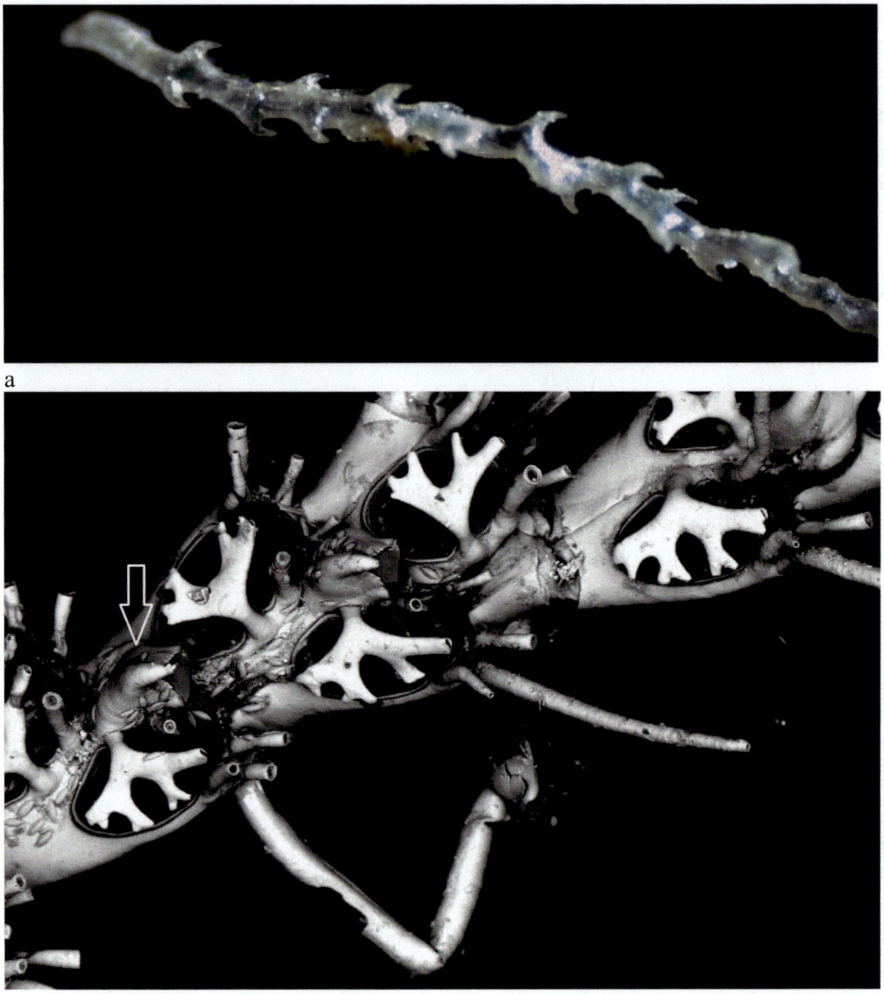

Fig. 4.70 *Cradoscrupocellaria reptans*. (**a**) Rhizoid with reverse hooks. (**b**) Part of a branch showing non-ovicellate zooids with scuta, spines, and frontal avicularia (one indicated by an *arrow*) (Heligoland, Germany)

Genus *Scrupocellaria* van Beneden 1845s. str. (Vieira et al. 2014b)

Scrupocellaria scrupea Busk, 1852

Description Colony erect, branched. The zooid bears three to four spines on the outer distal corner, and one or two on the inner distal corner. The zooid in the axillary branch has five distal spines, one of which is medial. The scutum is round or oval with the proximal portion larger than the distal. Vibracula are present basally.

Fig. 4.71 *Scrupocellaria scrupea.* (**a**) Colony on beached plastic at Bredene (Belgium). (**b**) Frontal side of the branch showing zooids with large scuta, colony grew on beached plastic, Zeebrugge (Belgium). (**c**) SEM image of the colony branches, basal view showing lateral and two axial (*arrows*) vibracula, colony beached at Westende (Belgium)

Two axial vibracula at a bifurcation. Lateral avicularia are large, small frontal avicularia are only present together with ooecia. Ovicells prominent, spherical, ectooecium with small fenestra (Fig. 4.71).

Occurrence

Belgium Occasionally on beached plastic objects and nets and on an empty oyster doublet between Oostduinkerke and De Panne in October 2004 (De Blauwe 2005a).

The Netherlands Bloklander and Leenhouts (1948) report washing ashore in the autumn of 1947. Five colonies beached on a piece of lobster trap near Ouddorp (de Ruijter 2015c).

Germany Not reported.

Scrupocellaria scruposa (Linnaeus, 1758)

Description Colony erect, branched. The zooid bears two to three spines on the outer distal corner, and one to two on the inner distal corner. Spines are often broken off. No scutum.

Vibracula occur on the basal surface and there are two axial vibracula at the bifurcation of branches. Lateral avicularia are large, frontal avicularia small and only present with ovicells. Ovicells are nearly spherical, often with teardrop-shaped fenestra on the inner side (Fig. 4.72).

Occurrence On algae, rocks, shells, Bryozoa (Hayward and Ryland 1998). In Belgium and the Netherlands, this species occasionally washes up on plastic objects.

Belgium Loppens (1906) described this species as fairly common on rocks, shells, hydroids, and *Flustra foliacea*. On several shipwrecks (Zintzen 2007). In June 2005, one colony was found on a rock between the Hinder Banks (De Blauwe et al. 2006).

The Netherlands Native in Zeeland throughout the Oosterschelde, in the Goes Canal near the lock and in the Western Scheldt near Vlissingen (Faasse and De Blauwe 2004) and on the Cleaver Bank (pers. comm. G. van Moorsel). Also mentioned on stones on West-Terschelling (Slager 1983), Den Helder (van der Sleen 1920), Dreischor (van Moorsel and Waardenburg 1999), and before the closure in the Grevelingen (Heerebout 1970). Collected on the Dogger Bank in 2016. On several shipwrecks and nets clinging to them (Lengkeek et al. 2013; van Leeuwen et al. 2017; van Leeuwen 2023).

Germany Collected in the German North Sea (Zettler et al. 2018) and at Heligoland (pers. comm. Britta Kind).

Genus *Tricellaria*

1 (a) Ovicell with multiple pseudopores, scutum usually lobed. .... *Tricellaria inopinata* (p. 235)
  (b) Ovicell with no or few pseudopores or a fenestra, scutum not lobed......... .... *Tricellaria ternata* (p. 237)

a

b

Fig. 4.72 *Scrupocellaria scruposa*. (**a**) Basal side with uncalcified joints, rhizoids, and two axial vibracula; the colony grew on washed-up plastic Westende (Belgium). (**b**) Frontal (left) and basal (right) sides of the colony branches

Tricellaria inopinata d'Hondt and Occhipinti Ambrogi, 1985

Description Colony erect, branched, attached by rhizoids. The branches consist of alternating zooids in two series. The first zooids after a branch have a chitinous joint. Zooids elongated, narrow proximally, distally an oval frontal membrane. Three spines on the outer distal angle, the most proximal of which is usually forked at the end. Two or three spines on the inner distal angle. The zooid in the axil of the branch bears one or more large spines. A scutum, often lobed, covers only the proximal half of the frontal membrane (entirely in *Cradoscrupocellaria*). However, the scutum is variable: it can be a smooth plate, or lobed in two, three, or four parts.

Fig. 4.73 *Tricellaria inopinata*. (**a**) General view of a colony from Goese Meer (the Netherlands). (**c**, **d**) Branches showing ovicellate and non-ovicellate zooids with scuta, spines, and avicularia

Polypide with 13 tentacles. Avicularia large, as wide as a zooid, lateral, triangular, with a mandible on the distal side. No frontal avicularia. Ovicells prominent, spherical, with pseudopores. Pink embryos. Released larva yellow with some pink-red dots. Ancestrula a small cup with ten long spines surrounding a frontal membrane (Fig. 4.73).

Occurrence *Tricellaria inopinata* is suspected to originate from the eastern Pacific Ocean. Indeed, the species matches the description of *Menipea* (*Tricellaria*) *occidentalis catalinensis* described from the west coast of the United States by Alice Robertson in 1905. The first European sighting comes from Venice where the species was described as new. Later, observations from the Atlantic Ocean (northern Spain), southern England, and finally the Netherlands, Belgium, and France follow. *Tricellaria inopinata* is rapidly colonizing the marinas of Europe (De Blauwe and Faasse 2001).

Belgium, The Netherlands, and Germany
This introduced species is very common in marinas, especially in August and September, where it grows on pontoons, ship hulls, *Sargassum*, *Codium*, rope, mussel shells, etc. Highly spreading species that can be expected on all coasts, mainly in marinas, but also on the infrastructure. Washes up on plastic and algae (De Blauwe 2002, 2003a; de Ruijter 2008, 2020b) and on goose barnacles (Heligoland 2023, De Blauwe unpubl.)

This species was first noticed in the Netherlands in the Goese Meer in August 2000. It was already the most common bryozoan species in the marina of Breskens.

Tricellaria ternata (Ellis and Solander, 1786)

Description A delicate white tuft, up to 3 cm high, attached by means of rhizoids. The zooid has three spines on the distal margin, two if there is an ovicell. A small undivided scutum, partially covering the frontal membrane. Lateral avicularia are triangular and large. Frontal avicularia are small and pointed proximally, usually absent and in the proximal part of an internode. Ovicells prominent, with ooecium are longer than wide, ectooecium with a single, few, or no pseudopores (Fig. 4.74).

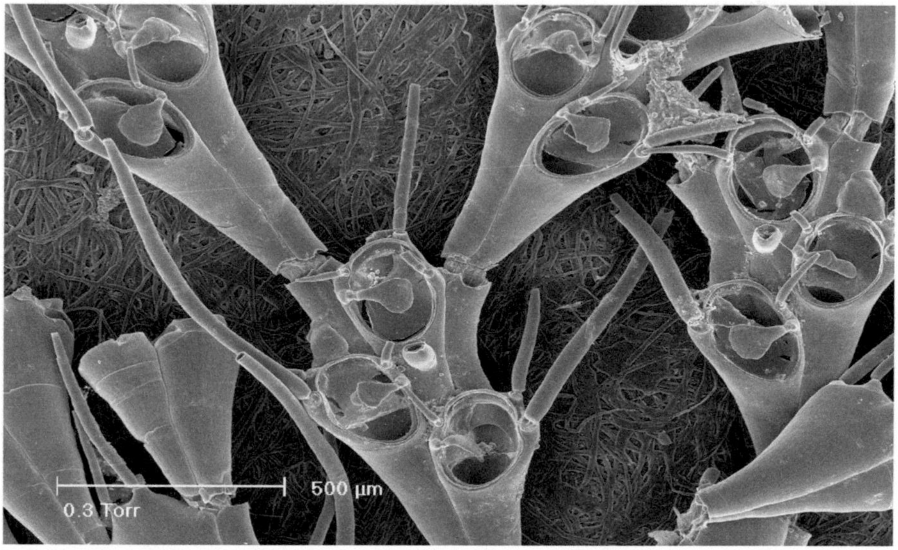

Fig. 4.74 *Tricellaria ternata.* SEM image of three partial branches, collected in Norway, by M. Faasse in 2004

Occurrence A northern species. For Belgium, this species is mentioned by Loppens (1906). He considered *T. ternata* to be rare on shells, algae, and especially on hydroids. His attached drawing bears more resemblance to *T. ternata* than to *Scrupocellaria scrupea*, a species with a similar scutum that he does not name. However, no material was found in the RBINS. Maitland (1851) mentions a washed-up colony from Katwijk.

Family Setosellidae

Genus *Setosella*

Setosella vulnerata (Busk, 1860)

Description The colony forms a small crust. Zooids oval and small. Below the frontal membrane is an extensive cryptocystal shelf with two narrow openings proximal to the near-semicircular opesia/orifice with its operculum. An interzooidal avicularium distal to the zooid with an oval body and rounded opening. This avicularium bears a slender mandible that is twice as long as a zooid. Ovicell subimmersed, closed by autozooidal operculum. Ancestrula smaller than autozooids with a much smaller cryptocyst (Fig. 4.75).

Occurrence

Belgium In 2007 two colonies were found on the inside of shells that Gilson collected in 1905 on the Hinder Banks. In May 2004, six dead colonies were found on shells from the Kwinte Bank and in June 2005 another two on shells from the Hinder Banks.

The Netherlands De Ruijter (2013c) reports a colony on washed-up plastic.

Family Microporidae

Genus *Steraechmella*

Autozooids lacking a gymnocyst. Cryptocyst occupying more than half the frontal area, opesia semielliptical, about one-third or less than autozooid length. Spines and avicularia absent. Ovicells immersed, closed by autozooidal operculum.

Steraechmella buski Lagaaij, 1952

Description The colony forms a small patch. Autozooids oval distally, truncate, or tapered proximally. Zooid boundaries marked by a raised lateral wall. Vertical walls thickened and crenellate. Cryptocyst slightly concave and crenellate. Opesia semielliptical with a slightly concave proximal edge, or trifoliate in the case in the presence of an ovicell. Ovicell immersed, ooecium is visible as a small, indistinct,

a

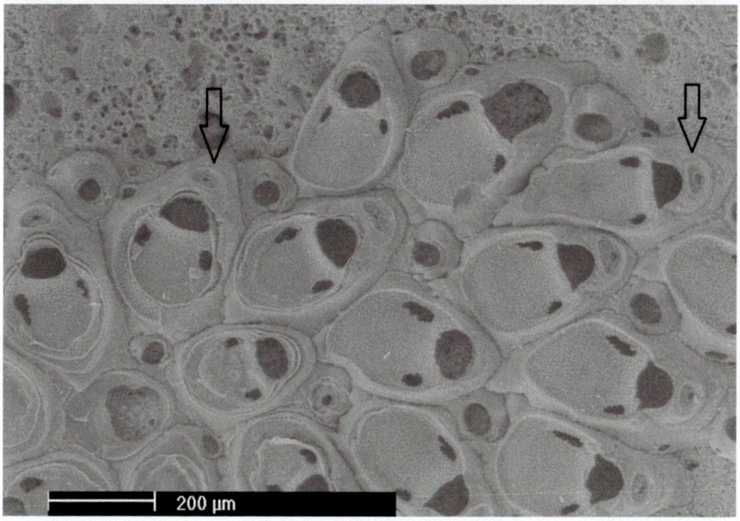

b

Fig. 4.75 *Setosella vulnerata.* (**a**) Small colony on a dead shell from the Kwinte Bank (Belgium). (**b**) Colony area showing ovicellate (two *arrows* pointing to an ovicell) and non-ovicellate zooids, ancestrula (on the left), and avicularia

shallow crescentic cap at the distal end of the maternal autozooid. Kenozooids with a reduced oval opesia randomnly present in the colony here described (Fig. 4.76).

Occurrence Described as a Pliocene fossil from the Low Countries, but found alive at greater depths than 100 m from the western Mediterranean to Faroe (Hayward and Ryland 1998).

The Netherlands De Ruijter (2014b) reports a colony on plastic washed ashore in 2013 at Egmond aan Zee, as *Rosseliana rosselii*. Further investigation revealed that is in fact *Steraechmella buski*.

Fig. 4.76 *Steraechmella buski.* (**a**) Part of the colony beached at Egmond aan Zee (the Netherlands). (**b**) Colony area showing non-ovicellate and one ovicellate zooids (left from the center)

Superfamily Cellarioidea

Family Cellariidae

Colony erect, jointed. Autozooids hexagonal, in longitudinal series, alternating around the branch. No gymnocyst. Cryptocyst depressed, covering frontal surface entirely. No spines, avicularia vicarious. Endotoichal ovicells, with brood cavity completely immersed in the distal zooid, and an independent opening in its cryptocyst, distal to the orifice of the maternal zooid.

Genus *Cellaria*

1 (a) Zooids in strictly alternating series, not in contact with the distal and proximal zooids of the same series. Avicularia as big as an autozooid and replacing it. .... *Cellaria salicornioides* (p. 242)
(b) Zooids in the same series in contact with their proximal and distal neighbors. Avicularia smaller than autozooids. 2

2 (a) Stout colony with internodes about 1 mm thick. Zooids hexagonal with broad ends. Avicularia half the size of an autozooid, squared, with a semicircular or roughly triangular mandible oblique to the axis of the internode. .... *Cellaria sinuosa* (p. 243)
(b) Internodes more slender, about 0.6 mm thick. Autozooids diamond-shaped or hexagonal, with pointed ends. Avicularia only one-third of autozooid length, quadrangular, with a narrow crescentic mandible, oriented transversely to the axis of the internode. .... *Cellaria fistulosa* (p. 241) (Fig. 4.77)

Cellaria fistulosa (Linnaeus, 1758)

Description Frequently branched tuft up to 6 cm high. Internodes cylindrical, up to 8 mm long. The internodes are not club-shaped. Usually five or six longitudinal zooid series. Zooids hexagonal, often terminally pointed or truncated, such that each zooid is in limited contact with its distal and proximal neighbor. The width of

Fig. 4.77 *Cellaria* colonies showing the width of the internodes: *C. salicornioides* (left), *C. fistulosa* (middle), *C. sinuosa* (right). Fishing harbor at Le Conquet (France)

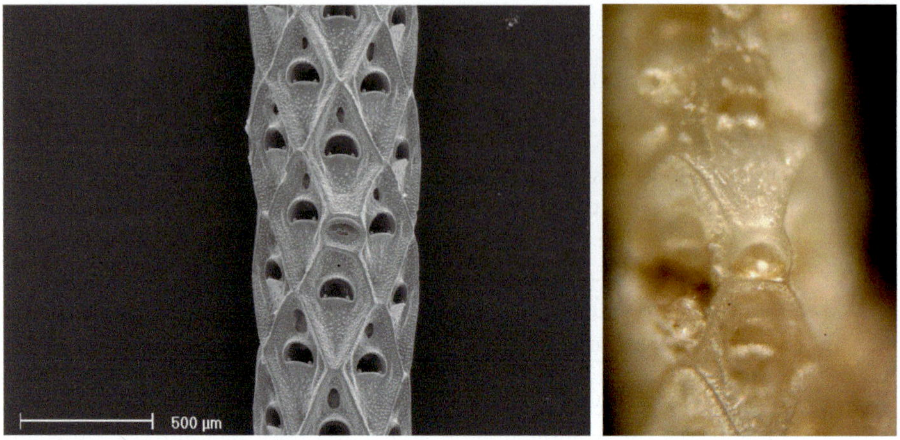

Fig. 4.78 *Cellaria fistulosa*. (**a**) Part of branch showing zooids and avicularium (center). (**b**) Magnified surface of the branch showing avicularium. Fishing harbor at Le Conquet (France)

an internode is 0.4–0.8 mm. The lateral walls are elevated around the sunken frontal surface. Opesia semicircular, with an erect margin. Two small teeth on the proximal edge of the opening. Avicularia between two consecutive zooids, one-third zooid length with a transverse elliptical opesia. Mandible less than a semicircle, only slightly angled to the axis of the internode. The opening of the endotoichal ovicell is small and rounded, just distal to the opesia (Fig. 4.78).

Occurrence On various hard substrates in the deeper sublittoral.

Belgium Occurs in the English Channel and two samples from the Flemish Banks from 1900 and 1911 are also preserved in the RBINS. Loppens (1906) considers this species to be quite rare.

The Netherlands De Ruijter (2015a) reports the washing-up of *Cellaria* cf. *fistulosa* on a plastic net and ten colonies on a piece of lobster trap (2015c).

Cellaria salicornioides Lamouroux, 1816

Description The smallest of the three species. Forms a tuft up to 5 cm high. Internodes 5–10 mm long, cylindrical, sometimes thickened to club-shaped where fertile zooids occur. Internodes 0.2–0.5 mm wide with two to five longitudinal zooid rows. Zooids implanted alternately. Consecutive zooids not in contact with each other. However, this arrangement is not followed where the internode is thickened. Zooids are hexagonal to oval, sometimes almost rectangular where the internode is at its widest. Opesia in the distal third of the frontal surface, semicircular with a straight or slightly convex proximal edge, with a short tooth at each corner. Operculum light brown with a granular surface and a noticeable marginal thickening.

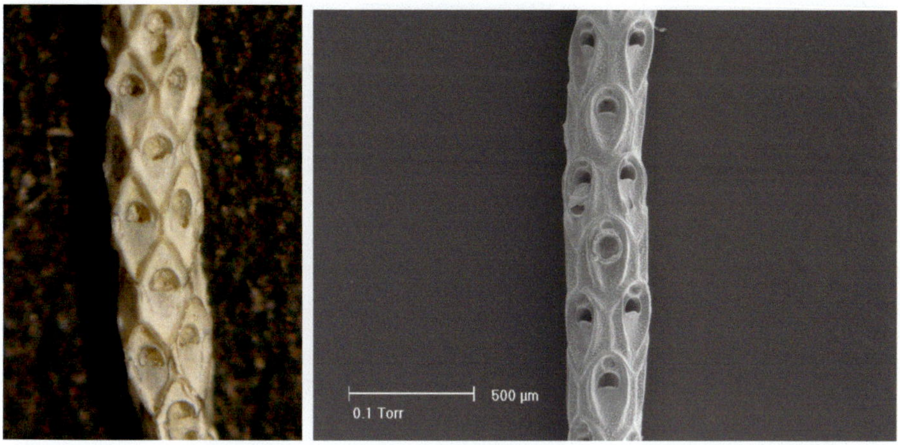

a b

Fig. 4.79 *Cellaria salicornioides*. (**a**) Fossil fragment, Oostende (Belgium). (**b**) Colony fragment with a secondarily calcified autozooid after damage (center), showing that consecutive zooids are not in contact with each other. Le Conquet (France)

Avicularia between two zooids and as large as a zooid, not common. An enlarged chitinous mandible stands on a hemispherical calcified rostrum, at an oblique angle to the frontal surface. The round opening of the endotoichal ovicell is located distal to the opesia, in older zooids it is transversely oval due to further calcification (Fig. 4.79).

Occurrence In Belgium, only known from fragments adhering to recently washed-up plastic (De Blauwe 2003a). These fragments were almost certainly brought in from outside the area. Such fragments have also been found in sub-fossil grit from Oostende.

Cellaria sinuosa (Hassall, 1840)

Description A dense bush, usually 4–5 cm high, sometimes up to 10 cm. Internodes cylindrical, club-shaped at the tips of the colony. Internodes usually 4–10 mm long, diameter 0.4–1.6 mm. The sturdiest species with up to ten longitudinal zooid rows. Zooids hexagonal with prominent transverse walls. Each zooid is in broad contact with its proximal and distal neighbors. Opesia semicircular, about the size of the operculum, its a proximal border with an erect, square lip and a hinge point on each side of it. A few smaller hinge points may be visible deeper, within the distal edge of the opesia.

Avicularia between two consecutive zooids, half as long as a zooid, rounded quadrangular, with elliptic opesia. Mandible semicircular or rounded-triangular, oriented obliquely to the axis of the internode. Ovicell endotoichal, with elongated opening, the proximal edge with an upright, angular lip (Fig. 4.80).

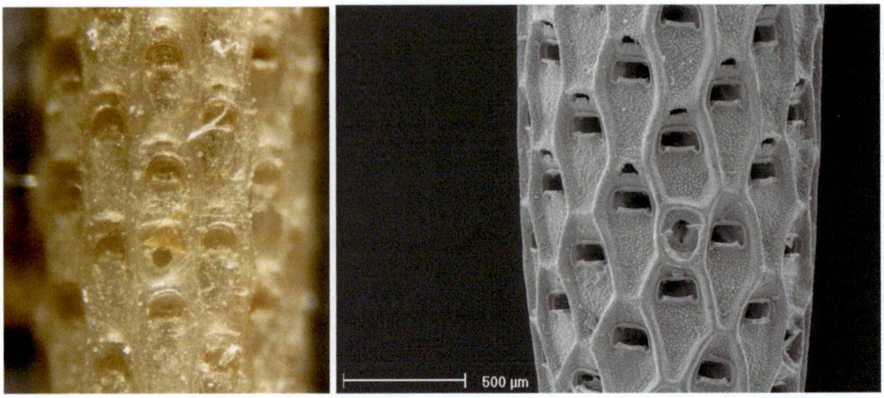

a b

Fig. 4.80 *Cellaria sinuosa.* (**a**) Colony surface showing zooids and avicularium with rounded-triangular mandible oriented obliquely to the axis of the internode. (**b**) Ovicell openings are seen above opesia. Fishing harbor Le Conquet (France)

Occurrence In the English Channel, deep sublittoral, even on sandy bottoms, attached to rough fragments.

Belgium Washed up in De Haan in September 2001 on plastic.

The Netherlands Beached in November 2023 at Texel (pers. comm. Rob Dekker).

Family Cribrilinidae

Genus *Collarina*

Collarina gautieri Harmelin, 2019

Description Colony encrusting, unilaminar, multiserial, small- to medium-sized, early mature. Autozooids with narrow marginal gymnocyst, costate shield comprising 7 to 12 costae bordered by a collar of large, prominent pelmatidia, one at the top of the steep base of each costa, pelmatidia often at the tip of a conical pillar; orifice with small L/W ratio; apertural bar with raised, massive umbo bearing two apical pelmatidia and two large basal pelmata; three to five oral spines; adventitious avicularia paired or single, associated with apertural bar, directed laterally; ovicells terminal, globular, associated with a pair of thick oral spines typically arched inwardly and distal avicularium directed proximally or distally, ectooecium punctured with many pseudopores with prominent edge, the smaller at the tip of conical processes. Ancestrula with a cribrimorph shield of six to seven costae, four or five oral spines, and without avicularium (Fig. 4.81).

a

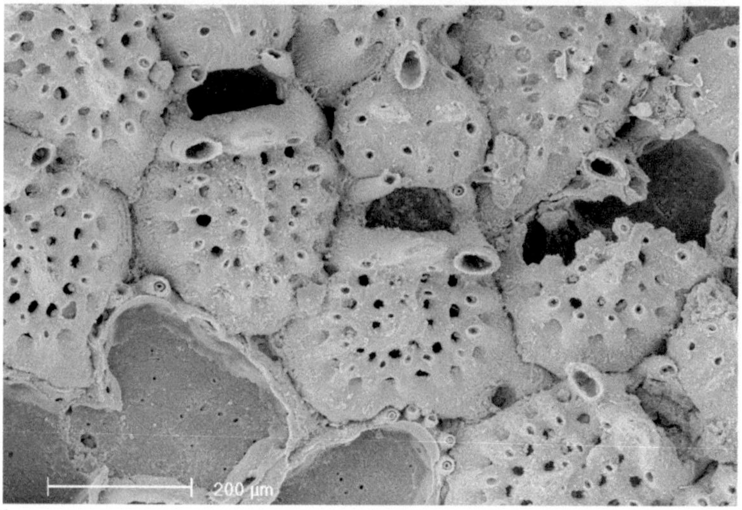

b

Fig. 4.81 *Collarina gautieri.* (**a**) Small colony on a dead shell. (**b**) Ovicellate and non-ovicellate zooids, avicularia, and spines, Westhinderbank (Belgium)

Occurrence Covers both the NE Atlantic and the Mediterranean. In the Atlantic, it is widely distributed from the Faeroes and the Shetlands to the southern coasts of the Iberian Peninsula. On the inner side of bivalves in Brittany, France at Roscoff and Pléneuf-Val-André, in the Rias in Spain (Ria de Ferol, Ria de Vigo, Ria de Pontevedra), the English Channel at Plymouth, in the Raasay Sound in Scotland and in the North Sea (Harmelin et al. 2019). Additional records from the author include Brittany, France: St-Jacut de la Mer 2014, Ile Callot 2015.

Belgium Common on the Vlaamse banken/Westhinderbank in 2004–2005.

The Netherlands De Ruijter 2014b reports a colony on beached plastic near Egmond aan Zee in 2013.

Genus *Cribrilina*

1 (a) Ovicells subimmersed or terminal, spherical ooecium, with or without pseudopores. 2
(b) Ovicells endozooidal, ooecium with transverse frontal ridge. Two to four oral spines in non-ovicellate zooids. ....
.... *Cribrilina cryptooecium* (p. 246)

2 (a) Ooecium with distinct pseudopores. Two to five oral spines ....
.... *Cribrilina punctata* (p. 246)
(b) Oooecium without pseudopores. A tooth projects over the opening from the first pair of costae. Oral spines 4; two forked spines in ovicellate zooids.....
....*Cribrilina belgica* nov. sp. (p. 249)

Cribrilina cryptooecium Norman, 1903

Description Colonies encrusting, multiserial, unilaminar, often extensive. Autozooids oval, small, convex, the boundaries marked by deep grooves. Frontal shield comprising 5 to 12 costae with one or two irregular lacunae between adjacent costae. Midline of shield not thickened. A single lacuna on a prominent tubercle may occur on each costa. The pair of distal costae form an apertural bar with a low, sometimes bifid, median mucro. Two short, thick, distal oral spines on zooids without an ovicell. Avicularia short, triangular on one or both sides of the aperture, obliquely directed distolaterally. Ovicells endozooidal, ooecium with a thick proximal transverse ridge and distal mucro. Embryos orange. Ancestrula tatiform with nine periopesial spines (Fig. 4.82).

Occurrence In the Netherlands found in Zeeland in the “Diepe Gat van Ouwekerk” and presumably in the Grevelingen at Ouddorp or den Osse (listed as *C. punctata*) (Faasse and De Blauwe 2004). Not in Belgium. Collected in Germany in the Tiefe Rinne, on empty shells, about 50 m deep (Kuhlenkamp and Kind 2012).

Cribrilina punctata (Hassall, 1841)

Description Colonies encrusting, multiserial, unilaminar, often extensive. Autozooids oval, convex, the boundaries marked by deep grooves. Frontal shield comprising 6 to 14 costae with two to four lacunae between adjacent costae. Lacunae may be present on the costal shield, the marginal lacunae often tuberculate. A carina develops along the midline of the costal shield in later ontogeny. The pair of distal

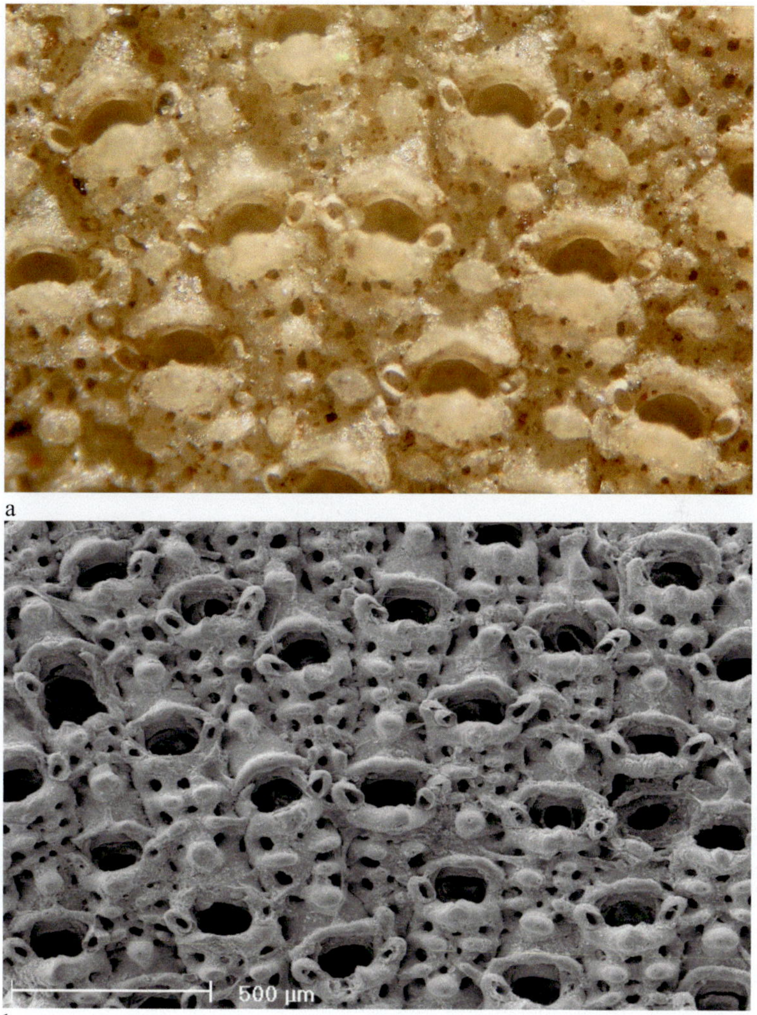

Fig. 4.82 *Cribrilina cryptooecium*. (**a**, **b**) Colony areas showing numerous ovicellate zooids with elevated ooecial ridge and distal mucro. Pointe du Rozel (Brittany, France)

costae form a thickened apertural bar, often mucronate. Three to five slender oral spines on young zooids, two oral spines on ovicellate zooids.

Avicularia at proximolateral corners of the orifice, often one or both absent. Avicularia directed slightly obliquely distally, not directly laterally as in *Collarina gautieri*. Ovicells subimmersed, some terminal, ooecium smooth, round, with distinct pseudopores. Pink embryos. Ancestrula tatiform, with periopesial spines. This species can be confused with *Collarina gautieri* (Fig. 4.83).

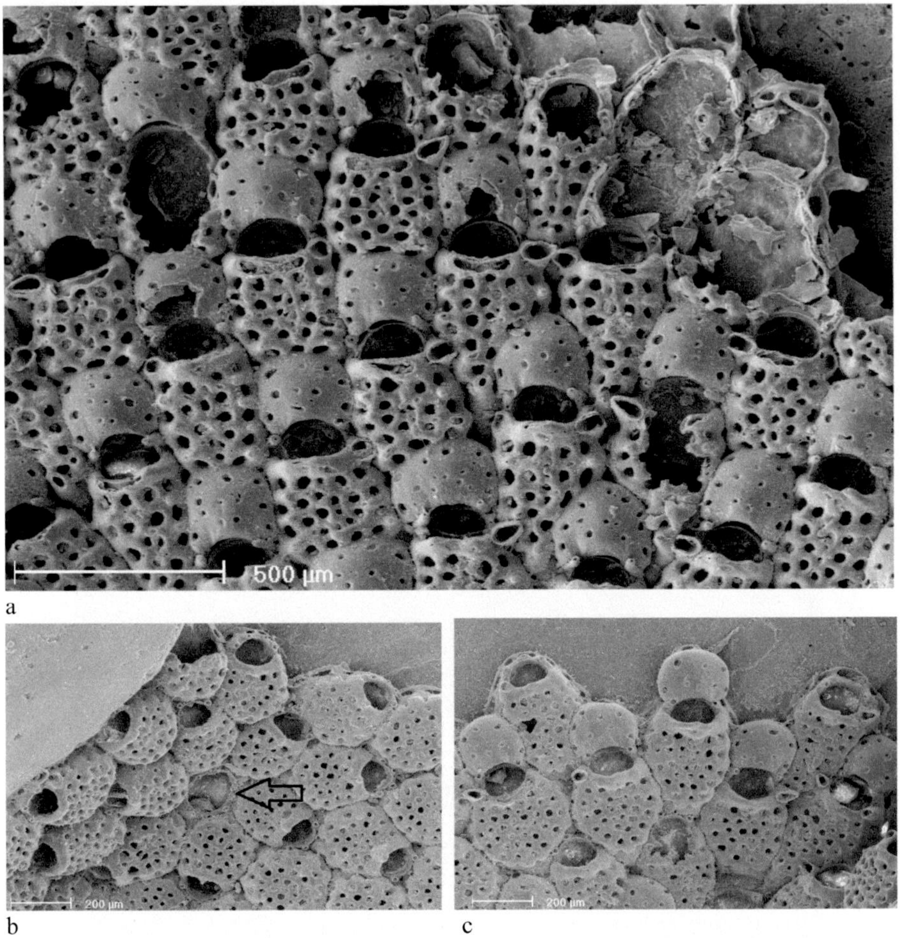

Fig. 4.83 *Cribrilina punctata.* (**a**) Part of the colony showing ovicellate zooids with ooecia formed by the distal autozooid. (**b**) Colony area with tatiform ancestrula (*arrow*). (**c**) Colony periphery showing terminal ovicells with ooecia formed by the distal kenozooid non-visible from above. Colonies on dead shells, Westhinderbank (Belgium)

Occurrence

Belgium Once on beached plastic at Zeebrugge. In 2005, a few colonies were found on a stone and on shells from the Westhinder Bank. In 2023, a single colony was collected on a shell between 51°30.511′ N, 2°35.273′ E and 51°29.559′ N, 2°36.378′ E.

The Netherlands The report in Maitland (1897) is unreliable. A dead colony was found on an old cockle near Westkapelle (de Ruijter 2013b). De Ruijter (2014b) reported a colony on beached plastic in 2013 near Egmond aan Zee. Five young colonies of *Cribrilina punctata* were found on three scour-protection rocks of platform L10-G consisting of 10 to approximately 30 non-ovicellate zooids (Beukhof et al. 2016).

From Friesland (Leiden ZMABRYO 02073) three fresh beautiful colonies on the inner side of a bivalve shell (coll. J.H.Stock, no date). On a shell from a bottom sample on the Cleaverbank (54.021019 N, 3.165225 E, DDNZS, June 18, 2017). In 2022, R. de Ruijter found colony fragments in grit from Texel.

Germany Collected in the Tiefe Rinne (Kuhlenkamp and Kind 2012).

Cribrilina belgica nov. sp.

Type Material

I.G. 34,864 / BRYT.230 holotype on shell fragment, gold coated for SEM, collected by Hans De Blauwe during replenishment of the beach with sand from the Kwinte Bank in 2004.

I.G. 34,864 / BRYT.230/2 paratype on shell fragment, gold coated for SEM, two colonies with ancestrula, collected by J.S. Houziaux by Gilson dredge (JSH-38) on the Hinder Banken in 2005.

I.G. 34,864 / BRYT.230/3 paratype on shell fragment, gold coated for SEM, colony with ancestrula, collected by J.S. Houziaux by Gilson dredge (JSH-38) on the Hinder Banken in 2005.

Description Colony encrusting, unilaminar, multiserial, small. Autozooids oval, convex, getting gradually larger distally in the colony, length 250–371 µm ($n = 8$), width 157–242 µm ($n = 5$). Frontal shield comprising 10 to 12 costae, pinnate except for suboral pair, each tapering to the center of the shield, with three or four irregularly shaped lacunae between them. The pair of distal costae form an apertural bar, which in ovicellate zooids bear a medial projection pointing distally and upward over the aperture. Zooids without an ovicell bear four oral spines. If there is an ovicell, the zooid has two oral spines that are flattened, triangular in shape with distal digitations, the distal digit resting in a groove on the ovicell. The colony soon produces ovicells (spot colony). Ovicells terminal, ooecium smooth or with few tubercles, without pseudopores; one to three pore chambers occur along its distal edge. Ooecium is produced by a distal kenozooid sometimes non-visible from the above or visible as a rudimentary frontal shield of one or two small costae between the distal end of the ectoooecium and uncalcified window of the distal pore chamber. Avicularia not observed. Ancestrula tatiform, 200–314 µm long ($n = 2$), with eight ($n = 2$) or ten ($n = 2$) periopesial spines (Fig. 4.84).

Remarks *Cribrilina belgica* nov. sp. forms a typical spot colony, with only a dozen autozooids and reproducing already at a very small colony size, characteristic of a species adapted to life on unstable, short-lived substrates. Ovicells lack pseudopores or proximal transverse ridge; two oral spines leaning on the ovicell are forked. *Cribrilina cryptooecium* forms larger sheets and has a transverse ridge on the ooecium.

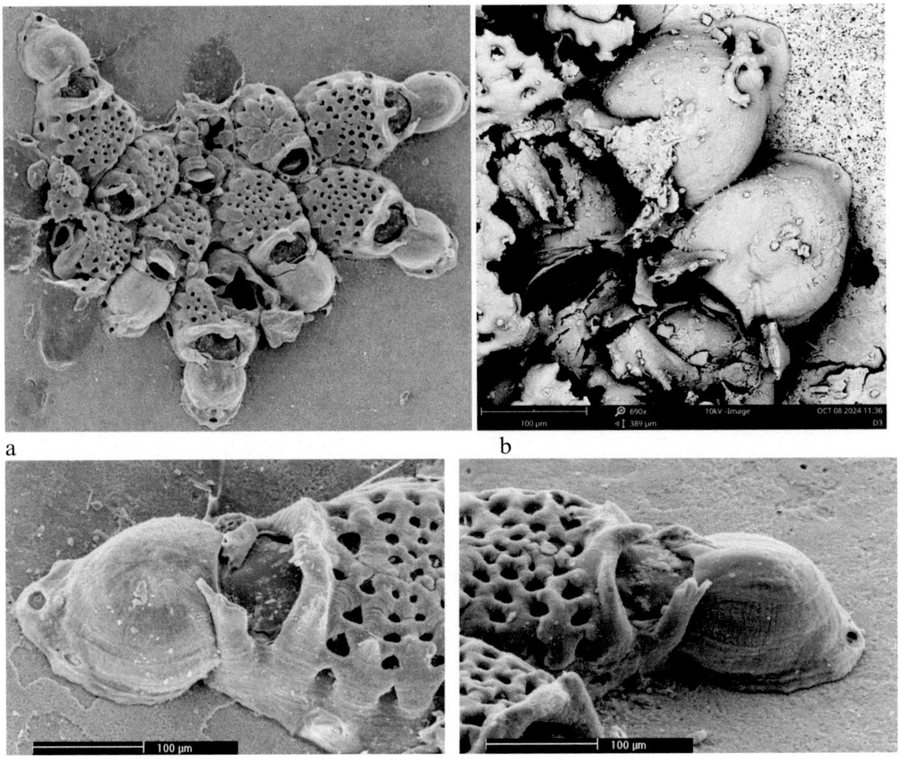

a b c d

Fig. 4.84 *Cribrilina belgica* nov sp. (**a**) Holotype colony on a dead shell. (**c**, **d**) Autozooids with terminal ovicells, forked oral spines, and small pore chamber distal to ooecium, Kwinte Bank (Belgium). (**b**) Two ovicellate autozooids; left ooecium is formed by the distal kenozooid visible as a rudimentary costate frontal shield from above, paratype BRYT.230/3

C. punctata has pseudoporous ectooecium. Ovicellate autozooids of *C. spitsbergensis* Norman, 1903, have similar flattened spines but have a pseudoporous ectooecium.

Occurrence Only known from Belgium and uncommon on the Flemish Banks. Found for the first time in 2004 on two shell valves on the Kwinte Bank. Collected again in 2005 on shell valves at the Westhinder Bank. The area where they were sampled is about 20 m deep, with gentle slopes up to 5 m deep, consisting of sand, shell valves, and pebbles.

Genus *Juxtacribrilina*

Diagnosis Zooids with variably extensive frontal shield of pinnate costal spines, with rows of intercostal lacunae, and at least one pseudopore, usually at tip; suboral costae stoutest, typically with frontally elevated tips. Articulated oral spines present. Ovicellate zooids in quincunx with autozooids and/or adventitiously borne and small or dwarfed; ooecium from well developed to vestigial, costa-like or flap-like, typically kenozooidal from maternal zooid, less commonly produced from distal autozooid, intimately associated with a pair of transversely orientated proximofrontal costal spines, each typically with one or more pseudopores. Interzooidal communications via basal pore chambers. Ancestrula costate (Yang et al. 2018; Dick et al. 2021).

Juxtacribrilina mutabilis (Ito, Onishi and Dick, 2015)

Description

Forms light pink, flat, small (5–7 mm across) circular encrusting colonies. Avicularia absent. Zooids oval, with two to three distal spines; spinocyst convex, composed of hollow, non-articulated costae resembling ribs with degree of lateral fusion dependent on zooid type. Zooids with costate frontal shields of three types:

R (rib) type: Spinocyst with costae widely spaced, large basal openings, fused at midline, appearing like "vertebrate rib cage." No intercostal pores.

S (shield) type: Spinocyst appearing as shield from distally fused costae; intercostal pores with radial arrangement; basal openings narrower than costae, size reduced; costae of apertural bar thickest, separate from adjacent costae by gap from basal opening and at least one lumen pore.

I (intermediate) type: Spinocyst with costae more numerous and narrower than R-type, with varying degree of distal and lateral fusion creating elliptical shield of variable size with scattered intercostal pores.

Only R- and I-type zooids reproducing. Reproductive zooids bear a vestigial compound (tripartite) ooecium and brood internally; live embryos yellow-orange. Dwarf-reproducing autozooids of R-type appear in some colonies (Fig. 4.85).

Occurrence Non-indigenous species, described from Japan as an obligate eelgrass epibiont.

Belgium Wilfried Laforce found a colony on a mussel shell in the Spuikom Oostende, a center of shell fish culture, in March 2025.

The Netherlands On beached plastic with *Schizoporella* cf. *japonica* in January 2017 at Bergen aan Zee (de Ruijter 2017b). Two colonies in a shell in the Oosterschelde on August 2023 by GIMARIS.

Germany On a mussel on pontoons in the south harbor of Helgoland by F. Hoffmann (Landesamt für Umwelt des Landes), in June 2022 and in November 2023.

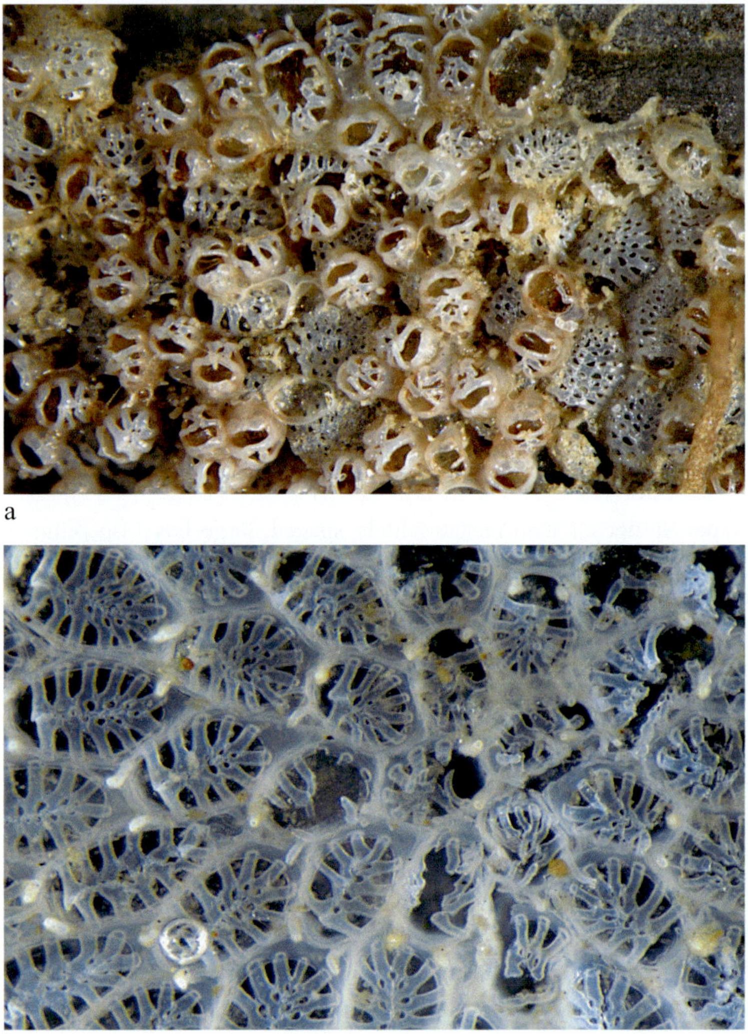

Fig. 4.85 *Juxtacribrilina mutabilis*. (**a**) Colony area showing zooids of R and I types and dwarfed frontal zooids, on plastic, Bergen aan Zee (the Netherlands). (**b**) Ancestrular region, Norway

Genera *Puellina—Cribrilaria—Glabrilaria*

Based on revision by Rosso et al. (2018, 2021):

Puellina Jullien (type species *Lepralia gattyae* Landsborough, 1852):

- Ovicells subimmersed or terminal; pseudoporous ooecium produced either by distal autozooid or kenozooid
- The absence of avicularia

- Comparably large gymnocyst and small costal shield formed by a few costae
- The presence of setiform papillae from the two distalmost papilla pores
- Articulated oral spines, which lack apophyses, reduced to two on ovicellate zooids
- Lateral walls exposing relatively small uncalcified windows of pore chambers
- The presence of a median proximal spine in the simple tatiform ancestrula
- Ancestrula regenerated as a miniature zooid or a kenozooid

Cribrilaria Canu and Bassler, 1929

- Ovicells prominent or subimmersed, with non-pseudoporous ooecium produced by either distal autozooid or kenozooid visible from the above
- Interzooidal avicularia, more or less frequent and variable in size and shape
- Gymnocyst narrow and frontal shield extensive, often formed by numerous costae
- The oral spines may have lateral pointed apophyses
- Oral spines four to eight (usually five) and decrease to four in ovicellate zooids
- The uncalcified windows of pore chambers are relatively large
- The ancestrula has a median proximal spine and is regenerated as a kenozooid

Glabrilaria Bishop and Househam, 1987

- Ovicells prominent, non-pseudoporous ooecium produced by either distal autozooid or a kenozooid visible or non-visible from the above
- The presence of pedunculate or semierect avicularia born from a pore chamber on zooidal lateral walls, consequently not in contact with the substratum
- Six to seven (rarely five) oral spines reduced to four in ovicellate zooids
- A relatively narrow gymnocyst and a relatively large frontal shield
- Tiny to moderately sized uncalcified windows of pore chambers
- A pointed median proximal spine in the ancestrula
- Ancestrula regenerated as a kenozooid

1 (a) Frontal shield with more than 12 costae. .... *Cribrilaria venusta* (p. 259)
(b) Frontal shield composed of 12 or less costae. 2

2 (a) Five oral spines in non-ovicellate zooids. 3
(b) Six of seven oral spines in non-ovicellate zooids. 7

3 (a) Ovicellate zooids with two oral spines. 4
(b) Ovicellate zooids with four oral spines. 5

4 (a) Four to six costae, ooecium with few pseudopores. .... *Puellina modica* (p. 262)
(b) Six to ten costae, the first pair parallel to each other. Ooecium with three to five short ridges fanning out from the center. .... *Cribrilaria directa* (p. 256)

(c) Five costae, zooids in uniserial rows, ooecium with a median ridge. .... *"Puellina" monostachys* nov. sp. (p. 264)

5 (a) Colony small, up to 20 zooids. Ovicells terminal, ooecium with five prominent tubercles, often with uncalcified top (pseudopore). .... *Puellina praecox* (p. 267)
(b) Ooecium is formed by the daughter zooid. 6

6 (a) Colony forming an extensive sheet. Big, often triangular, single median suboral lacuna on the distal slope of a suboral mucro. .... *Cribrilaria innominata* (p. 258)
(b) Colony small. Median suboral lacuna as large as intercostal pores. Ooecium with four short radiating ridges. .... *Cribrilaria bifida* (p. 256)

7 (a) Six oral spines. 8
(b) Seven oral spines. .... *Glabrilaria corbula* (p. 260)

8 (a) One suboral lacuna, costae forming a raised median rim or umbo. .... *Cribrilaria arrecta* (p. 254)
(b) at least two suboral lacunae, the first pair of costae forming a high arc around the lacunae. .... *"Puellina" nana* (p. 264)

Cribrilaria arrecta Bishop and Househam, 1987

Description Colony encrusting, unilaminar, multiserial, small. Autozooids roundly subhexagonal to suboval. Frontal shield comprising eight to nine (rarely seven to ten) pinnate costae. Each costa with a prominent tubercle toward the periphery of the shield, often with a minute pelmatidium on top of the tubercle. Costae forming a suboral umbo or short carina in the midline of the shield. Two to three lacunae between adjacent costae, lacunae uniform in size, reniform owing to a denticle (ligula) adjacent to the intercostal suture. The most peripheral opening between costae is a lipped papilla pore, slightly larger than the adjacent lacuna. Orifice transversely D-shaped, wider than long, proximal edge straight. Six long oral spines, jointed some distance above the base, in non-ovicellate zooids, reduced to four in ovicellate zooids. Spines with inwardly directed half-circle of denticles (apophyses) above the joint. Apertural bar sloping upward from the proximal margin of the orifice to the first row of lacunae, with median suture and often a minute pelmatidium on a low tubercle on each side of the midline. The first pair of costae define a very small, sometimes large, median suboral lacuna; three to four moderately sized uncalified external windows of basal pore chambers on each distolateral side of autozooids at the colony margin. Avicularia interzooidal, half the size of an autozooid, with semicircular opesia and distally broadening rostrum. Ovicell prominent, ooecium formed by the distal autozooid or kenozooid visible from above, sometimes met in the same colony, without pseudopores, with two or three low ridges and sometimes a discernible median suture. Kenozooids, with cribrimorph frontal shield, may occur. Ancestrula tatiform, with 11 periopesial spines, median proximal spine bifid; sometimes regenerates as a kenozooid (Fig. 4.86).

Occurrence Present in the nearby English Channel.

Belgium In June 2005 four colonies were found on the inside of shell valves from the Westhinder Bank.

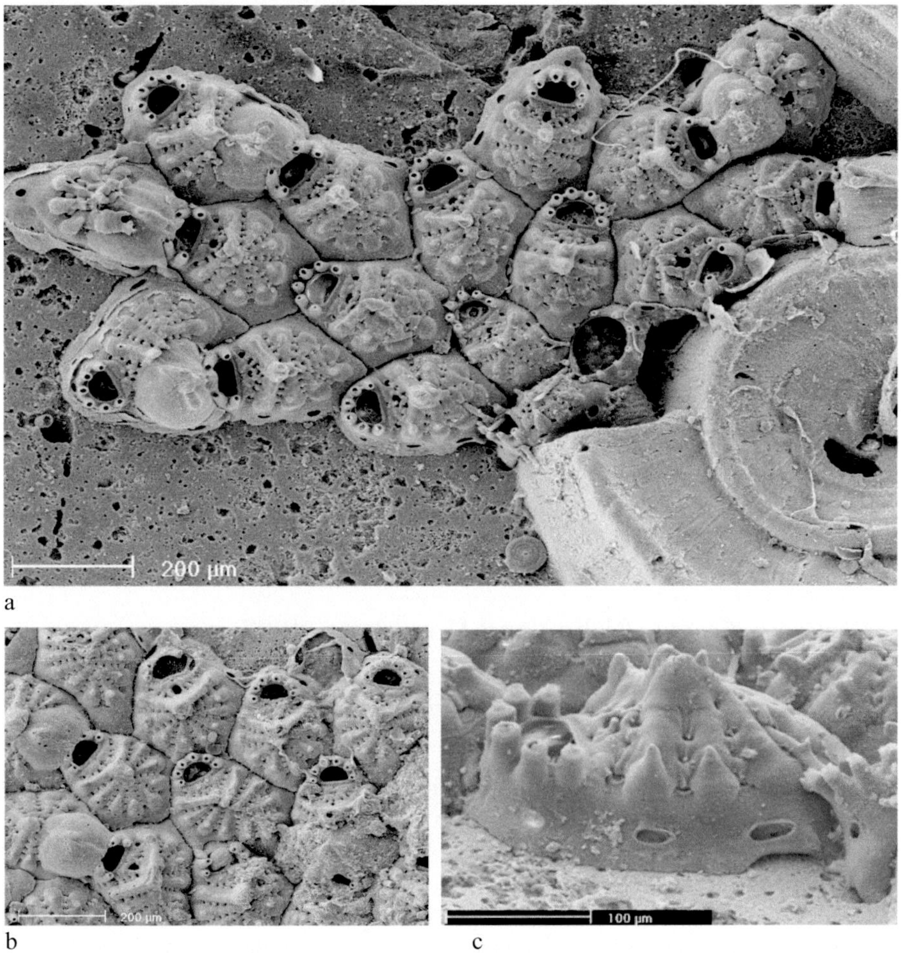

Fig. 4.86 *Cribrilaria arrecta*. (**a**) Small colony on a dead shell showing two ovicells with ooecia formed by the distal kenozooid and three ooecia formed by the distal autozooid. (**b**) Colony area showing six oral spines in non-ovicellate zooids and four in ovicellate zooids. (**c**) Lateral view of autozooid showing lipped papilla and reniform lacunae between costae, tubercles with minute pelmatidium on the periphery of the costal shield, base of jointed oral spines, oral spine with apophyses above its base leaning on the proximal side of its distal zooid and uncalcified windows of pore chambers. Westhinder Bank (Belgium)

Cribrilaria bifida (d'Hondt, 1970)

Description Colony encrusting, unilaminar, multiserial. Autozooids roughly suboval to irregularly polygonal, sometimes with proximal or lateral extensions between neighboring zooids. Frontal shield comprising 6 to 11 costae, mostly 7 to 9, sharply ridged and highest at the periphery of the shield, centrally rather flat without mucro nor carina. Four to five subrectangular to reniform lacunae between adjacent costae. Autozooidal orifice transversely D-shaped, wider than long, proximal margin straight. Five oral spines, reduced to four in ovicellate zooids. Oral spines basaly jointed, and generally with inwardly directed spiniform apophysis near base. Apertural bar with a tuberculate pelmatidium on each side of the median suture. A small suboral lacuna near the first row of intercostal lacunae. Five to eight large uncalcified windows of distolateral pore chambers visible in zooids on the edge of the colony. Avicularia (Fig. 4.87b) sparse, interzooidal, mostly near the edge of the colony. Ovicell prominent, formed by the distal autozooid; ectooecium without pseudopores, often with about four short ridges in a radiating pattern and a median suture. Kenozooids infrequent, smaller than autozooids, with a costate shield, lacking an orifice. Ancestrula tatiform, with 11 periopesial spines, median proximal spine bifid. The ancestrula sometimes regenerates as a kenozooid (Fig. 4.87).

Occurrence Offshore on shell substrata.

Belgium Subtidal on the inner side of bivalve shells on the Kwinte Bank and the Hinder Banken. Consequently found on beaches that are replenished with sand from these localities.

The Netherlands Found on the replenished beach of Nieuwvliet. De Ruijter (2013c) reports a colony on beached plastic.

Cribrilaria directa (Bishop and Househam, 1987)

Here reassigned from *Puellina* to *Cribrilaria* due to the ovicell type, ooecia lacking pseudopores, despite the lack of avicularia.

Description Colony encrusting, unilaminar, multiserial, small. Autozooids roughly suboval, sometimes with short proximal or lateral extensions between neighboring zooids. Gymnocyst surrounding a frontal shield comprising seven to eight (rarely nine or ten) costae. Distalmost pair of costae parallel, forming a double straight ridge across the shield; more proximal costae smaller and radiating from the middle of the shield to the edge. Two to four lacunae and a lipped papilla pore of the same size between adjacent costae. Autozooidal orifice transversely D-shaped, wider than long, proximal margin straight. Apertural bar sloping upward from the proximal margin of the orifice to the first row of lacunae, a minute pelmatidium on a low tubercle on each side of the suboral lacuna. The oval suboral lacuna is sometimes divided by a small ligula. Six oral spines, jointed some distance above the base, in non-ovicellate zooids, reduced to four in ovicellate zooids. Avicularia are not observed.

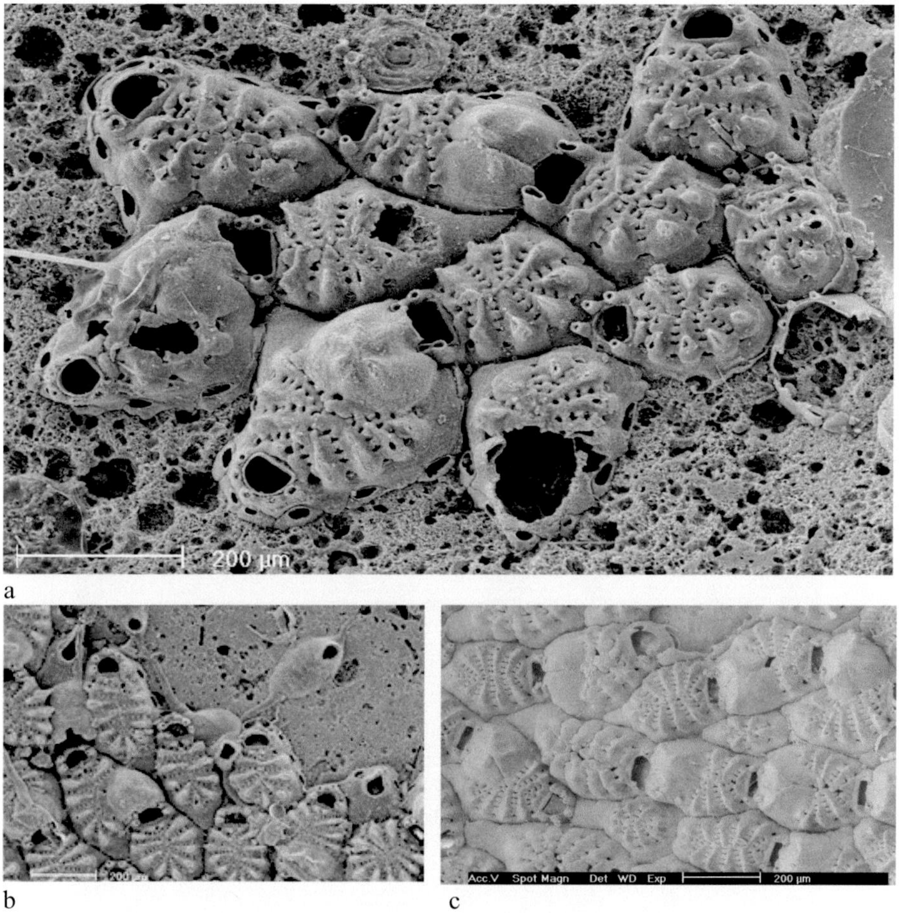

Fig. 4.87 *Cribrilaria bifida*. (**a**) Young colony with two ovicells formed by the distal zooids; ooecia with radiating ridges and median suture; broken ancestrula to the right. (**b**) Edge of colony with two interzooidal avicularia and zooids of *Hippothoa flagellum*, Westhinder Bank (Belgium). (**c**) Image showing part of the colony with ovicellate and non-ovicellate autozooids and two kenozooids; ooecia are formed by the distal autozooid, Kwinte Bank (Belgium)

Ovicell subimmersed, ooecium formed by the distal autozooid; ectooecium without pseudopores, often with a median ridge and a few short ridges in a radiating pattern. Kenozooids infrequent, with a costate shield, lacking an orifice. Ancestrula tatiform, with 11 periopesial spines without apophyses, median proximal spine bifid (Fig. 4.88).

Occurrence Only known from the English Channel off Normandy and Brittany (Bishop and Househam 1987) and off Belgium since 2005 when several colonies were sampled on the Westhinder Bank (De Blauwe 2009a).

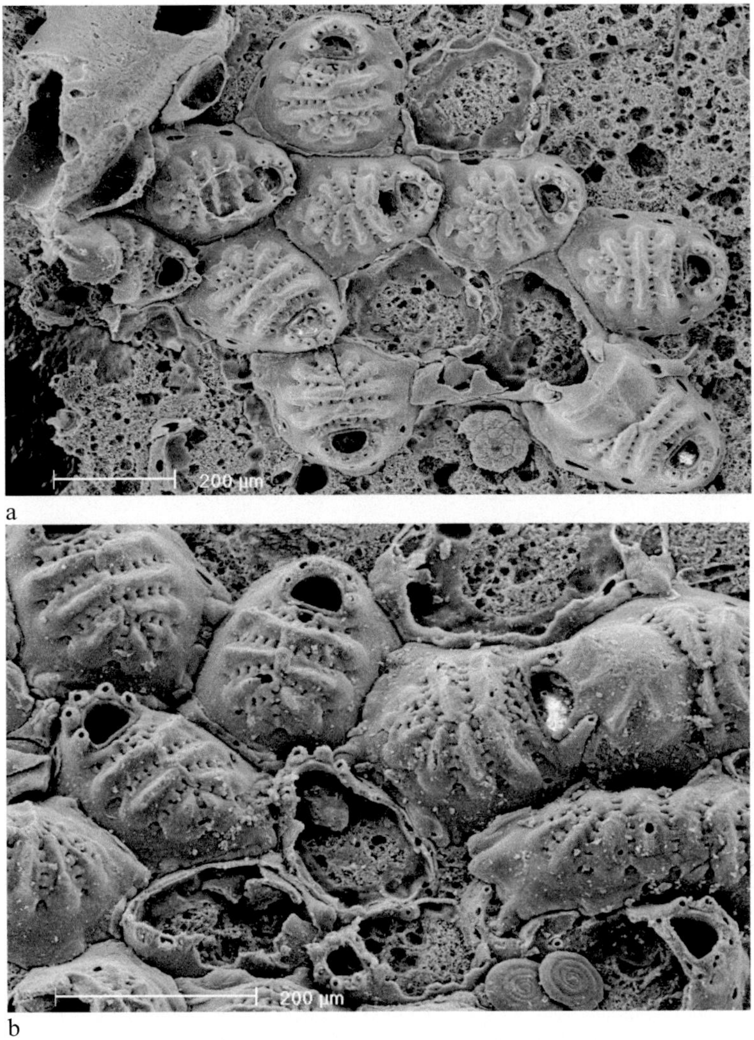

Fig. 4.88 *Cribrilaria directa*. (**a**, **b**) Partial view of two colonies on a dead shell. (**b**) Broken ancestrula is seen in the center, Westhinderbank (Belgium)

Cribrilaria innominata (Couch, 1844)

Description Colony encrusting, unilaminar, multiserial, extensive. Autozooids suboval, bifid proximally when passing to both sides of the proximal zooid, sometimes with short proximal or lateral extensions between neighboring zooids. Frontal shield comprising 11 to 12 (rarely 9 to 14) costae, with prominent suboral umbo and variable carina proximal to umbo. Each costa with often a pelmatidium or tubercle near its base.

One to five lacunae and near their base a larger, lipped papilla pore between adjacent costae. Autozooidal orifice transversely D-shaped, wider than long, proximal margin straight. Apertural bar forming a triangular area, with minute, non-tuberculate pelmatidium at both sides of the median suture. Very large suboral lacuna, often rounded-triangular.

Five oral spines, jointed just above base, in non-ovicellate zooids, reduced to four in ovicellate zooids. Seven to nine distolateral pore chambers, uncalcified external windows, visible in zooids at the colony edge, as large as or larger than suboral lacuna. Interzooidal avicularia often present, half als long as an autozooid.

Ovicell prominent, ooecium formed by the distal autozooid, ectooecium without pseudopores, often with four tubercles or short ridges in a radiating pattern. Kenozooids infrequent, with a costate shield, lacking an orifice. Ancestrula tatiform, with 11 periopesial spines with apophyses, median proximal spine bifid (Fig. 4.89).

Occurrence

Belgium In 1904 and 1906, single colonies had already been collected at the Flemish Banks (RBINS collection). More recently present in Belgium on pebbles and especially on shell valves on the Hinder Banken. Also on beached plastic (De Blauwe 2009a).

The Netherlands On beached plastic (de Ruijter 2014b, d, 2018a) and on 18/01/2022 at Westerschouwen on plastic.

Cribrilaria venusta (Canu and Bassler, 1925)

Description Colony encrusting, unilaminar, multiserial, extensive with several hundreds of autozooids. Autozooids suboval, bifid proximally when passing to both sides of proximal zooid, sometimes with short proximal or lateral extensions between neighboring zooids. Frontal shield comprising 12 to 23 (mostly 15 to 18) costae, without suboral umbo nor carina. Each costa with up to 3-minute pelmatidia along its length. Three to four lacunae, and near their base a lipped papilla pore of the same size, between adjacent costae. Autozooidal orifice transversely D-shaped, wider than long. Five oral spines, jointed just above base, in non-ovicellate zooids, reduced to four in ovicellate zooids. Spines above the joint constricted and lacking apophyses. Apertural bar with a very short triangular area, in front of the suboral mucro, lacking a suboral lacuna. Nine distolateral pore chambers, uncalcified external windows, visible in zooids at the colony edge. Interzooidal avicularia often present, half als long as an autozooid.

Ovicell prominent, ooecium formed by the distal autozooid, ectooecium without pseudopores, often variably ridged. Kenozooids infrequent, with a costate shield. Ancestrula tatiform, with 11 periopesial spines (Fig. 4.90).

Occurrence The species occurs in the English Channel and more to the south and west but not in the North Sea. Settles occasionally on plastic, that, when detached

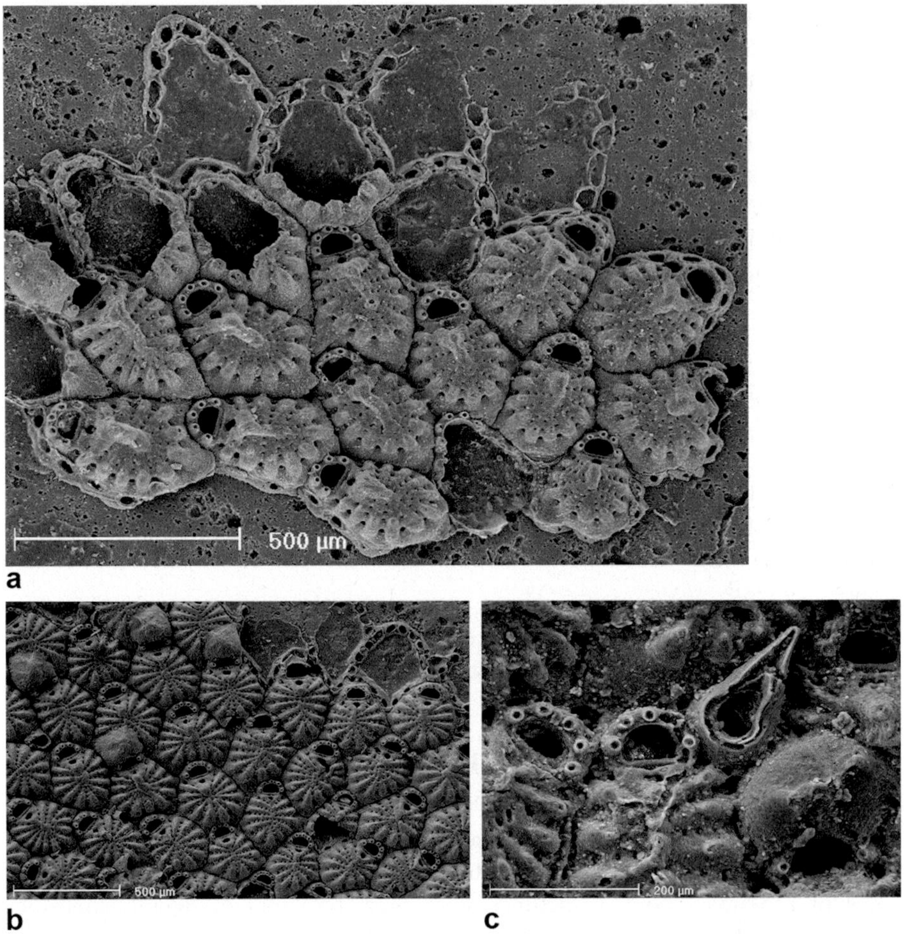

Fig. 4.89 *Cribrilaria innominata*. (**a**) General view of young colony. (**b**) Partial view of a colony showing ovicellate and non-ovicellate zooids. (**c**) Close-up of the colony periphery showing two non-ovicellate zooids, ooecium, and interzoidal avicularium, Westhinderbank (Belgium)

from the bottom, drifts to the north into the North Sea. A colony beached on plastic at Egmond aan Zee (the Netherlands) in 2013, was identified by present author and published by de Ruijter (2014b).

Genus Glabrilaria Bishop and Househam, 1987

Glabrilaria has non-pseudoporous ooecia that are exclusively produced by the distal kenozooid, erect or semierect avicularia, six to seven (rarely five) oral spines, small to moderately sized uncalcified windows of pore chambers (Rosso et al. 2018).

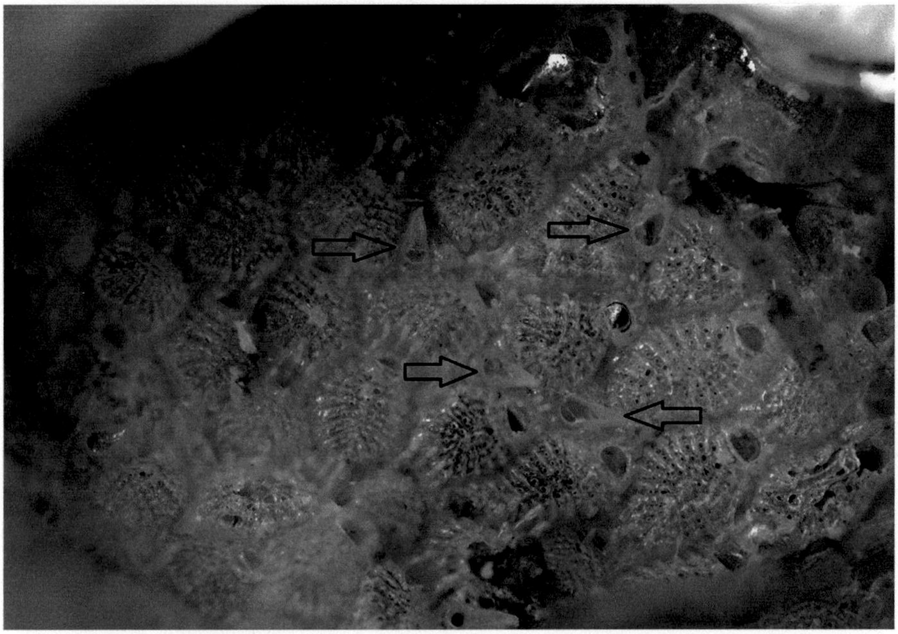

Fig. 4.90 *Cribrilaria venusta,* large colony with interzooidal avicularia (*arrows*) on beached plastic, Newgale, Wales (United Kingdom)

Glabrilaria corbula Bishop and Househam, 1987

Description Colony encrusting, unilaminar, multiserial. Zooids irregularly oval, rather flat. Frontal shield comprising 9 to 14 (usually 10 to 12) costae. Costae with broad base tapering to a narrow nearly horizontal portion. The central part of the costal shield is flat. Six to eight rectangular to reniform lacunae between adjacent costae. Papilla pore between bases of adjacent costae with same-sized intercostal lacunae.

Orifice D-shaped with one or two (rarely three) suboral lacunae between the orifice and the first row of intercostal lacunae. Seven oral spines, four in ovicellate zooids. Avicularia very small. Ovicells with four to six low ridges, ooecium formed by the distal kenozooid not visible from above. Kenozooids are uncommon, each with a costate shield. Ancestrula tatiform, with 12 slender periopesial spines (Fig. 4.91).

Occurrence

Belgium In 2005, a damaged colony was collected on the Westhinder Bank, which corresponds to the aforementioned description, given by Bishop and Househam.

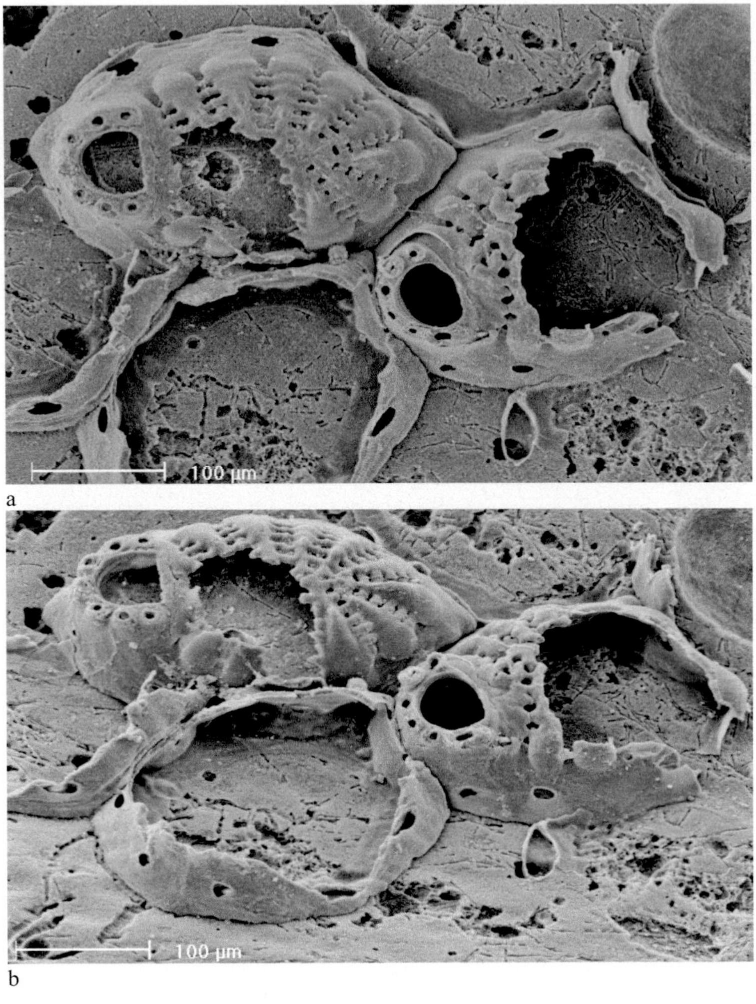

Fig. 4.91 *Glabrilaria corbula*. (**a**, **b**) Small destroyed colony, Westhinder Bank (Belgium)

Puellina modica Bishop and Househam, 1987

Description Colony encrusting, unilaminar, multiserial, small. Autozooids variable, often suboval. Frontal shield comprising four to six short, tapering, pinnate costae. Costae ridged, decreasing in height from the periphery of the frontal shield toward the center. A lipped papilla pore at the base and one or two lacunae between adjacent costae.

Autozooidal orifice transversely D-shaped, slightly wider than long, proximal margin straight. Suboral lacuna larger than intercostal lacunae. Five oral spines in non-ovicellate zooids, with a brown chitinous joint just above their base, reduced to two in ovicellate zooids. Ovicells terminal, ooecium formed by the distal kenozooid not visible from above, with medial ridge and pseudopores, sometimes with tubercles. Avicularia not observed. Kenozooids rare, with cribrimorph shield, without an orifice. Ancestrula tatiform with 11 evenly distributed spines, might regenerate as autozooid or kenozooid (Fig. 4.92).

Occurrence Known from the English Channel, Irish Sea, and "off Antrim" (Bishop and Househam 1987).

Belgium A colony was dredged in 1905 on the Hinder Banken (RBINS collection). Not common on shell valves on the Kwinte Bank and on the Hinder Banks.

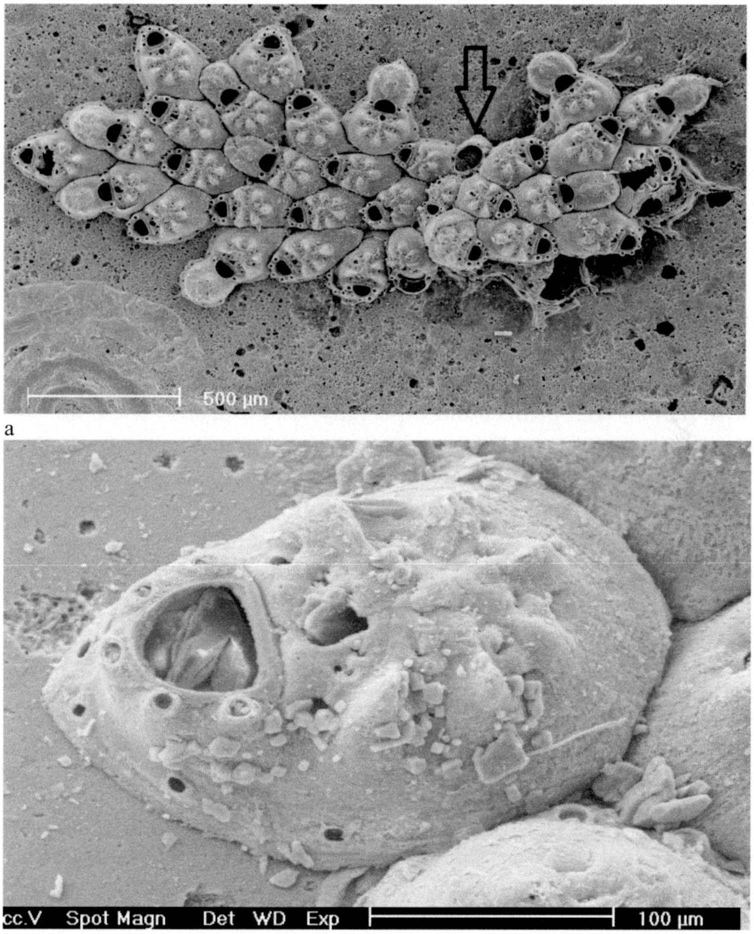

Fig. 4.92 *Puellina modica*. (**a**) General view of the colony showing ancestrula (*arrow*) and ovicellate zooids, Westhinder Bank (Belgium). (**b**) Lateral view of autozooid, Kwinte Bank (Belgium)

"Puellina" monostachys nov. sp.

Type Material

I.G. 34,864/BRYT.229 holotype on shell fragment, gold coated for SEM, sampled by J.S Houziaux (JSH 52) on the Westhinder Bank with Gilson dredge in 2005.

Description Colony very small, encrusting, uniserial, and branching. Autozooids roughly suboval with tapering proximal part. Zooids very small, about 0.2 mm long unless the narrow proximal part is elongated. Frontal shield comprising five costae. Distal pair of costae with a broad base and highly ridged near the center of the shield. A papilla pore and 2-minute lacunae between adjacent costae. Pelmatidia are absent. Autozooidal orifice transversely D-shaped, wider than long. Apertural bar without suboral lacuna but with a prominent, proximolateral tuberculate papilla pore that resembles a broken spine. Oral spines five, four in periancestrular zooids and reduced to two in ovicellate zooids. Avicularia not observed. Ovicell terminal, swollen, ooecium formed by the distal kenozooid not visible from above, with low median ridge, without pseudopores. Ancestrula tatiform, with ten periopesial spines, appears transformed into a kenozooid with a calcified frontal plate (Fig. 4.93).

Remarks This colony is very small (an ancestrula, ten autozooids, two ovicells). The colony is uniserial, a primitive character recalling some fossil material. It is not yet clear to what genus this species belongs but it can not be allocated to *Puellina*, *Cribrilaria*, or *Glabrilaria* (pers. comm. A. Rosso). The absence of pseudopores excludes *Puellina*.

Glabrilaria is characterized by avicularia (pedunculate or sessile) that are associated with autozooidal orifice, and especially with ovicells. These are missing in this colony. No trace of avicularia (typically occurring in *Cribrilaria*) occurs in this colony that is however very small. On the other hand, it is a typical spot colony, reproducing at a very small colony size. It will not be easy to find more specimens of this material because this was the only one found during the examination of thousands of shell valves. A detailed evaluation of the literature is necessary before going in the direction of a new genus.

Occurrence Known from only one colony on the inner side of the valve of *Striarca lactea* dredged on the Westhinder Bank (Belgium) in 2005.

"Puellina" nana Reverter-Gil and Fernandez Pulpeiro, 1996

Description Colony encrusting, unilaminar, very small. Autozooids suboval. Frontal shield comprising seven to eight costae, the distal pair forming a high suboral peristome. The upper edge of this peristome has a number of triangular points. The next pair of costae also has an upright pointed edge, but much less high. Three lacunae and a larger papilla pore between adjacent costae. Autozooidal orifice transversely D-shaped, wider than long, proximal margin straight. Six long oral spines, jointed some distance above the base, in non-ovicellate zooids, reduced to four in

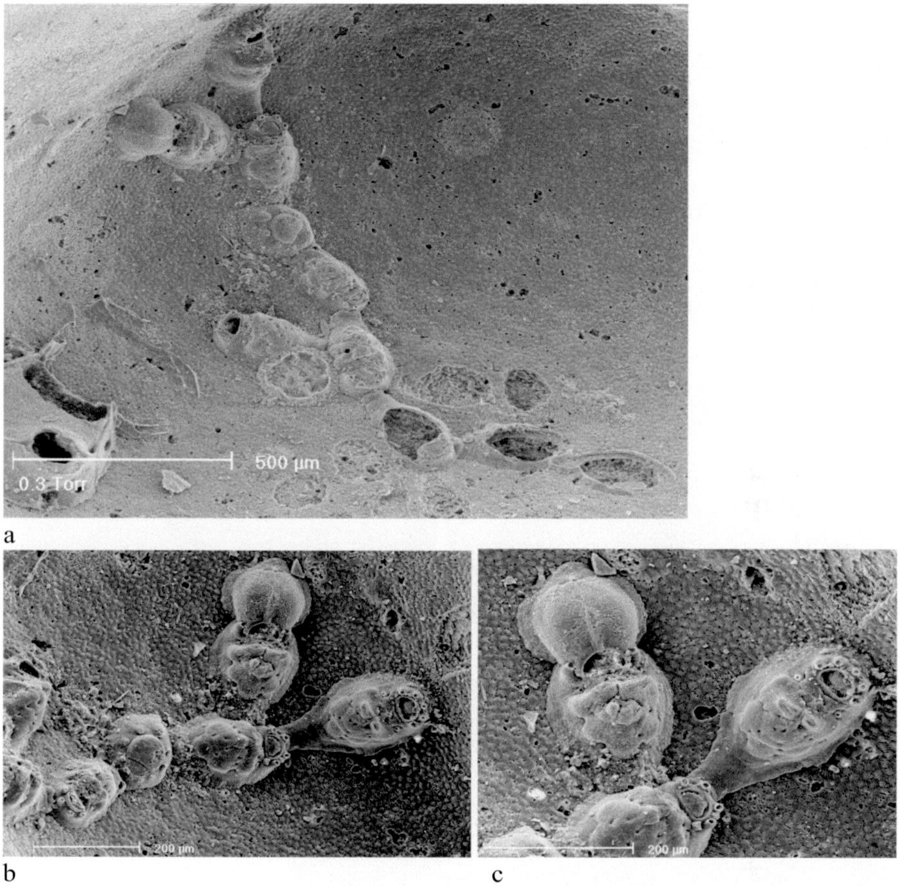

Fig. 4.93 *"Puellina" monostachys* nov. sp. (**a**, **b**) Small colony on a dead shell. (**c**) Distal part of the branch showing ovicellate and two non-ovicellate zooids, Westhinderbank (Belgium).

ovicellate zooids. Spines with inwardly directed half-circle of denticles or nodules (apophyses) above the joint. Two large papilla pores distolateral in the suboral area enclosed by the peristome; papilla setiform, seen in unbleached material. Medio-proximal in the same area lies another large lacuna that can be divided into three smaller lacunae. Ovicell terminal, globular, ooecium formed by the distal kenozooid not visible from above, smooth with median ridge. Avicularia and kenozooids not observed. Ancestrula tatiform, with 11 periopesial spines, that have apophyses near their base (Fig. 4.94).

Occurrence Described in 1996 from a damaged colony collected in Ria de Ferrol (Northern Spain). In 2005, four colonies were collected on the Westhinder Bank, of which at least one was alive.

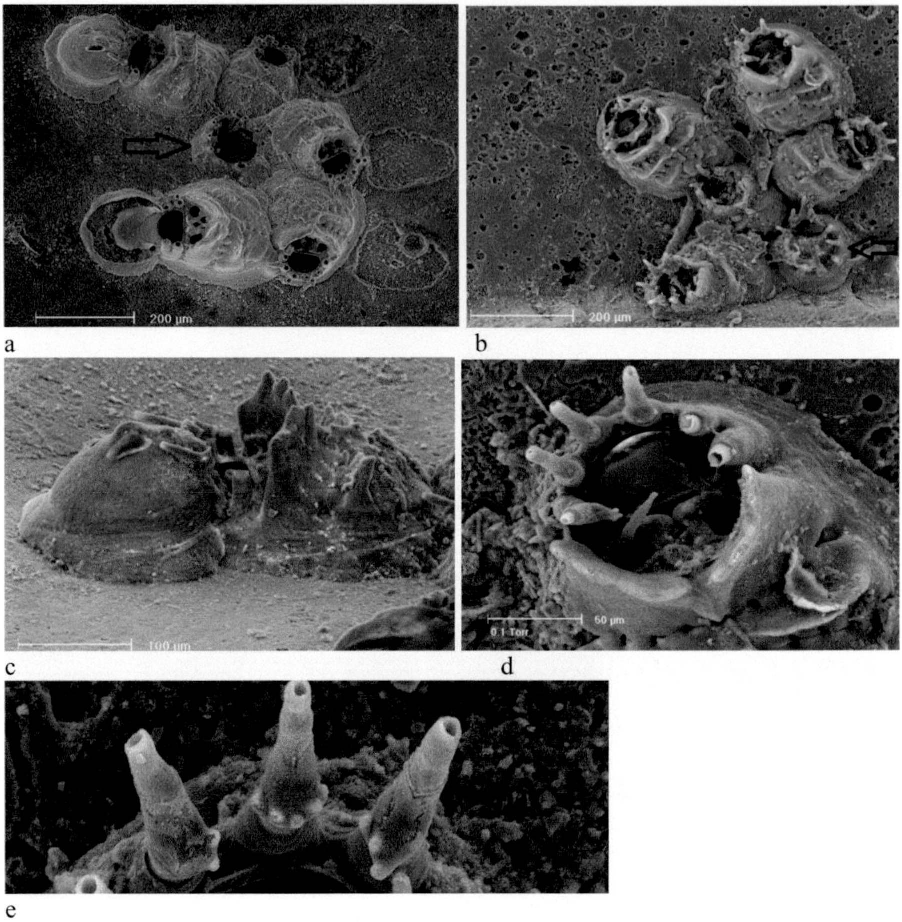

Fig. 4.94 *"Puellina" nana*. (**a**) General view of small colony with two ovicellate zooids and ancestrula (*arrow*). (**b**) Colony of non-ovicellate zooids and ancestrula (*arrow*). (**c**) Lateral view of ovicellate zooid; ooecium formed by the distal kenozooid. (**d**) Zooidal orifice showing oral spines and two setiform papillae (*arrows*) protruding from lacuna pores. (**e**) Oral spines with inwardly directed half-circle of apophyses above the joint, Westhinderbank (Belgium)

Remarks

Because the material from Spain is damaged, the original description is not optimal (pers. comm. Reverter). The description given above has been adapted on the basis of the material collected on the Westhinder Bank.

The original name (*Puellina parva*) was already in use in the subgenus *Cribrilaria*, namely *Cribrilaria parva* Winston and Hakansson 1986 of Florida. Therefore, the name of *Puellina parva* Reverter-Gil and Fernandez Pulpeiro 1996 has been changed to *Puellina nana* (Reverter and Fernandez 2007).

It is not yet clear to what genus this species belongs: ooecium formed by the distal kenozooid not visible from above suggests *Puellina* or *Glabrilaria* but the absence of pseudopores excludes *Puellina*. On the other hand, the presence of setiform papillae is a character of *Puellina* and excludes *Glabrilaria*. If *Cribrilaria* have ovicells with ooecia formed by the distal kenozooid, not visible from the above, then the absence of pseudopores and the spines with apophyses are in favor of this genus, although the papilla pores are not.

Puellina praecox Bishop and Househam, 1987

Description Colony encrusting, unilaminar, small, up to 20 zooids. Autozooids roughly suboval. Frontal shield comprising eight to ten costae, without umbo or carina. Costae ridged, decreasing in height from periphery to center of shield. Costa with a tubercle near its base, occasionally with a pelmatidium. Three to four lacunae and a papilla pore between adjacent costae. Lacunae between distal costae larger than between proximal costae.

Autozooidal orifice transversely D-shaped, slightly wider than long, proximal margin straight. Apertural bar with a pelmatidium on a tubercle on both sides of the median suture. Median suboral lacuna larger than the intercostal lacunae.

Suboral lacuna larger than intercostal lacunae. Oral spines five in non-ovicellate zooids, jointed just above their base, reduced to four in ovicellate zooids. Spines without apophyses. Ovicell terminal, globular, cleithral; ooecium formed by the distal kenozooid not visible from above, with two to five tubercles, sometimes with a pseudopore on their top. Avicularia and kenozooids not observed. Ancestrula tatiform with 11 spines, can regenerate as a kenozooid.

Very similar in frontal shield morphology to *Cribrilaria bifida*, but immediately distinguishable if ovicells are present (in *C. bifida*, the ovicell is formed by the distal autozooid) (Fig. 4.95).

Occurrence

Belgium In 1906, a colony was collected at the Flemish Banks (RBINS collection). Nowadays, quite common on shell valves on the Kwinte Bank and the Hinder Banken.

The Netherlands Not reported.

Genus *Distansescharella*

Extant species are distributed in the NE Atlantic and the Mediterranean on bathyal hard substrates (Harmelin 2024), but curiously enough, also on empty shells collected at low tide on shores of the English Channel (*D. seguenzai* Cipolla, 1921: De Blauwe 2019).

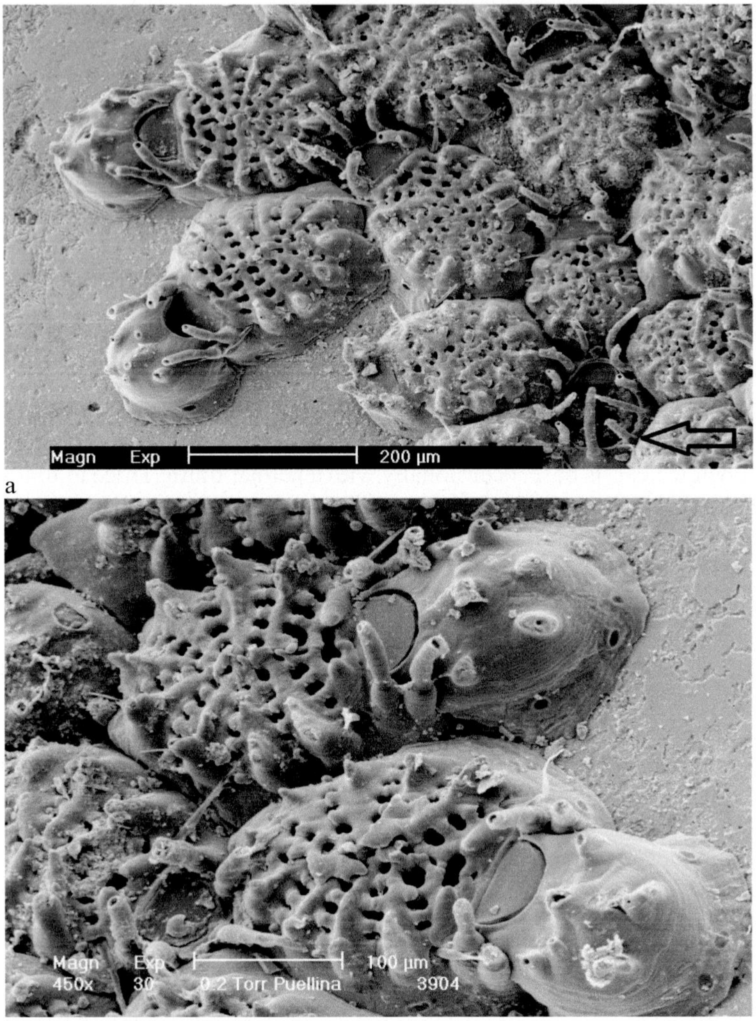

Fig. 4.95 *Puellina praecox*. (**a**) Part of the colony showing two ovicellate zooids and ancestrula (*arrow*). (**b**) Two ovicellate zooids showing terminal cleithral ovicells, Kwinte Bank (Belgium)

1 (a) Autozooids disjunct, with tubular connections (Fig. 4.16e), interzooidal kenozooids and avicularia present, ovicellate zooids have four spines. ………..…………………………….. *Distanseschareila seguenzai* (p. 269)

(b) Autozooids uniserial, not connected by tubes, no kenozooids or avicularia, ovicellate zooids have two spines. ………………………………...…….. ………..…....................… *Distanseschareila spirorbicola* nov. sp. (p. 270)

Distanseescharella seguenzai Cipolla, 1921

Description Colony encrusting, unilaminar, consisting of zooids linked by tubular connexions. Autozooids oval. Frontal shield comprising 12 to 15 costae (usually 13), and a poorly developed carina. Four lacunae between adjacent costae. The first pair of costae are slightly directed proximally, making the proximal edge of the orifice slightly V-shaped. Four distal spines, also present in ovicellate zooids. Ovicells terminal, spherical, ooecium is formed by the distal kenozooid not visible from above, with smooth ectooecium without pseudopores and with small distal pore chamber. Avicularium oval, interzooidal, with a rounded rostrum. Kenozooids elliptical, with a central circular area and a central pore, surrounded by three to five conical spikes. Ancestrula tatiform with periopesial spines (Fig. 4.96).

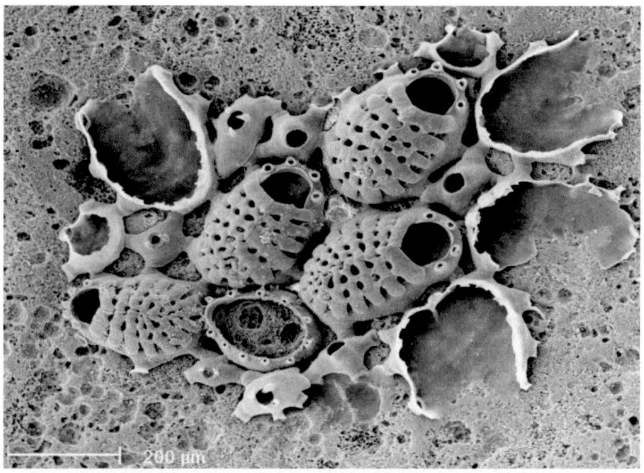

a

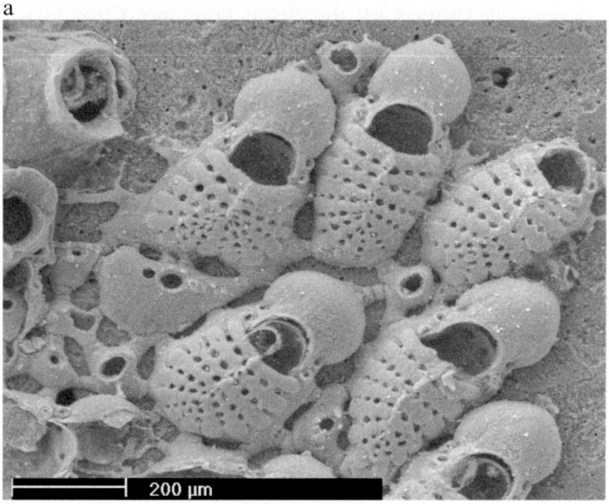

b

Fig. 4.96 *Distanseescharella seguenzai*. (**a**) Young colony showing central ancestrula, periancestrular zooids, and forming zooidal buds. (**b**) Colony area showing ovicellate autozooids and kenozooids, Westhinderbank (Belgium)

Occurrence Known from fossil (Pliocene) and recent material (Harmelin et al. 1989) in the Mediterranean area, also from Ria de Ferrol in Spain (Reverter and Fernandez 1996).

Belgium In 2005 dredged on the Westhinder Bank (De Blauwe et al. 2006). Due to the known distribution area, a recent introduction was considered, but during a review of shell material collected in 1905 at 51°29′15″N, 2°32′45″E, four colonies were found on a shell valve, proving the usefulness of historical collections.

The Netherlands de Ruijter (2014b) reports the washing-up in 2013 of a colony on plastic near Egmond aan Zee.

France Pleneuf-Val-André on the lower beach, in total 25 colonies on 5 shells (De Blauwe 2019).

United Kingdom Eastern Channel and Thames Estuary (De Blauwe 2019).

Distanseschareila spirorbicola nov. sp.

Type Material

I.G. 34,864 / BRYT.228 holotype on shell fragment, gold coated for SEM, sampled by J.S Houziaux (JSH 52) on the Flemish Banks with Gilson dredge in 2005.

Description The colony forms a small encrustation. Zooids irregularly oval, convex. Frontal shield composed of eight to ten costae with three or four irregularly shaped lacunae between them. The orifice is proximally bordered by the first pair of costae, forming an angle of about 150°. The base of some costae shows a knob-shaped protrusion. Four oral spines, two in the presence of an ovicell. Avicularia not observed. The colony soon produces ovicells. Ovicells terminal, somewhat elongated, ooecium is formed by the distal kenozooid not visible from above, with smooth ectooecium lacking pseudopores. Ancestrula tatiform with nine periopesial spines (Fig. 4.97).

Remarks Its generic affiliation might be not convincing, especially because of the lack of kenozooids, avicularia, and tubular connections between colony members. But this species is very similar to *Distanseschareila seguenzai* in the shape of the costal shield, costae smooth, without pelmatida, although some autozooids show at the base of some costae a knob-shaped protrusion. The proximal edge of the orifices and the ovicells in both species are identical. They differ in the number of spines in ovicellate zooids: constantly four in *D. seguenzai* and constantly two in the new species. The species was first mentioned in De Blauwe (2009a) as *Cribrilina* sp. 2.

Etymology Seems to like to settle next to *Spirorbis*. It can be a coincidence, but all three colonies collected were located against a *Spirorbis* tube.

Occurrence Only known from Belgium and uncommon on the Flemish Banks.

Found for the first time in 2004 on two shell valves on the Kwinte Bank. Collected again in 2005 on shell valves at the Westhinder Bank.

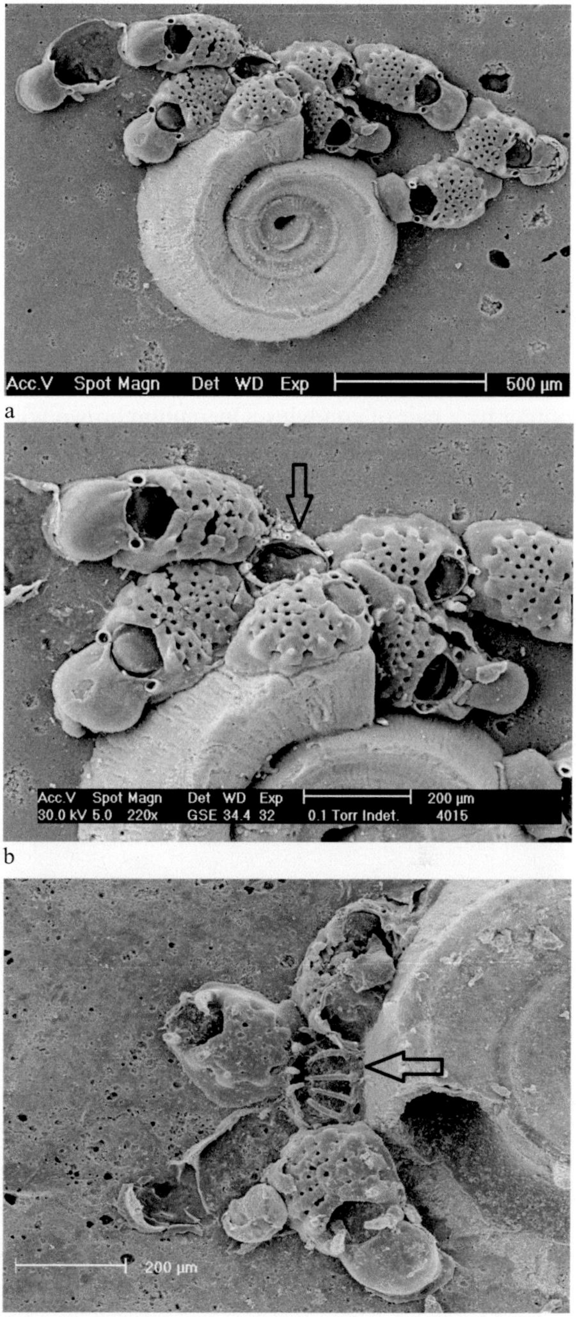

Fig. 4.97 *Distanseescharella spirorbicola* nov. sp. (**a**, **b**) Colony showing ancestrula (*arrow*), and ovicellate and non-ovicellate zooids, next to *Spirorbis* tube, on a dead shell, Kwinte Bank (Belgium). (**c**) Early stage of a second colony formation with ancestrula (*arrow*) in the center, again next to *Spirorbis* tube, Westhinder Bank (Belgium)

Genus *Figularia*

Figularia figularis (Johnston, 1847)

Description Colony encrusting, unilaminar, dull white to light brown. Autozooids large, up to 1 mm long, oval to rectangular. Gymnocyst present as broadband and proximally comprising up to one-fourth of the zooid length. Frontal shield comprising 10 to 14 tapered costae. Apertural bar formed by the distal pair of costae, broad and thick. Each costa with a prominent tuberculate pseudopore toward the periphery of the frontal shield. Three to four lacunae between adjacent costae, those at the periphery of the shield large, slit-shaped. Orifice large, distal portion rounded quadrangular, separated from a proximal portion with a curved and arched proximal border by a pair of condyles. Operculum brown with thick marginal sclerite. Vicarious avicularia rare (not seen in material from the North Sea), rostrum spoon-shaped distal to the thick pivot bar. Ovicell prominent, cleithral, formed by the distal autozooid, ectooecium with a longitudinal suture and a large drop-shaped, uncalcified fenestra on each side. Ancestrula tatiform (Fig. 4.98).

Occurrence Southern species occurring as far as the western half of the English Channel and absent from the North Sea.

Belgium On a beached fishing net at Oostende on November 4, 2000.

The Netherlands On beached plastic (de Ruijter 2013a, 2014b, 2015a, 2017c, 2018a, 2019a).

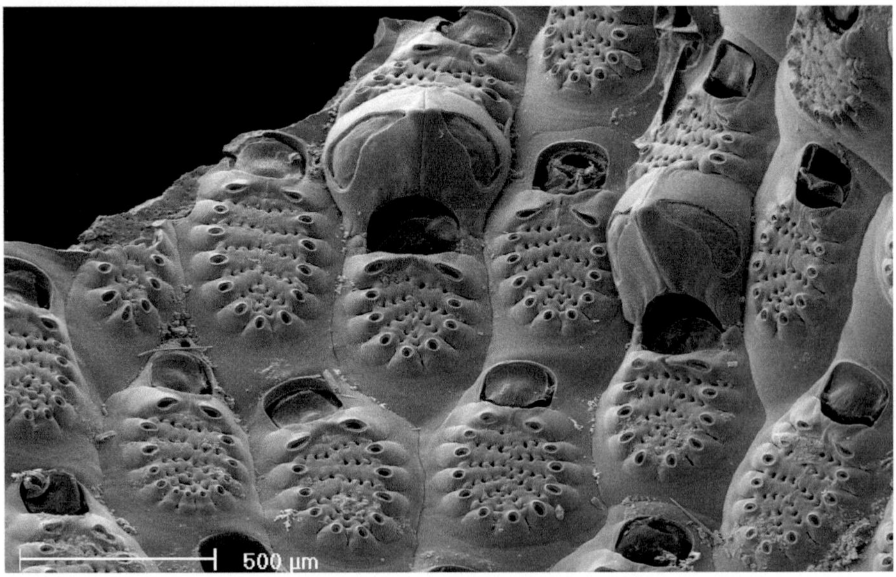

Fig. 4.98 *Figularia figularis,* colony area with ovicellate and non-ovicellate zooids, on a shell fragment from Erqui (France)

Genus *Membraniporella*

Membraniporella nitida (Johnston, 1838)

Description Colony encrusting, unilaminar, multiserial. Autozooids elongate oval. Frontal membrane obscured by a rib-cage-like shield comprising eight to ten pairs of flattened, rib-like spines. Spines fused in the central part of the shield, one slit-like lacuna between neighboring spines. Four slender, erect oral spines, the distal pair obscured by the proximal part of an ovicell in ovicellate zooids. Avicularia small, with a pointed triangular rostrum, acute to the frontal plane; one or two in the proximal corners of the autozooid, sometimes one more proximolateral. Ovicell hyperstomial, cleithral; the ooecium is formed by the distal autozooid, with an ectooecium membranous and an entooecium granularly calcified. Embryos orange. Ancestrula tatiform, with four erect oral spines and ten acute marginal spines.

This species is most likely to be confused with *Callopora rylandi*, of which the frontal shield consists of fewer spines and its ovicell is not closed by the autozooidal operculum (Fig. 4.99).

Occurrence

Belgium Loppens (1906) reported that this species is rare, occurring on stones, shells, and algae. Occurs on washed-up plastic (De Blauwe 2009a).

The Netherlands Maitland (1897) reported this species. Lacourt (1949) listed finds in Schouwen, Scheveningen, and Den Helder. Lacourt (1978) considered this species to be very numerous, especially on seaweed, along the entire Dutch coast. Presumably most (or all) of that material is *Callopora rylandi*, a similar species not mentioned by Lacourt. A find in a valve of *Mya arenaria* on the beach of Schiermonnikoog in 1998 was reported in Holsteijn (1998). Reported on washed-up plastic (de Ruijter 2011c, 2012b, 2013a, 2014a, b, 2015d, 2018a, 2021c).

Germany Collected at Helgoland and in the Tiefe Rinne (Kuhlenkamp and Kind 2012).

Superfamily Hippothooidea

Family Hippothoidae

Genus *Hippothoa*

1 (a) Ovicellate zooid (zooid with distal ooecium) as long as an autozooid. Zooids usually with longitudinal keel.. *Hippothoa divaricata* (p. 274)
(b) Ovicellate zooid much shorter than an autozooid. Zooids without longitudinal keel. .... *Hippothoa flagellum* (p. 276)

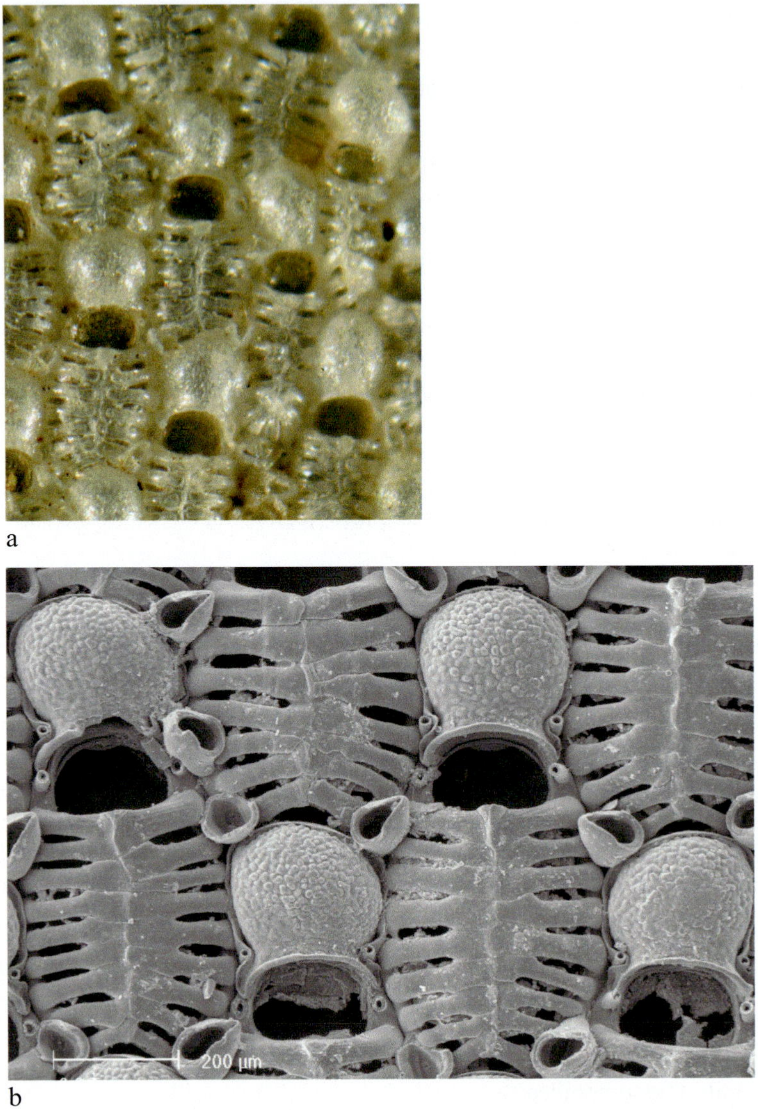

Fig. 4.99 *Membraniporella nitida.* (**a**) Colony area showing ovicellate zooids, Northumberland (United Kingdom). (**b**) Close-up of ovicellate zooids, Normandy (France)

Hippothoa divaricata Lamouroux, 1821

Description Colony adnate, uniserial, zooid rows branching. Zooids are of three types: autozooids, ovicellate female zooids, and zooeciules. Zooids with a distal dilatation, usually with a longitudinal central keel, tapering to a narrow tube-like proximal portion. At first hyaline, later white. Orifice longer than wide, sinus

V-shaped. Ovicellate female autozooids (including ovicell) are as large as an autozooid, with usually a shorter proximal portion. Orifice longer than wide, rounded distally, tapering proximally, with rounded V-shaped sinus between condyles. Ovicells terminal, globular, ooecium smooth, umbonate or ridged, without pseudopores, sometimes with distal budding. Opening of the ovicell combined with the zooidal orifice and closed by the zooidal operculum. Very small zooeciules with minute orifices may arise from a normal budding point (Fig. 4.100).

Occurrence

Belgium Found between 1903 and 1908 (RBINS collection) and in 2004–2005 on boulders and shell valves on the Hinder Banks. Small stones were collected by F. Kerckhof (RBINS) in the extreme NW part of the Belgian continental shelf (BCS), on the border with the United Kingdom on July 12, 2018, during a haul with a Gilson dredge with starting coordinates 51°35′26″ N, 2°15′14″ E and ending point 51°35′36″ N, 2°15′28″ E, 100 stones were examined on epibionts and 3 of them contained living colonies of *H. divaricata*. Sometimes on washed-up plastic, usually inconspicuous. In 2023, one small colony was collected on a shell by beam trawl between 51°30.511′ N, 2°35.273′ E and 51°29.559′ N, 2°36.378′ E.

The Netherlands De Ruijter (2013c, 2014b, 2015d) reports a colony on washed-up plastic in the Netherlands.

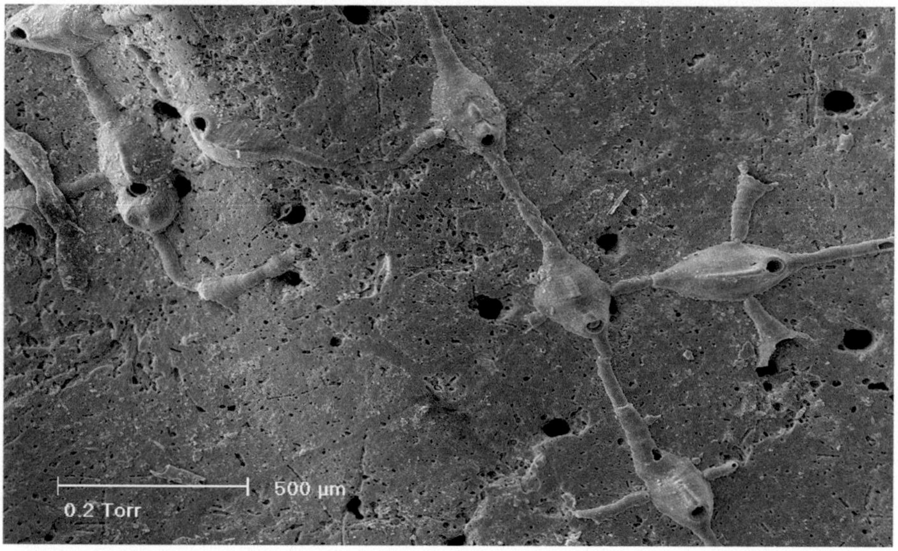

Fig. 4.100 *Hippothoa divaricata*. Autozooids, zooeciules, and ovicellate zooids with tube-like parts of the distally budding zooid, on a dead shell from the Westhinder Bank

Hippothoa flagellum Manzoni, 1870

Description Colony adnate, uniserial, zooid rows branching. Zooids are of three types: autozooids, ovicellate female zooids, and zooeciules. Zooids, with an unkeeled distal dilatation, tapering to a very long thread-like proximal portion. At first hyaline, later white. Orifice longer than wide, U-shaped sinus between small condyles. Ovicellate female autozooids wedge-shaped with a short proximal portion. Ovicells terminal, globular, with smooth ooecium without pseudopores. Opening of the ovicell combined with the zooidal orifice, wider than long and closed by the zooidal operculum. Very small zooeciules with minute orifices may arise from a normal budding point (Fig. 4.101).

Occurrence

Belgium Dredged in 1905 on shell valves on the Hinder Banks (RBINS collection). Dredged again in 2004 on the Kwinte Bank and in 2005 on the Westhinder Bank on shell valves. In 2023, one small colony was collected on a shell by beam trawl between 51°30.511′ N, 2°35.273′ E ° and 51°29.559′ N, 2°36.378′ E. Much more common than *H. divaricata*. Sometimes on washed-up plastic, usually inconspicuous.

The Netherlands No records.

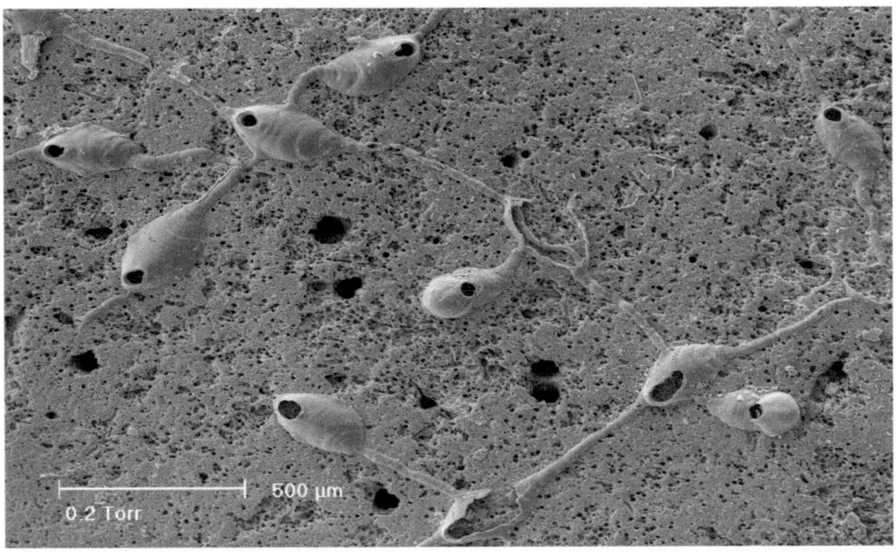

Fig. 4.101 *Hippothoa flagellum*. Colony with two ovicellate zooids on dead shell from the Westhinder Bank (Belgium)

Genus *Celleporella*

Celleporella hyalina (Linnaeus, 1767)

Description Colony encrusting, often on algae. Zooids 0.3–0.4 mm long, sausage-shaped, transparent, later opaque. Zooids contiguous with narrow interstices, which are later hidden by frontally budded dwarf zooids—male and female, the latter with ovicells. Pore chambers tubular. Autozooids and males elongate, the wholly gymnocystal frontal wall strongly convex and striated, often with an umbo proximal to the orifice. Orifice round with wide shallow sinus, without condyles. Polypide with 12 tentacles. Female zooids are short and triangular and associated with a globular terminal ovicell. Ooecium with frontal pseudopores. Embryos pale yellow. The common orifice of the female zooid and ovicell is much wider than long. Male zooids like autozooids but with a smaller orifice. No avicularia. Ancestrula oval with an orifice like the autozooids (Fig. 4.102).

Occurrence

Belgium Loppens (1906) found this species on washed-up *Flustra foliacea*. In more recent years often on washed-up plastic, *Himanthalia* buttons and also on *Abietinaria abietina*. It is the dominant epiphyte on washed-up *Himanthalia* buttons (De Blauwe 2000c, 2005a). *Himanthalia* probably broke loose on the coasts of the English Channel and drifted northeastward. Subtidally on stones of the erosion protection in Belwind wind farm in August 2018 and on mussels at Belwind wind farm, at a depth of 15 m (pers. comm. Francis Kerckhof, RBINS).

The Netherlands Washed up along the entire coast in the Netherlands on *Flustra foliacea*, *Patella*, algae, and cork (Lacourt 1978) and more recently on plastic.

Celleporella hyalina has been found on hard substrates of the offshore Prinses Amalia windpark near IJmuiden (Vanagt et al. 2013). M. Faasse found it in the Eastern Scheldt and on Ensis (sample 15,004–26.2). On fishing nets on the wreck, Niponia nr 2305 was collected by DDNZS in 2016 (52°44.520′ N, 3°58.702′ E) (van Leeuwen et al. 2017).

Germany Collected at Heligoland and in the Tiefe Rinne (Kuhlenkamp and Kind 2012) and also on mussels on pontoons in the south harbor of Heligoland in 2023.

Two main parapatric lineages for NE Atlantic *Celleporella hyalina*. Range expansions have resulted in two contact zones in Spain and W. Ireland. Lineage 1 is found from Ireland to Spain, with recent population expansion into the Irish Sea, S. Ireland, S. England, and Spain. Lineage 2 is found from Iceland to Spain (Gomez et al. 2007).

a

b

Fig. 4.102 *Celleporella hyalina.* (**a**) Colony showing a basal layer of autozooids and frontally budded male (with small orifice) and female dwarfs with ovicells, Prinses Amalia windpark (the Netherlands, photo by M. Faasse). (**b**) Colony on beached *Himanthalia*, Oostduinkerke (Belgium)

Genus *Plesiothoa*

Plesiothoa gigerium (Ryland and Gordon, 1977)

Description Colony adnate, consisting of partially aggregated chains of zooids, branching on either side on the distal or distolateral part of zooids. Zooids of three types: autozooids, ovicellate female zooids, and zooeciules. Autozooids about 0.4 mm long, teardrop-shaped or club-shaped, narrowest proximally, but wider than in *Hippothoa*. Frontal surface at first translucent, becoming opaque, without longitudinal keel but variably striated or hump-ribbed, with an umbo proximal to the orifice. Orifice longer than wide with a U-shaped sinus. Female zooids are shorter than autozooids, with a wider, but similar orifice with a sinus. Zooeciules variable, usually smaller than zooids, variable in shape and size. Ovicells terminal, globular, striated in variable directions and sometimes with a few frontal pseudopores. Ancestrula oval with serrated edge and a suboval opening (Fig. 4.103).

Occurrence

English Channel In south-west England and the Channel Islands on red and brown algae (Hayward and Ryland 1999). Davoult et al. (1999) report three colonies on hydroids off Cap Gris-Nez in June 1992.

Belgium A colony on beached plastic washed ashore at Zeebrugge in December 1999.

The Netherlands On beached plastic in 2011 (de Ruijter 2012b).

Germany Collected in the Tiefe Rinne (Kuhlenkamp and Kind 2012).

Fig. 4.103 *Plesiothoa gigerium*, zooids of the colony growing on plastic beached, Zeebrugge (Belgium)

Family Trypostegidae

Genus *Trypostega*

Trypostega venusta (Norman, 1864)

Description Colony encrusting, unilaminar. Zooids 0.4–0.6 mm long, irregular in outline. Frontal wall smooth, and evenly pseudoporous, with a small umbo proximal to the orifice. Orifice keyhole-shaped, distally rounded, separated from a wide triangular proximal part by short, pointed condyles. Operculum brown. Kenozooids of variable shape and size occur distal to almost every zooid and ooecia. Their orifice is small and also has an operculum. Ovicells subimmersed, ooecium oval, longer than wide, evenly pseudoporous, often with a frontal umbo and small distal kenozooid, closed by a zooidal operculum (Fig. 4.104).

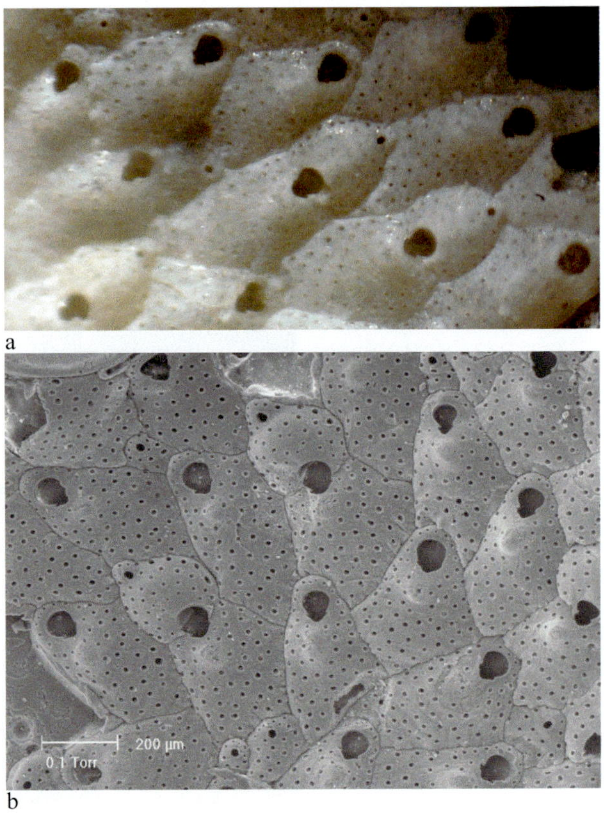

Fig. 4.104 *Trypostega venusta*. (**a**) Part of a colony showing ovicellate and non-ovicellate zooids, with small kenozooids in between, Westhinder Bank (Belgium). (**b**) SEM image of the same colony

Occurrence

Belgium Offshore encrusting shell (Hayward and Ryland 1999). In 1906 a colony was collected at the Flemish Banks (RBINS collection). In 2005, several colonies were found on shell valves on the Westhinder Bank.

Family Chorizoporidae

Genus *Chorizopora*

Chorizopora brongniartii s.l.

Description Colony encrusting. Zooids 0.4–0.6 mm long, oval to pear-shaped. Zooids are connected to each other by short tubes and separated by the narrow slits. The substrate is often visible between the tubes and the zooids. Frontal surface glassy and translucent, later white. Transverse frontal ridges may appear, as well as a conical umbo. Orifice D-shaped. Basal walls are not calcified. Avicularia adventitious, small, elliptic, mandible triangular, and directed distally. Almost always in line with an autozooid or distal to ooecia. Ovicell prominent, terminal, elongate globular and without pseudopores, closed by the zooidal operculum. Embryos red. Small kenozooids with a membranous window between autozooids. Ancestrula oval with a D-shaped orifice, with four frontal and four oral spines. Small kenozooids (Fig. 4.105).

Remark It is apparent that *Chorizopora brongniartii* s.s. is not cosmopolitan—there is no type material, it is uncertain whether the species was described from the Red Sea or the Mediterranean, the original drawing differs from Mediterranean congeners in its large suboral umbo and populations can appear " different everywhere!" (pers. comm. Bjorn Berning).

Occurrence *Chorizopora brongniartii* s.l. encrusts many types of organic and inorganic substrates.

Belgium Lives on pebbles on the Hinder Banks, dead colonies were found there on shell valves. Collected in 2016 on the Dogger Bank. Small stones were collected by Francis Kerckhof (RBINS) in the extreme north-western part of the BCS, on the border with the United Kingdom on July 12, 2018, during a haul with a Gilson dredge with starting coordinates 51°35′26″ N, 2°15′14″ E and ending point 51°35′36″ N, 2°15′28″ E, 100 stones were examined on epibionts and 8 of them contained living colonies of *C. brongniartii*. In 2023, one small colony was collected on a shell by beam trawl between 51°30.511′ N, 2°35.273′ E and 51°29.559′ N, 2°36.378′ E. Common on washed-up plastic and on *Himanthalia* buttons.

Germany Collected at Heligoland and in the Tiefe Rinne (Kuhlenkamp and Kind 2012).

Fig. 4.105 *Chorizopora brongniartii* s.l. (**a**) Colony on beached plastic, Wales. (**b**) Ovicellate and non-ovicellate zooids, avicularia, and kenozooids

Family Haplopomidae Gordon in De Blauwe (2009a)

Zooids convex, translucent, pseudoporous by scattered or areolae. An ascopore is present and is sometimes connected to the orifice by means of a suture. Orifice D-shaped, spineless. No avicularia. Ovicells prominent, ooecium pseudoporous, formed by the distal kenozooid; its opening closed by the operculum of the maternal zooid. Zooids and ooecia surrounded by a string of pore chambers.

Genus *Haplopoma*

Haplopoma bimucronatum (Moll) is not in the key below but can be expected (not yet found) on washed-up red seaweeds from the English Channel. The zooids are not longer than 0.5 mm, have prominent distally directed projections and the basal surface has an uncalcified window.

1 (a) Zoooid twice as long as wide, more or less oval with many frontal pores. .... *Haplopoma graniferum* (p. 283)
(b) Zooid one and a half times as long as wide, rather hexagonal with fewer frontal pores. .... *Haplopoma impressum* (p. 283)

Haplopoma graniferum (Johnston, 1847)

Description Colony encrusting, vitreous or white. Zooids 0.6–0.9 mm long, elongate oval to hexagonal, convex, and translucent. Many scattered frontal pores. Orifice D-shaped. An ascopore proximal to the orifice, joined to it by a suture. Sometimes a suboral umbo is present. The orifice of ovicellate zooids is not larger than the orifice of non-ovicellate zooids. Ovicells prominent, with pseudoporous ooecium formed by the distal kenozooid non-visible from above; its opening closed by the operculum of the maternal zooid. Embryos pink. Ancestrula oval with a D-shaped orifice with a very small sinus or an ascopore (Fig. 4.106).

Belgium Occasionally found on washed-up plastic or *Himanthalia* buttons originating from the English Channel (De Blauwe 2000c, 2005a; Vanhaelen et al. 2006).

The Netherlands First report on a beached *Himanthalia*, November 2009 near Kwade Hoek (de Ruijter 2010). Subsequently regularly reported on washed-up *Himanthalia* buttons and plastic (de Ruijter 2011c, 2012a, 2013c, 2017c, 2018d, 2020a).

Germany Collected in the Tiefe Rinne (Kuhlenkamp and Kind 2012) and on beached *Himanthalia* buttons on Heligoland (pers. obs. October 2023).

Haplopoma impressum (Audouin, 1826)

Description Colony encrusting, vitreous. Zooids 0.5–0.6 mm long, hexagonal–oval. The thin frontal calcification is translucent (not when dried), with some scattered frontal pores. Proximal to the D-shaped opening is an ascopore, often with a small proximal umbo. Ovicellate zooids have a larger orifice. Ovicells prominent, globular, ooecium with a few scattered pseudopores; its opening closable by the operculum of the zooid. Embryos pink. Ancestrula oval, without pores; its orifice keyhole-shaped. Ancestrula budding three periancestrular zooids (Fig. 4.107).

a

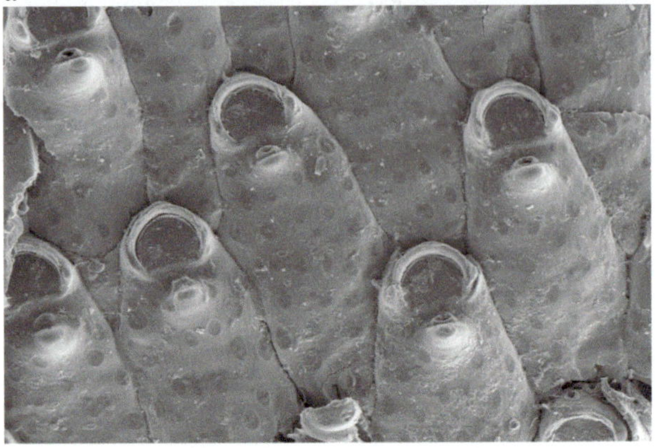

b

Fig. 4.106 *Haplopoma graniferum*. (**a**) Part of colony with two ovicellate zooids on blue plastic beached at Newgale (UK). (**b**) Ascopores between orifice and proximal umbo. On plastic at St-Malo (France)

Occurrence

Not native in the North Sea. Rarely found in Belgium and the Netherlands on washed-up plastic or *Himanthalia* buttons (Bloklander and Leenhouts 1948; De Blauwe 2003a; Vanhaelen et al. 2006; de Ruijter 2020a). *Himanthalia elongata* as substrate suggests an origin from the English Channel or further south, the loose seaweed has been observed floating as far as the German island of Heligoland.

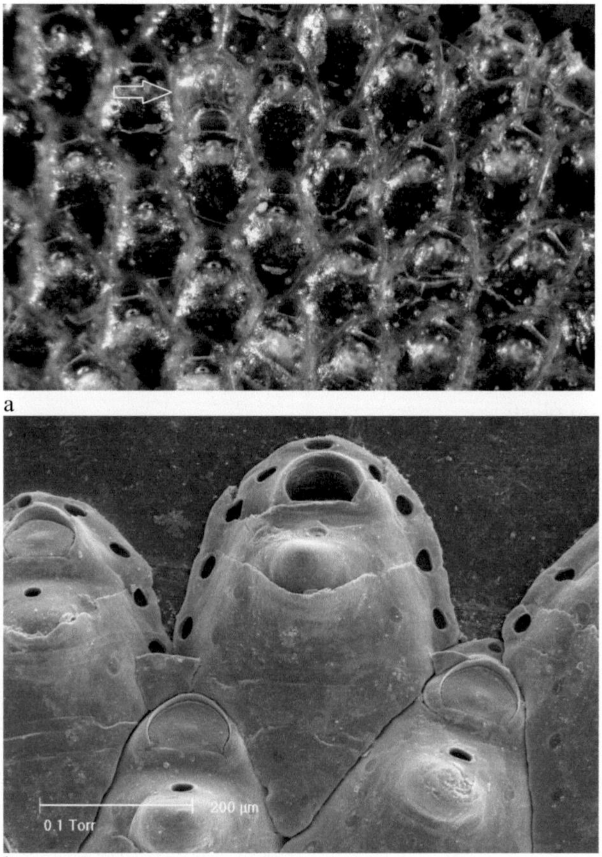

a

b

Fig. 4.107 *Haplopoma impressum.* (**a**) Part of colony on beached *Himanthalia*, Oostduinkerke (ooecium shown by *arrow*). (**b**) Zooids with ascopores at the colony periphery

Superfamily Adeonoidea

Family Adeonidae

Genus *Reptadeonella*

1 (a) Frontal shield with areolae and a single central perforation. ....
.... *Reptadeonella violacea* (p. 287)
(b) Frontal shield with areolae and a double central perforation. ……… ....
.... *Reptadeonella insidiosa* (p. 285)

Reptadeonella insidiosa (Jullien, 1903)

Description Colony encrusting, white or yellowish-white. Frontal membrane glistening in dried colonies. Zooids 0.5–0.6 mm long. Frontal surface of autozooids flat with areolae (areolae) and with a central depression. Two pores (spiramen) in the central depression. Orifice semicircular with a short peristome. Avicularia small and triangular, directed distally, between the orifice and the spiramen. A tubular kenozooid at the proximal extremity of each autozooid, additional kenozooids may be present on the proximolateral walls. Ovicells are absent; embryos are incubated in the internal brood sac in enlarged modified autozooids (gonozooids). Gonozooids are twice as broad as autozooids, their orifice is wider and their avicularium is shorter. The settled larva immediately develops into six ancestrular autozooids arranged in a star shape (Fig. 4.108).

Occurrence

Belgium No observations.

The Netherlands De Ruijter (2015c, 2015d, 2017c) reports a colony washing-up on plastic.

a

b

Fig. 4.108 *Reptadeonella insidiosa.* (**a**) Colony growing on plastic, Bergen aan Zee (the Netherlands). (**b**) Colony area showing zooids with frontal avicularium and paired spiramen

Reptadeonella violacea (Johnston, 1847)

Description Colony encrusting, often extensive, purple when living. Zooids 0.5–0.7 mm long. Frontal surface of autozooids slightly concave with areolae and with a central depression. A single pore (spiramen) in the central depression. Orifice semicircular with a short peristome. Avicularia small and triangular, usually directed distally, between the orifice and the spiramen. Ovicells are absent; brown embryos brooded in enlarged gonozooids. Gonozooids are up to twice as broad as autozooids, their orifice ~~is~~ wider and somewhat more frontal. Their spiramen is often transversely elongate and the avicularium may be absent. The settled larva immediately develops in to six ancestrular autozooids arranged in a star shape (Fig. 4.109).

Occurrence

Belgium Between 1902 and 1911 several colonies were collected on the Flemish Banks (RBINS collection). Also found this century in the anchorage area before Zeebrugge and at least 130 colonies at the Westhinder Bank on silex boulders and shells. Small stones were collected by Francis Kerckhof (RBINS) in the extreme north-western part of the Belgian continental shelf, on the border with the United Kingdom on July 12, 2018, during a haul with a Gilson dredge with starting coordinates 51°35′26″ N, 2°15′14″ E and ending point 51°35′36″ N, 2°15′28″ E. 100 stones were examined on epibionts and only one of them contained a colony of *R. violacea*. Found once on washed-up plastic at Ostend.

The Netherlands No native population discovered so far, four times on washed-up plastic (de Ruijter 2014b, 2015d, 2019b, 2021b).

Superfamily Umbonuloidea

Family Bryocryptellidae

Genus *Palmiskenea*

Palmiskenea skenei (Ellis and Solander, 1786)

Description Colony forms an erect lobed or branched colony up to 3 cm high, on an encrusting base. Autozooids 0.5–0.7 mm long. Frontal shield thickly calcified, convex, with widely spaced areolae. Orifice longer than wide with a weakly convex proximal margin, condyles are small and blunt. No oral spines. Peristome obscuring the orifice, developed as lateral flanges and a mid-proximal umbo. Avicularium suboral, at the base of the umbo, proximally directed. Additional small avicularia may occur on umbonate processes developed on the flanges of the peristome. Ovicell prominent, ooecium somewhat longer than wide and flattened frontally. The ectooecium with irregular pseudopores. Ancestrula tatiform, opesia surrounded by spines (Fig. 4.110).

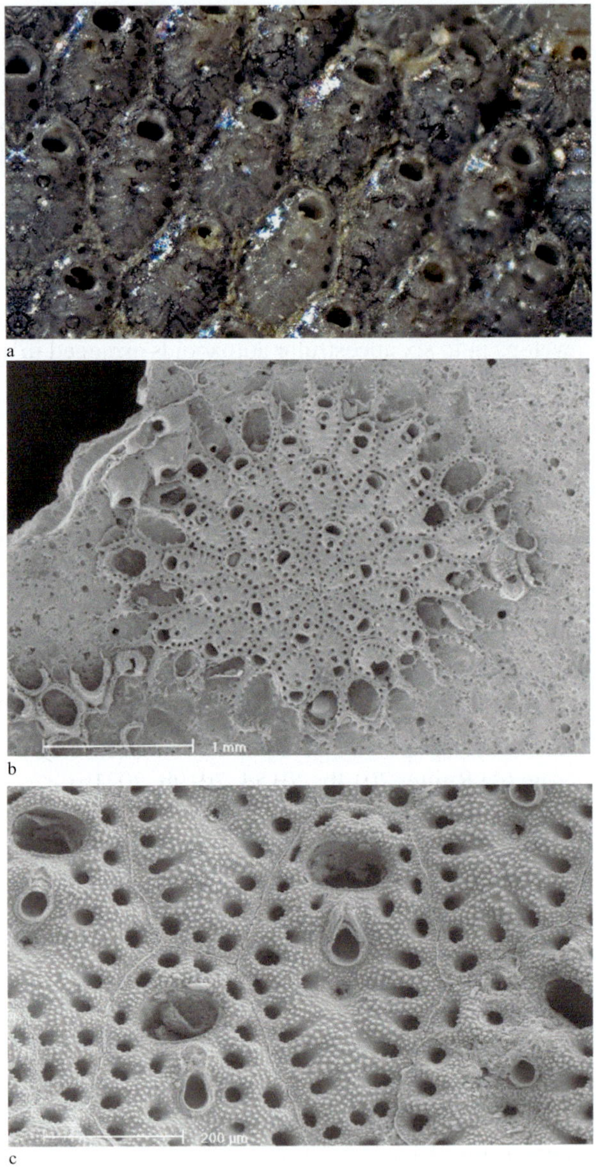

Fig. 4.109 *Reptadeonella violacea*. (**a**) Living colony with a typical purplish color, Brittany (France). (**b**) Young colony showing ancestrular complex of six autozooids arranged in a star-like pattern, Westhinder Bank (Belgium). (**c**) Enlarged autozooids with frontal avicularium and non-paired spiramen

Occurrence

Offshore, occurring off all British coasts, especially Scotland and Shetland, and ranging south to the Bay of Biscay (Hayward and Ryland 1999).

Belgium No records.

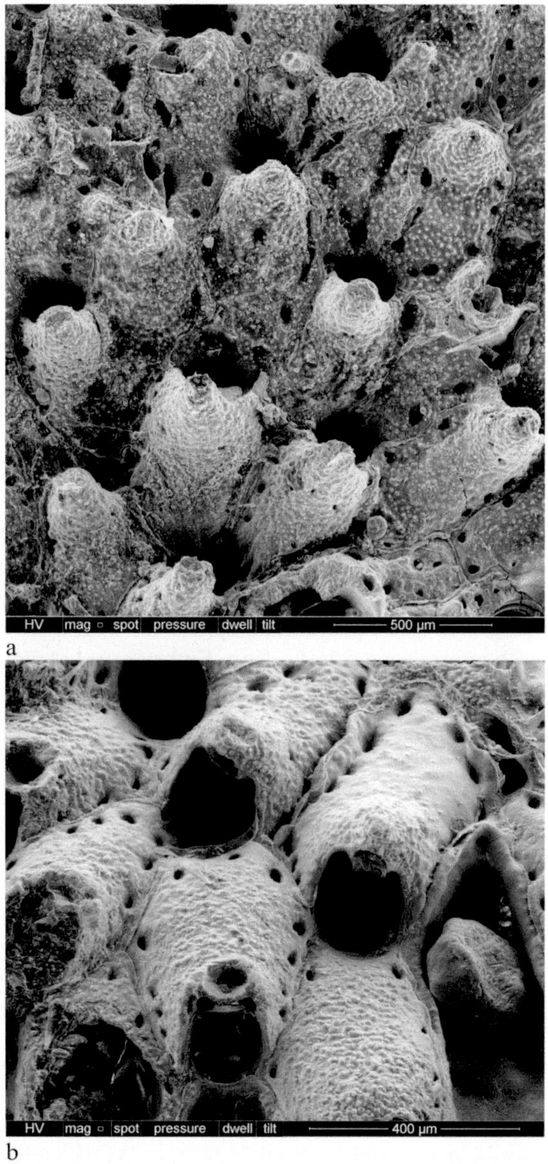

Fig. 4.110 *Palmiskenea skenei*. (**a**) Part of the colony that grew on plastic beached at Texel (The Netherlands). (**b**) Magnified zooids, view from a distal position showing the suboral avicularium

The Netherlands Not native. Known from one colony on beached plastic at Texel, reported in de Ruijter (2013c). It concerns a young encrusting colony, where the typical erect colony form has not yet developed. The peristome is not as well developed laterally as described, but the mid-proximal umbo is clearly present, although broken off in many zooids. Ovicells are missing in the studied colony.

Genus *Porella*

1 (a) Extensive patch, zooids sinuous in outline. .... *Porella concinna* (p. 290)
(b) Small patch, zooids oval or polygonal in outline. ....…. .... *Porella chadleyi* nov. sp. (p. 291)

Porella concinna (Busk, 1854)

Description Colony encrusting, unilaminar. Zooids 0.4–0.5 mm long, hexagonal, often irregular or sinuous in outline, separated by deep grooves. Frontal surface thick, from glistening glassy to grainy and dull, with areolae. Orifice roughly quadrangular with a low, broad, barely visible lyrula. The orifice has a low peristome which quickly thickens and partly hides the orifice. Rarely two distal oral spines on zooids at the periphery of the colony. Avicularia with oval rostrum, proximal to the orifice, enclosed within the peristome and perpendicular to the orifice. This avicularium may be replaced by a large one with a spatulate rostrum, frontal on the zooidal surface. In later ontogeny, additional avicularia may develop on the zooidal margin. Ovicell prominent, spherical, with a single pseudopore, which disappears with progressive calcification. Embryos red. Ancestrula tatiform, opesia surrounded by nine spines (Fig. 4.111).

Occurrence

Belgium At least 11 colonies were collected on the Flemish Banks between 1899 and 1907 (RBINS collection); 32 deceased colonies in 2004 and 2005 on stones and shells on the Hinder Banks. Rarely on plastic objects: at Zeebrugge in 1999 and 2000, and at Ostend in 2000. Small stones were collected by F. Kerckhof (RBINS) in the extreme north-western part of the Belgian continental shelf, on the border with the United Kingdom on July 12, 2018, during a haul with a Gilson dredge with starting coordinates 51°35′26″ N, 2°15′14″ E and ending point 51°35′36″ N, 2°15′28″ E, 100 stones were examined on epibionts and 33 of them contained living colonies of *P. concinna*. In 2023, 2 colonies were collected on a shell by beam trawl between 51°30.511′ N, 2°35.273′ E and 51° 29.559′ N, 2°36.378′ E.

The Netherlands Only known from beached substrata such as plastic, in 1947 (Bloklander and Leenhouts 1948), in Westkapelle in 2004 (De Blauwe 2009a), and on plastic (de Ruijter 2014b, 2015d, 2019a).

Germany Collected in the Tiefe Rinne (Kuhlenkamp and Kind 2012).

Fig. 4.111 *Porella concinna.* (**a**, **b**) Colony growing in a dead shell, Westhinder Bank (Belgium). (**b**) SEM image showing the irregular outline of non-ovicellate autozooids

Porella chadleyi nov. sp.

Type Material

I.G. 34,864 / BRYT.222/1a: holotype on shell, gold coated for SEM. Sampled by J.S Houziaux (JSH 36) on the Flemish Banks with Gilson dredge in 2005.

I.G. 34,864 / BRYT.222/2: paratype, two colonies, on a shell. Sampled by J.S Houziaux (JSH 37) on the Flemish Banks with Gilson dredge in 2005.

Description Colony encrusting, unilaminar, forming a small flaring patch comprising up to 30 autozooids. These spot colonies produce ovicells as early as the five-zooid stage. Autozooids oval or polygonal, 305–340 μm long, 125–200 μm wide ($n = 6$). Frontal shield with conspicuous areolae. Vertical walls with multiparous septula. Orifice wider than long, with a low, broad lyrula. Orifice about 120 μm wide, lyrula 52–56 μm wide. Condyles indistinct or absent. Traces of two oral spines in non-ovicellate zooids. A peristome encloses the orifice proximally and laterally and has an umbo mid-proximal to the orifice. Avicularium perpendicular to

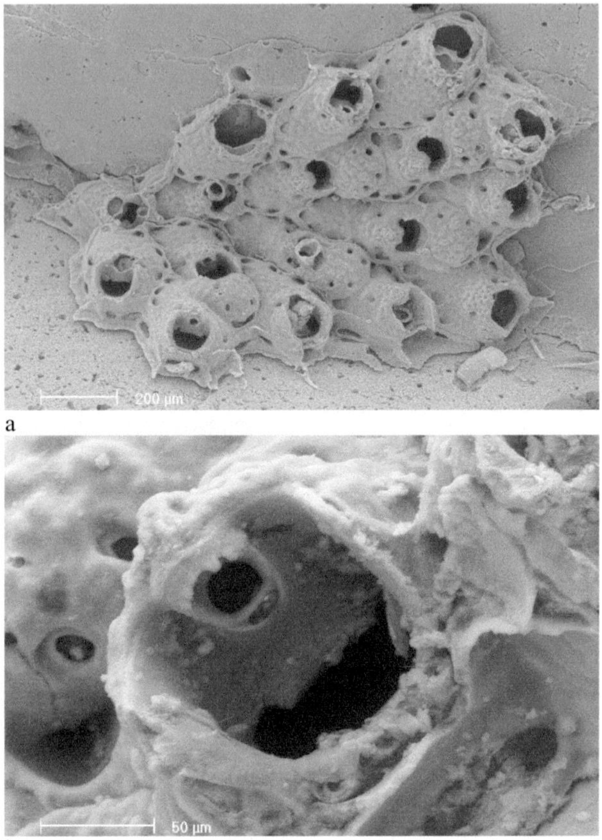

Fig. 4.112 *Porella chadleyi* nov. sp. (**a**, **b**) Colony from the Westhinder Bank (Belgium). (**a**) Colony, the tubes in some orifices are sheaths of epibiotic ciliophorans. (**b**) Proximal side of the orifice with lyrula and suboral avicularium

the peristome, obscuring the proximal border of the aperture making the lyrula difficult to see. Avicularium oval with a thin, complete crossbar. Distal margin of avicularian rostrum with a row of blunt teeth. Ovicell prominent, not closed by autozooidal operculum, ooecium formed by the distal autozooid, ectooecium coarsely granular, 213–225 µm wide ($n = 3$), with two to five conspicuous round or oval pseudopores, 13–16 µm wide; one 27-µm long teardrop-shaped pseudopore also observed. Ancestrulae severely damaged in the studied specimens; they probably had four oral spines (Fig. 4.112).

Etymology
Named after Chadley, our grandson.

Occurrence In June 2005, at least 12 colonies were dredged on shell valves on the Hinder Banks.

Remarks Largely similar to *Porella minuta* but the ovicells have pseudopores, and there are traces of two distal oral spines. *Porella minuta* has ooecia without pseudopores and is spineless. Another *Porella* sp. with ooecia without pseudopores, but with three oral spines, beached between Blankenberge and Wenduine on an object composed of plastic and wood, was found on September 15, 2001.

Family Romancheinidae

Genus *Escharella*

1 (a) Six oral spines (four in ovicellate zooids). 2
(b) Two to four oral spines (six in periancestrular zooids). 3

2 (a) Large encrustation, older zooids with white, lateral, secondary calcification, accentuating the zooid margins. Brood chamber rests on the distal zooid. .... *Escharella immersa* (p. 294)
(b) Small patches on the inner surface of lamellibranch valves. Ovicells globular, terminal. .... *Escharella gilsoni* (p. 293)

3 (a) Four oral spines. 4
(b) Two or three oral spines. Zooids up to 0.8 mm long. Lyrula slender, anvil-shaped. Mucro often bifid. Zooids flat with conspicuous areolae. .... *Escharella variolosa* (p. 296)

3 (a) Zooids small, up to 0.5 mm long. Lyrula longer than wide. .... *Escharella labiosa* (p. 296)
(b) Zooids larger than 0.5 mm. Lyrula wider than long. .... *Escharella ventricosa* (p. 299)

Escharella gilsoni De Blauwe, 2006

Description Colony encrusting, unilaminar, small, on the inner surface of lamellibranch valves. Zooids oval, convex. Frontal surface finely granulated, areolae not crowded and not separated by ridges. Orifice with a short wide, concave lyrula. Condyles blunt. Six oral spines, four in ovicellate zooids. Peristome thin and often damaged. Ovicells large, terminal, already present in colonies of a dozen zooids; ectooecium membranous, not present in dry colonies (Fig. 4.113).

Occurrence

Belgium Collected on the Kwinte Bank (24 colonies in 2004) and the Westhinder Bank (15 colonies in 2005).

Germany Known from Heligoland (Zettler et al. 2018).

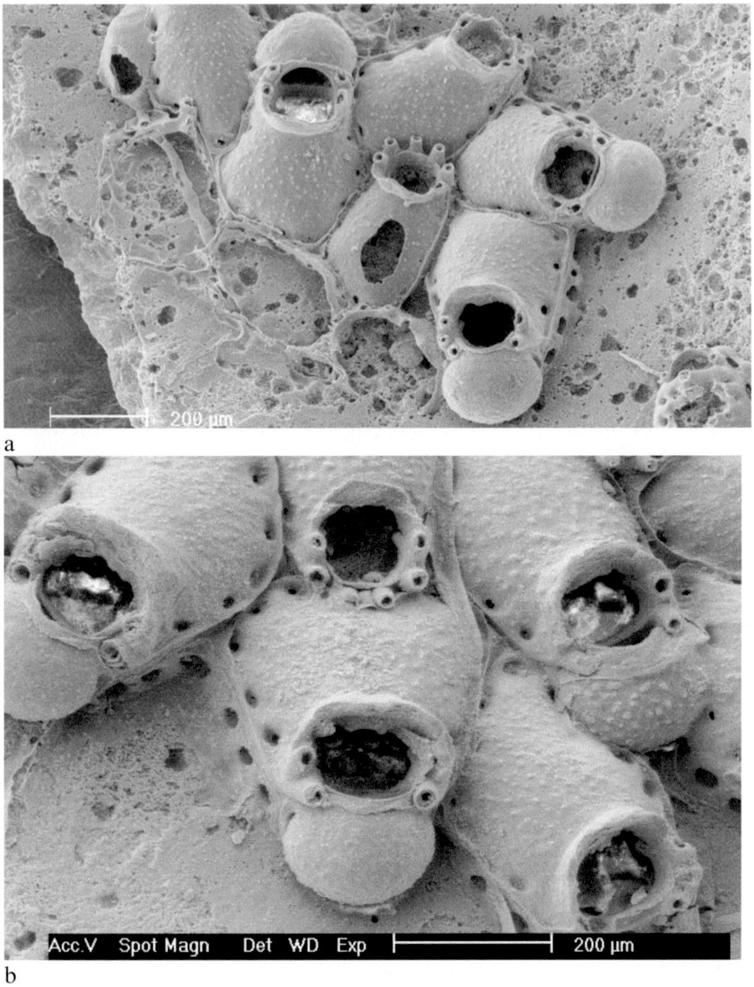

Fig. 4.113 *Escharella gilsoni*. (**a**) Young colony growing in a dead shell, Kwinte Bank (Belgium). (**b**) Colony edge showing ovicellate zooid with terminal ovicells, oral spines, areolae, and pore chambers, Westhinder Bank (Belgium)

Escharella immersa (Fleming, 1828)

Description Colony encrusting, unilaminar, becoming an extensive sheet. Zooids only 0.3–0.4 mm long, oval to hexagonal, and convex. Zooids are initially glassy, due to further calcification a white chalk line appears on the zooid edge.

Frontal surface with areolae that are inconspicuous at first and become more pronounced in older zooids. The broad lyrula is anvil-shaped with often a concave distal margin. Orifice with blunt condyles. Six distal oral spines are characteristic of zooids without an ovicell, reduced to four spines in ovicellate zooids. Peristome

Fig. 4.114 *Escharella immersa.* (**a–c**) Colonies on beached plastic from the Belgian coast. (**a–b**) Frontal view showing colony areas with secondarily calcified zooids. (**c**) Autozooids with six oral spines and an ovicellate zooid with four oral spines remaining

with a stout, pointed mucro proximal to the orifice. Polypide with 11 to 12 tentacles. Ovicells subimmersed, ooecium formed by the distal zooid, ectooecium membranous, not present in dry colonies. Embryos pink. Ancestrula tatiform, surrounded by 10 to 11 spines. The proximal part of its opesia is underlain by an extensive cryptocyst (Fig. 4.114).

Occurrence Very common on washed-up plastic.

Belgium On dredged boulders of the Thornton Bank (2004), locally very common on boulders between the Hinder Banks (June 2005). Small stones were collected by Francis Kerckhof (RBINS) in the extreme north-western part of the Belgian continental shelf, on the border with the United Kingdom on July 12, 2018, during a haul with a Gilson dredge with starting coordinates 51°35′26″ N, 2°15′14″ E and ending point 51°35′36″ N, 2°15′28″ E, 100 stones were examined on epibionts and 25 of them contained living colonies of *Escharella immersa*.

The Netherlands In Zeeland growing on stones and oysters in Westkapelle, Zoutelande, the Outer Harbor of Vlissingen and the western dam of the Sloehavens (Faasse and De Blauwe 2004). Found on the Cleaver Bank in 2002 (pers. comm. van Moorsel G) and in 2019 (pers. comm. Faasse M.) and collected in 2016 on the Dogger Bank.

Germany Collected at Heligoland and in the Tiefe Rinne (Kuhlenkamp and Kind 2012).

Escharella labiosa (Busk, 1856)

Description Colony encrusting, unilaminar, forming small patches. Zooids 0.4–0.5 mm long, oval, convex. Frontal surface with areolae only, small and inconspicuous. Lyrula slender, quadrangular with concave distal margin. Always four distal oral spines, even on ovicellate zooids (five or six spines on the first three zooids emerging from the ancestrula). Peristome stout, with a tall peak proximally. Ovicells prominent, ooecium formed by the distal zooid, ectooecium membranous, not present in dry colonies, endooecium with granulated surface. Ancestrula tatiform, surrounded by 11 to 12 spines. The edge of the opesia is underlain by a narrow cryptocyst (Fig. 4.115).

Occurrence Reported from the west and south of the United Kingdom and Guernsey as the closest location to the Low Countries (Hayward and Ryland 1999).

Belgium On washed-up plastic on December 28, 1999, in Zeebrugge, on January 16, 2000, in Westende, and in January 2024 at De Panne.

The Netherlands On washed-up plastic (de Ruijter 2012b, 2015a, 2017b, 2018a, 2019a, b).

Escharella variolosa (Johnston, 1838)

Description Colony encrusting, unilaminar, forming extensive sheets. Zooids elongate, 0.6–0.8 mm, rather flat, vitreous. Areolae closely spaced with intervening ridges. Orifice orbicular, lyrula narrowly anvil-shaped, no condyles. Two, rarely one or three distal oral spines. Proximal mucro pointed, sometimes bifid. Ovicells ini-

Fig. 4.115 *Escharella labiosa*. (**a**, **b**) Colony areas consisting of ovicellate and non-ovicellate zooids, on beached plastic, Zeebrugge (Belgium)

tially prominent, but becoming subimmersed because of secondary calcification, ooecium wider than long, formed by the distal autozooid, ectooecium membranous, not present in dry colonies, androecium granulated. Embryos orange. Ancestrula tatiform, frontal membrane surrounded by nine spines (Fig. 4.116).

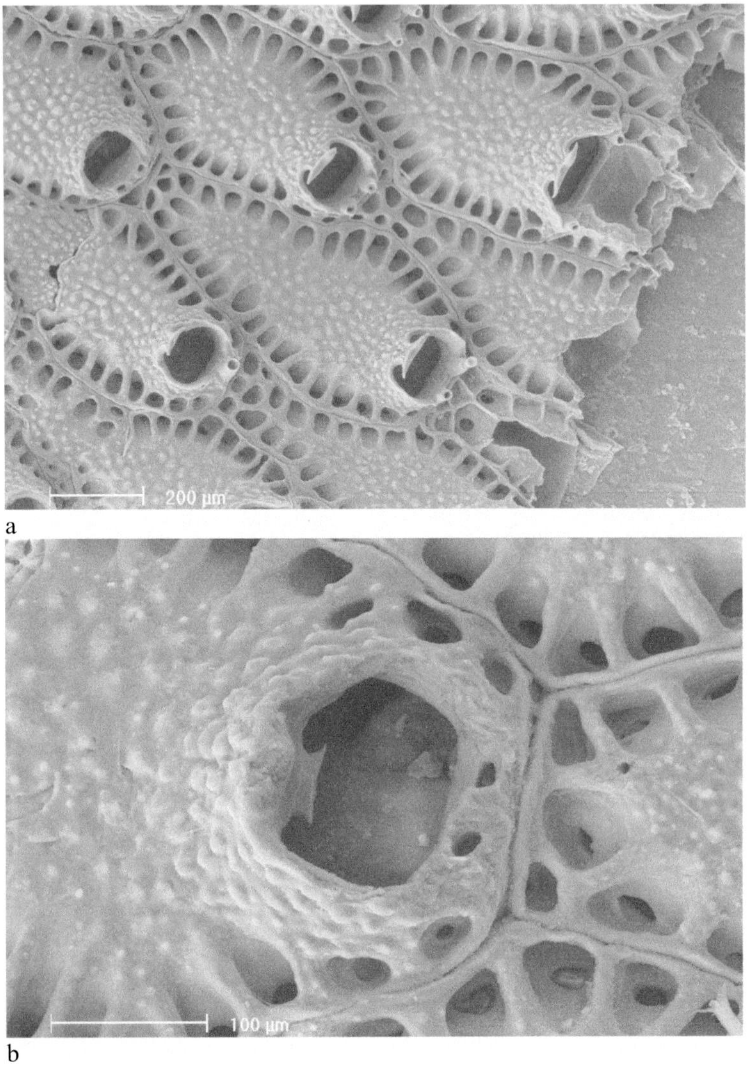

Fig. 4.116 *Escharella variolosa*. (**a**, **b**) Periphery of the colony growing on plastic beached at the Belgian coast; zooidal orifice showing the shape of lyrulae. (**b**) Orifice of older zooid showing lyrula

Occurrence

Belgium Two colonies were collected on the Flemish Banks in 1904, with a third in 1906. In June 2005, 35 deceased colonies were found on stones and shells on the Hinder Banks. Occasionally on washed-up plastic.

The Netherlands Occasionally on washed-up plastic.

Germany Collected at Heligoland and in the Tiefe Rinne (Kuhlenkamp and Kind 2012).

Escharella ventricosa (Hassall, 1842)

Description Colony encrusting, unilaminar, silvery white sheet. Autozooids oval to hexagonal, strongly convex. Zooid length between *E. immersa* and *E. variolosa*, i.e., 0.55 to 0.7 mm. Frontal surface vitreous and glittering, patterned with minute wrinkles. Very small, inconspicuous areolae. The lyrula is widest at its distal edge and has frontally deflected pointed corners. Condyles minute. Four thick distal oral spines, all remaining in ovicellate zooids (six spines on the first autozooids emerging from the ancestrula). Truncate mucro proximal to the orifice. Ovicells subimmersed, ooecium formed by the distal zooid, ectooecium membranous, not present in dry colonies, endooecium with granulated surface. Ancestrula tatiform, opesia surrounded by 11 spines, the proximal half of the membrane underlain by a cryptocyst (Fig. 4.117).

Occurrence

Belgium In June 2005, a living colony was discovered on a rock between the Hinder Banks. Small stones were collected by Francis Kerckhof (RBINS) in the extreme north-western part of the Belgian continental shelf, on the border with the United Kingdom on July 12, 2018, during a haul with a Gilson dredge with starting coordinates 51°35′26″ N, 2°15′14″ E and ending point 51°35′36″ N, 2°15′28″ E. 100 stones were examined on epibionts and 1 of them contained a living colony of *E. ventricosa.* Sometimes washes up on plastic.

The Netherlands De Ruijter (2014b) reports the washing-up of a colony on plastic near Egmond aan Zee in 2013. Washed up on plastic on Terschelling on June 19, 2022 (Sjoerd Groos, pers. comm.).

Germany Collected in the Tiefe Rinne (Kuhlenkamp and Kind 2012).

Genus *Escharoides*

1 (a) Avicularia absent. .... *Escharoides bishopi* (p. 300)
(b) Avicularia present. 2

2 (a) Bumps scattered over the frontal surface. Avicularia large, oriented laterally or distolaterally. .... *Escharoides coccinea* (p. 302)
(b) Bumps in a row pattern radiating from the suboral mucro to the zooid margin. Avicularia smaller, directed distally. ....
.... *Escharoides mamillata* (p. 303)

Fig. 4.117 *Escharella ventricosa*. (**a**) Colony growing on empty shell collected in the harbor of Le Conquet (France). (**b**) Non-ovicellate zooids growing on plastic beached at the Belgian coast

Escharoides bishopi De Blauwe, 2006

Description Colony encrusting, unilaminar, forming small patches of a dozen autozooids on the inner surface of lamellibranch valves. Zooids convex, frontal surface smooth or covered with indistinct bumps arranged in rows. Areolae large and prominent. Six hollow oral spines present, often broken, reduced to two very conspicuous oral spines in ovicellate zooids. Peristome widely flaring and lacking a mucro. Orifice with two condyles and a slender proximal tooth that has two tips at the end. Avicularia lacking in all material. Ovicell terminal, globular, smooth, or corrugated, with small peripheral pits. Ancestrula tatiform, opesia surrounded by 13 spines. All known colonies consist of 2 to 24 zooids, and the fourth zooid is often already ovicellate. This shows that the species is adapted to life on unstable substrate by resorting to sexual reproduction at an early stage (Fig. 4.118).

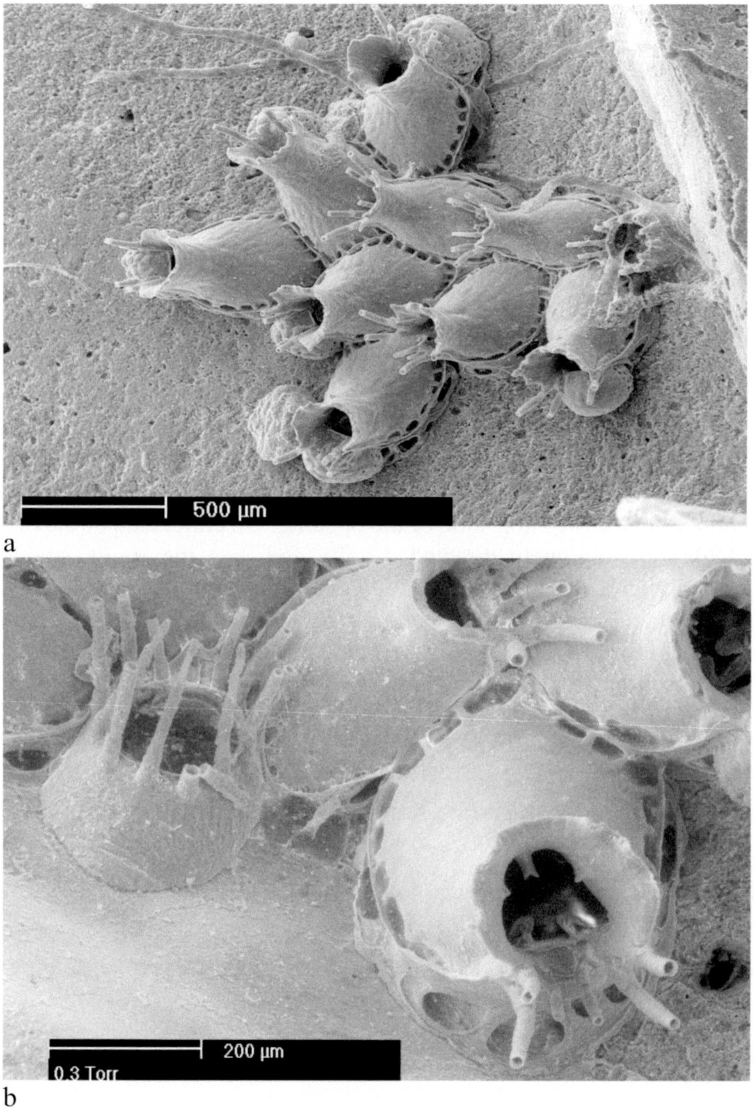

Fig. 4.118 *Escharoides bishopi*. (**a**, **b**) Two colonies growing in a dead shell, Le Val André (France). (**a**) Entire colony with ancestrula and ovicellate periancestrular zooid. (**b**) Ancestrular region of a colony showing the orifice of a periancestrular zooid with a flaring peristome, condyles, and a slender proximal tooth that has two tips at the end

Occurrence All known finds are listed below.

Belgium Five colonies were discovered on the inside of empty clams (*Cerastoderma edule*) from the Kwinte Bank (De Blauwe 2006b). In June 2005, another six colonies were collected on the Westhinder Bank (De Blauwe 2009b). One colony also from Buiten Ratel in March 2012 (De Blauwe 2012).

France 20 colonies were found in one shell valve of *Acanthocardia echinata* from Le Val André (Brittany). On a second quest, 11 colonies were in one valve at Le Val André (De Blauwe 2019).

United Kingdom One colony on the East Coast was found in a benthic sample collected in July 2013 at 53819′0.12″N 0840′21.87″E from a depth of 21.1 m, NHMUK 2013.10.28.2 (Wasson and De Blauwe 2014). A second colony was collected in the Thames Estuary in August 2013 (De Blauwe 2019).

Escharoides coccinea (Abildgaard, 1806)

Description Colony encrusting, unilaminar, forming extensive orange sheets. Zooids 0.6–0.7 mm long. The initially smooth frontal surface quickly calcifies. The irregularly distributed bumps are reminiscent of the spots of a ladybug. Frontally not pseudoporous, areolae are present but are hidden by progressive calcification. Initially six large spines on the distal edge of the orifice, which later break off. The proximal edge of the aperture is raised and serrated. A square tooth stands above the operculum within the proximal side of the opening. On either side of the orifice is a large avicularium, obliquely directed distally and, like the zooid, studded with bumps. Mandible elongated, triangular. Ovicell prominent, ooecium formed by the distal zooid, ectooecium membranous, not present in dry colonies, endooecium with a peripheral series of small pores, flattened frontally and with the same bumpy surface as the zooid. Embryos red. Ancestrula tatiform, surrounded by 13 to 14 spines (Fig. 4.119).

Occurrence

Belgium Common on washed-up plastic and other objects and *Himanthalia* buttons (De Blauwe 2000a, b, c, 2005a, b; Lacourt 1949; Vanhaelen et al. 2006).

The Netherlands Common on washed-up plastic and other objects and *Himanthalia* buttons (de Ruijter 2005b, 2012b, 2018a, c, 2020c, 2021b).

France Davoult et al. (1993) report this species on boulders off Cap Gris-Nez in 1989 at a depth of between 35 and 45 m.

Germany Collected in the Tiefe Rinne (Kuhlenkamp and Kind 2012).

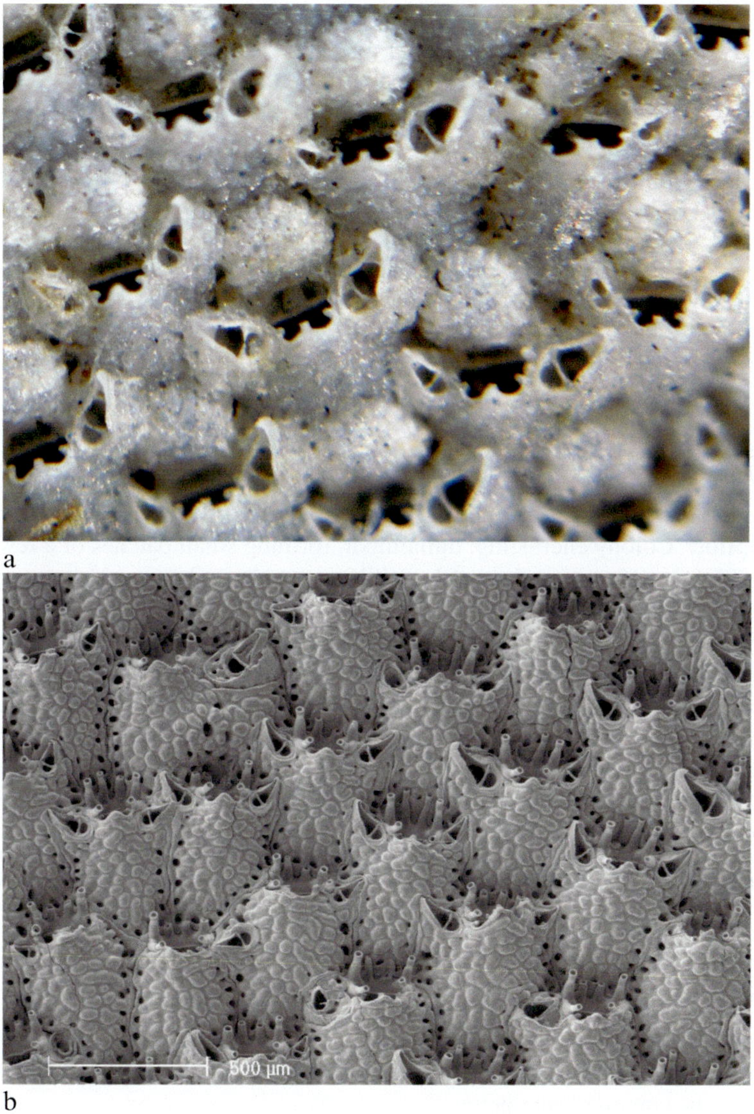

Fig. 4.119 *Escharoides coccinea.* (**a**) Part of colony of ovicellate zooids with avicularia growing on beached plastic, Zeebrugge (Belgium). (**b**) Non-ovicellate zooids with avicularia and oral spines growing on a lobster pot, St-Malo (France)

Escharoides mamillata (Wood, 1844)

Description Colony encrusting, unilaminar, forming orange sheets. Zooids 0.5–0.8 mm long. Young frontal surface smooth, later covered with bumps arranged in rows radiating from the large, blunt suboral mucro. Areolae small and only visible in zooids at the colony margin. Four or six oral spines on autozooids. Proximally inside the orifice is a short angular tooth flanked by a triangular tooth at both sides. On either side of the orifice is an avicularium, smaller than in *E. coccinea* and directed distally. Ovicells from prominent to subimmersed, ooecium formed by the distal zooid, ectooecium membranous, not present in dry colonies, endooecium with a peripheral series of small pores, surface quickly covered with bumps (Fig. 4.120).

Occurrence A rarely reported species (Hayward and Ryland 1999).

Belgium Washed up on cork in De Panne on October 5, 2004 (De Blauwe 2005a).

Genus *Neolagenipora*

Neolagenipora collaris (Norman, 1867)

Description Colony encrusting, unilaminar, forming small patches of a few dozen autozooids on the inner surface of lamellibranch valves. Autozooids strongly convex, 0.4–0.5 mm long, frontal surface smooth. Areolae widely spaced. Orifice slightly wider than long, proximal border shallowly concave, bordered by blunt condyles. No lyrula. A pair of distal oral spines may be present on some peripheral autozooids. Peristome forming a prominent flaring rim round the proximal and lateral borders of the orifice. Ovicells prominent, not closed by the zooidal operculum, ooecium formed by the distal zooid, surrounded by a "collar" of secondary calcification, ectooecium with small scattered pseudopores. No avicularia. Ancestrula tatiform, surrounded by short spines (Fig. 4.121).

Remark A large number of colonies of the Flemish Banks show linear growth (De Blauwe 2006b), possibly a different but related species.

Occurrence Often found in association with *Callopora discreta* and *Phylactella labrosa* (Hayward and Ryland 1999).

Belgium In 1906, colonies were collected on the Flemish Banks (RBINS collection), recently generally on shells on the Kwinte Bank and the Westhinder Bank. Rarely washes up on plastic items.

The Netherlands One colony on washed-up plastic (de Ruijter 2013c).

Germany Reported by Zettler et al. (2018) from Heligoland.

Fig. 4.120 *Escharoides mamillata.* (**a**) Part of colony with ovicellate zooids with avicularia, Ploubazlanec (France, photo by F. Gully). (**b**) SEM image of a colony area growing in a dead shell from the fishing harbor of Le Conquet (France)

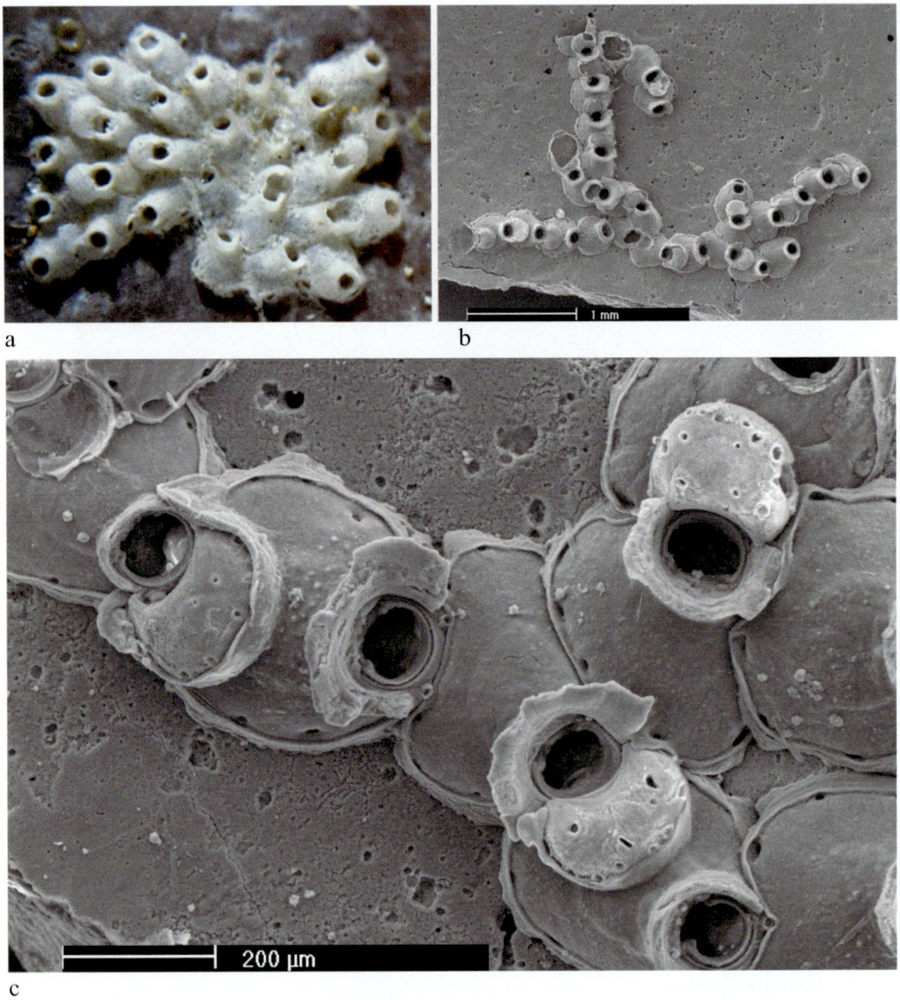

Fig. 4.121 *Neolagenipora collaris.* (**a**) Colony with typical growth form from the Westhinder Bank (Belgium). (**b**, **c**) Uniserial colony showing ovicellate zooids, Kwinte Bank (Belgium)

Family Umbonulidae

1 (a) Frontal shield with 6 to 12 marginal areolae, the pronounced interareolar ridges converging on the suboral umbo. Ovicells occur. .... *Umbonula ovicellata* (p. 307)

(b) Frontal shield with 14 to 20 marginal areolae, interareolar ridges feebly developed. No ovicells. .... *Oshurkovia littoralis* (p. 307)

Genus *Oshurkovia*

Diagnosis (Grischenko and Mawatari 2005): Colony encrusting, multiserial, sheet-like. Zooids with umbonuloid frontal shield, with marginal areolae, rising to prominent suboral umbo. Secondary orifice large, cormidial, i.e., comprising four lobes with distinct sutures between them, with thin, prominent marginal rim. Oral spines and ovicells absent, avicularia not always present. Multiparous septula present in vertical walls.

Oshurkovia littoralis (Hastings, 1944)

Description Colony encrusting, unilaminar, orange when alive. Secondary orifice longer than broad. Operculum with marginal sclerite in the distal region of the secondary orifice. 14 to 20 large areolae. The interareolar ridges converge toward the base of the umbo. Avicularia not always present, situated immediately proximal to the orifice, rostrum oval, perpendicular to the frontal plane. No ovicells. Red embryos are brooded in internal brood sacs, and two or more embryos recorded per brood sac (Ostrovsky 2013) (Fig. 4.122).

Occurrence Not native to Belgium or the Netherlands.

Belgium In July 2000, a colony washed ashore on a *Himanthalia* button between Zeebrugge and Blankenberge (De Blauwe 2000c), again in 2005 between Oostduinkerke and De Panne (Vanhaelen et al. 2006) and in 2019 at Raversijde (Nathalie Colpaert, pers. comm.)

The Netherlands Bloklander and Leenhouts (1948) mention the washing-up of a colony in the winter of 1947–48, possibly at Schouwen. On washed-up plastic in 2011 and 2013 (de Ruijter 2012a, 2013c).

Genus *Umbonula*

Revised diagnosis (Grischenko and Mawatari 2005): Colony encrusting, multiserial, sheet-like. Zooids with umbonuloid frontal shield. Orifice simple, with symmetrically placed suboral avicularium. Ovicells prominent, ectoecium with pseudopores. Uniparous or multiparous septula present in distal and lateral walls.

Umbonula ovicellata Hastings, 1944

Description Colony encrusting, unilaminar, orange when alive. Autozooids 0.6–0.8 mm long. Frontal shield vitreous at first, becoming coarse and nodular, with 6 to 12 marginal areolae. The interareolar ridges converge toward the proximal border of the orifice. Avicularia not always present, situated immediately proximal to the orifice, rostrum oval, perpendicular to the frontal plane. Ovicells prominent,

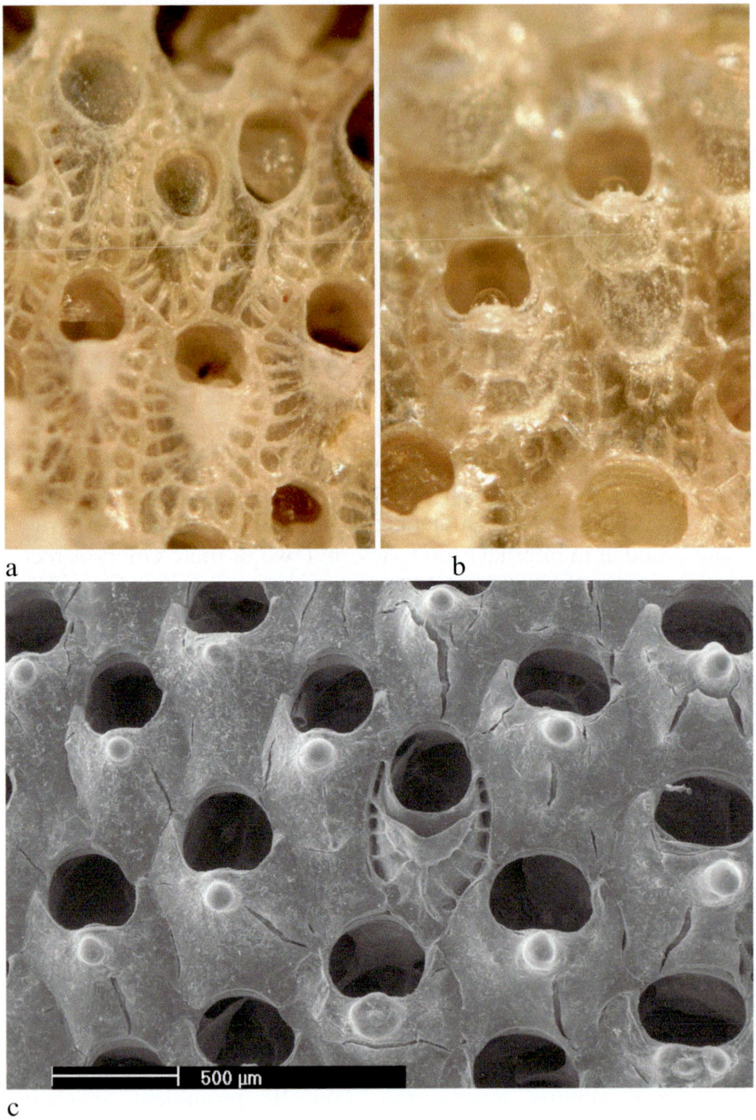

Fig. 4.122 *Oshurkovia littoralis*. (**a**, **b**) Parts of the same colony showing zooidal morphology, Pointe du Rozel (France). (**b**) Zooids showing suboral avicularia with raised mandible. (**c**) Autozooids covered by their frontal membrane, except one, on beached *Himanthalia*, Oostduinkerke (Belgium)

ectoecium with irregular frontal pseudopores, the largest at the periphery (Fig. 4.123).

Occurrence Not native to Belgium or the Netherlands.

The Netherlands Washed up on plastic in August 2017 at Texel (de Ruijter 2017c) and on a lobster pot in February 2024 near Scheveningen (pers. comm. Eric Wisse).

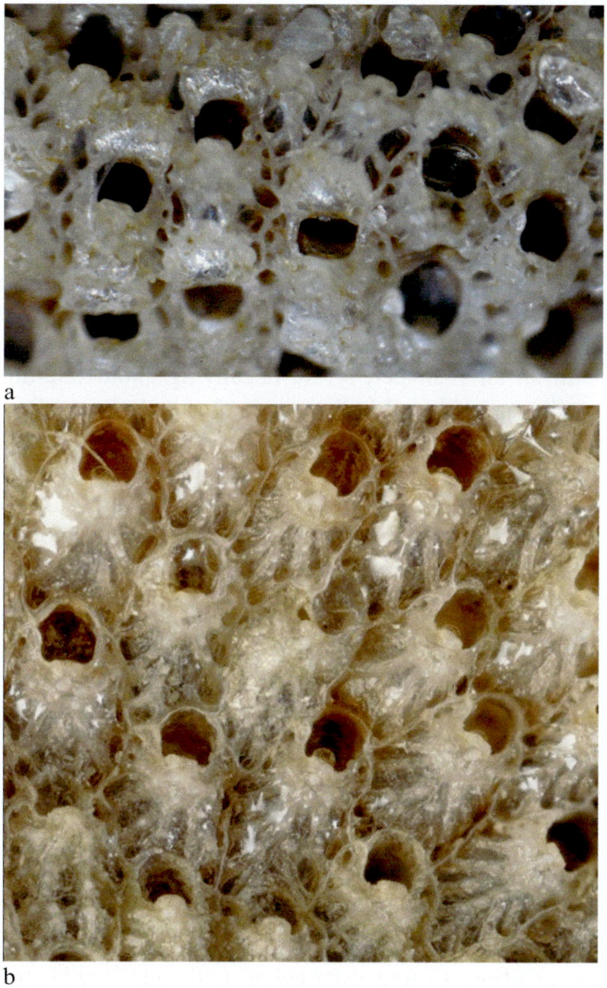

Fig. 4.123 *Umbonula ovicellata.* (**a**) Colony area with ovicells, beached at Texel (the Netherlands). (**b**) Part of colony showing non-ovicellate zooids, Verdelet (France, photo by F. Gully)

Superfamily Smittinoidea

Family Smittinidae

Genera *Prenantia* and *Smittina*

1 (a) Avicularium orientated transversely. .... *Smittina affinis* (p. 310)
(b) Avicularium in the longitudinal direction of the zooid. 2
(c) Avicularium absent. .... *Prenantia cheilostoma* (p. 310)

2 (a) Oral spines on young zooids, small avicularia immediately proximal to the gap of the peristome proximal to the orifice. .... *Smittina crystallina* (p. 311)

(b) Oral spines never present, small avicularia situated in the proximal gap of the peristome proximal to the orifice. Spatulate large avicularia may replace a small avicularium or occur on the frontal surface of an autozooid. .... *Smittina landsborovii* (p. 313)

Prenantia cheilostoma (Manzoni, 1869)

Description Colonies orange to light reddish purple, encrusting, multiserial, unilaminar. Autozooids hexagonal, gently convex in early ontogeny, later rather flat and separated by shallow grooves. Primary orifice wider than long; proximal border with an anvil-shaped lyrula occupying about one-third of total width; condyles small, rounded, and indistinct. Oral spines absent. Peristome forming a low thickened rim completely encircling the orifice, mid-proximal portion deeply U-shaped. Frontal shield thick, vitreous, densely pseudoporous with numerous small pores. Ovicell initially subimmersed, and finally almost totally covered by the progressive development of the secondary calcification, ooecium wider than long, with many small frontal pseudopores; secondary calcification of the ooecium is continuous with the peristome (Fig. 4.124).

Occurrence Shallow sublittoral and offshore, often encrusting shell. Common throughout the Mediterranean, ranging northward to the western English Channel and the Isle of Man (Hayward and McKinney 2002).

Belgium No observations.

The Netherlands On beached plastic at Petten, November 2023 (de Ruijter 2024a).

Smittina affinis (Hincks, 1862)

Description Colony encrusting, unilaminar, reddish-orange. Zooids 0.5–0.6 mm long, oval to hexagonal, slightly convex. Frontal shield evenly pseudoporous. Orifice round with small condyles and a slender quadrate lyrula. No oral spines. Peristome with a broad proximal gap. Avicularia small, in the proximal gap of the peristome, directed transversally. Mandible triangular and oblique to the frontal plane. Ovicells prominent, ooecium formed by the distal zooid, ectooecium with numerous small pseudopores (Fig. 4.125).

Occurrence

Belgium On plastic washed up in Zeebrugge in 1999. In March 2007, F. Kerckhof collected a living colony on a tube of *Lanice conchilega* on the Birkenfels buoy.

The Netherlands On washed-up plastic (de Ruijter 2013a, 2015d, 2021b).

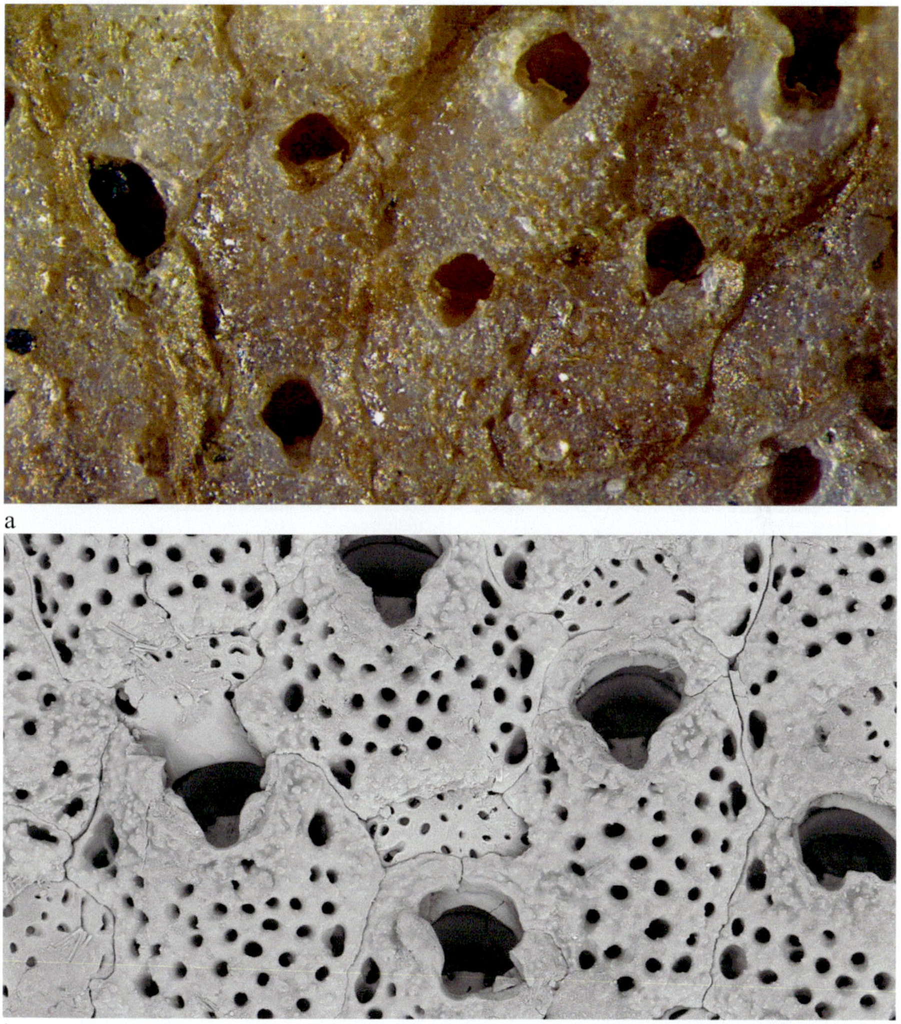

a

b

Fig. 4.124 *Prenantia cheilostoma.* (**a**, **b**) Part of colony with two ovicellate zooids (left), on beached plastic, Petten (the Netherlands). (**b**) Ovicellate zooids with ooecia covered by the secondary calcification

Smittina crystallina (Norman, 1867)

Description Colony encrusting, unilaminar, forming a small patch. Autozooids 0.4–0.5 mm long. Orifice wider than long, with a broad lyrula with longitudinal median ridge and large, pointed condyles. Young zooids with four slender oral spines, lost in later ontogeny. Peristome developed as lateral flaps, avicularium situ-

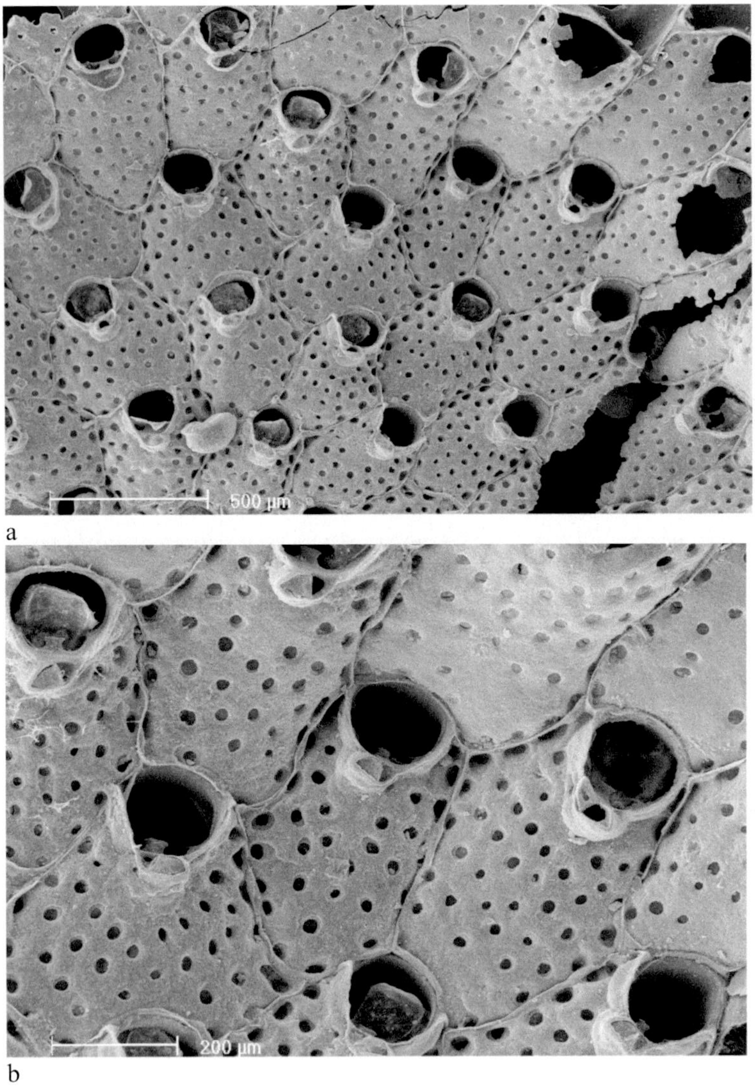

Fig. 4.125 *Smittina affinis.* (**a**) Colony growing on plastic beached at Zeebrugge (Belgium). (**b**) Non-ovicellate autozooids showing lyrulae and suboral avicularia

ated proximal to the proximal gap in the peristome. Frontal shield evenly pseudoporous. Ovicells prominent, ooecium formed by the distal zooid, ectooecium with numerous small pseudopores (Fig. 4.126).

Occurrence

The Netherlands Washed up on plastic at Egmond aan Zee in December 2014 (de Ruijter 2015c).

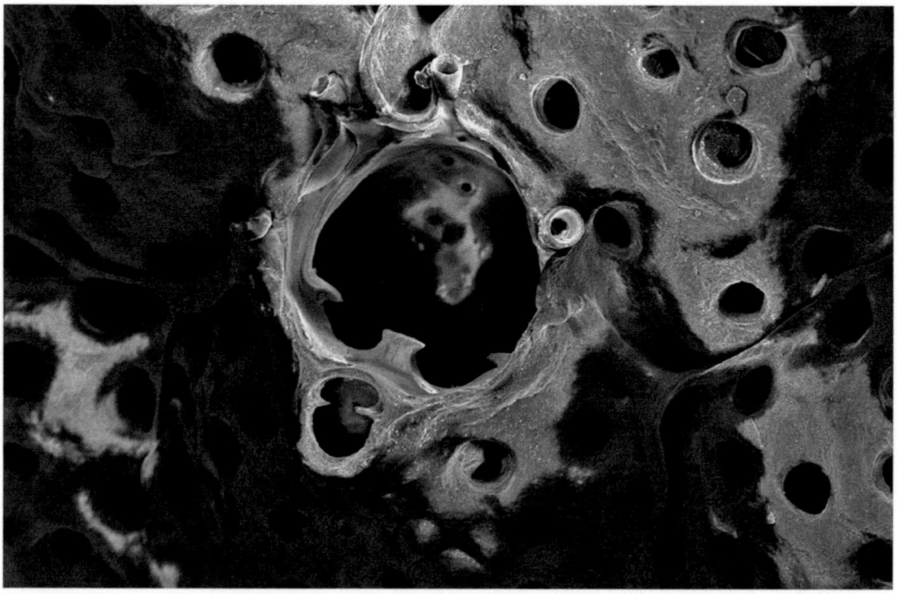

Fig. 4.126 *Smittina crystallina*. Zooidal orifice with spines, lyrula, condyles, and avicularium. Colony beached at Egmond aan Zee (the Netherlands)

Smittina landsborovii (Johnston, 1847)

Description Colony encrusting, unilaminar, able to form unilaminar or bilaminar erect foliaceous lobes. Zooids 0.5–0.7 mm long, slightly convex, frontal shield with closely placed large round pseudopores. Orifice broader than long, with short quadrate or anvil-shaped lyrula and sharp, basally deflected condyles. No oral spines. Peristome as lateral flaps with a proximal gap. Avicularium small with a semielliptical mandible, in the proximal gap of the peristome. Spatulate large avicularia may replace the small avicularium or occur on the frontal surface of an autozooid. Ovicells prominent, ooecium formed by the distal zooid, ectooecium with numerous small and irregular pseudopores (Fig. 4.127).

Occurrence

Belgium Present on washed-up plastic in Blankenberge on December 27, 1999, and in Zeebrugge on January 8, 2000.

The Netherlands A colony washed up on cork on January 16, 1949, near Scheveningen (Perk 2000). Reported once on washed-up plastic (de Ruijter 2014a).

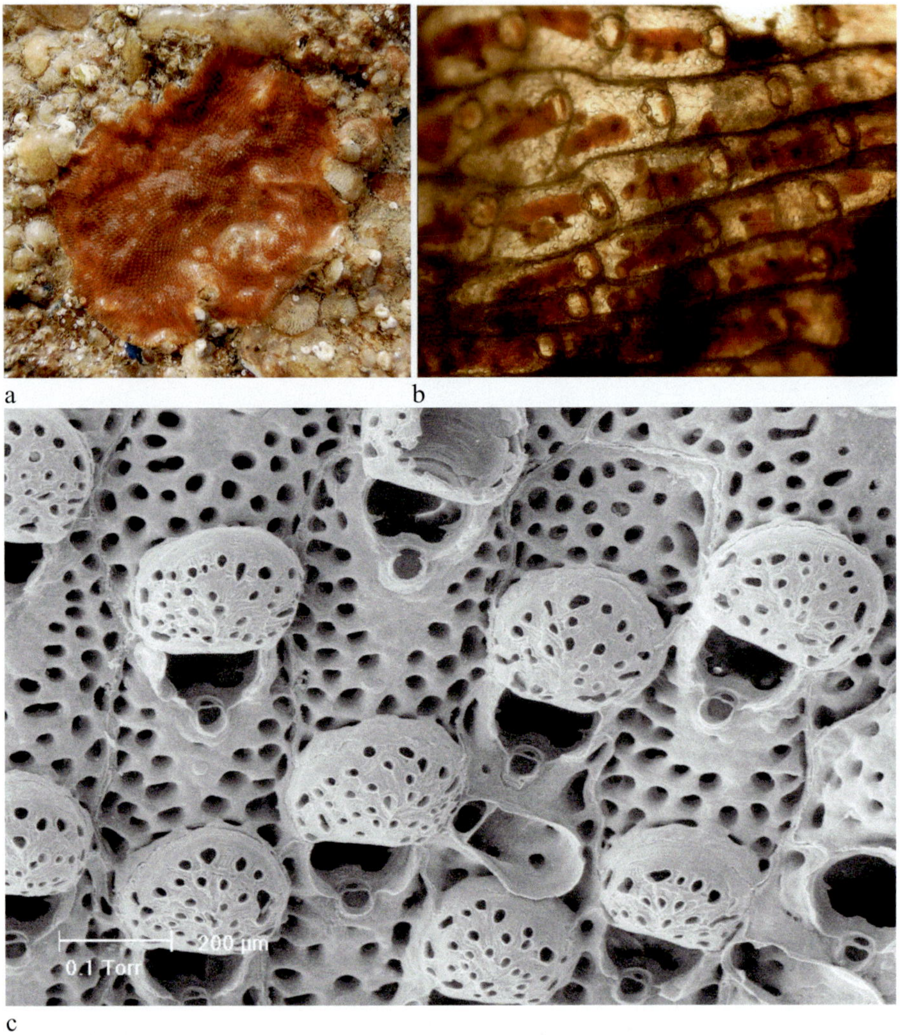

Fig. 4.127 *Smittina landsborovii.* (**a**) Colony growing on stone, Presqu'île de Quiberon (France). (**b**) Colony area of non-ovicellate zooids. (**c**) Colony area with ovicellate zooids having two types of avicularia, on beached plastic, Zeebrugge (Belgium)

Genus *Smittoidea*

1 (a) Suboral avicularium perpendicular to the frontal shield. .... *Smittoidea prolifica* (p. 315)
(b) Suboral avicularium in the plane of the frontal shield. .... *Smittoidea reticulata* (p. 315)

Smittoidea prolifica Osburn, 1952

Description Colony encrusting, unilaminar, forming white-pink patches. Zooids 0.5–0.7 mm long. Frontal shield smooth at first, later roughly calcified. Large areolae separated by solid ridges. Orifice with a lyrula and sharp condyles. Lyrula variable in width, one-fourth to one-half of the orifice width. Peristome only developed laterally. Young zooids bear two to four oral spines. Polypide with 12 tentacles. Avicularium in the distal side of a conspicuous umbo, mandible semicircular, and oriented perpendicular to the frontal surface. Ovicells prominent, progressively becoming subimmersed in the secondary calcification, ooecium formed by the distal autozooid, ectooecium with round or irregular pseudopores. Ovicells present in almost all zooids except in the youngest two or three rows. Twinned ovicells exceptionally occur (Markert et al. 2015; Kind and Kuhlenkamp 2016). Embryos orange (Fig. 4.128).

Occurrence A species originating from California, probably introduced to Europe with oyster imports to Zeeland (the Netherlands). Found on wood, stones, bivalves, and *Sargassum muticum* (De Blauwe and Faasse 2004). Present in Brittany (France) in 2022 (pers. comm. Florence Gully), and Castlemaine Harbour (Ireland) (pers. comm. 2022 Liz Hewitt Hebog Environmental Ltd).

Belgium Not yet reported!

The Netherlands Examination of reference material in the collection of Bureau Waardenburg revealed that *S. prolifica* had already been collected on October 26, 1995, at Schelphoek by Van Moorsel (1996), reported as *Smittoidea reticulata*. Collected in 2008 in the harbor region of Rotterdam (Faasse et al. 2013) and in the Prinses Amalia Wind Farm off Ijmuiden in 2011 (Vanagt et al. 2013). *Smittoidea prolifica* was first sighted in the Dutch Wadden Sea in 2008, east off Texel on a mussel culture plot on the inside of a dead Mya shell (Dekker and Drent 2013). Since then, the species was not found in the region anymore until 2014 when it was found to occur widespread in the western Wadden Sea off Terschelling and Texel on sublittoral mussel and oyster banks (Gittenberger et al. 2015a), still present in June 2023 (D. Oude Luttikhuis, NIOZ; WM 2023, MPA, PosKey 21,270, N 53°8.22′ E 5°8.17′).

Germany Collected in the Tiefe Rinne (Kuhlenkamp and Kind 2012; Kind and Kuhlenkamp 2016). In November 2014, several colonies were found in two harbors in the Jade estuary (Markert et al. 2015). In the German Wadden Sea, just north of Sylt, the species was first recorded in 2012 (Gittenberger et al. 2013).

Smittoidea reticulata (MacGillivray, 1842)

Description Colony encrusting, unilaminar, forming an extensive sheet. Zooids 0.5–0.8 mm long, oval to polygonal, separated by thin raised sutures. Frontal shield with large areolae. Orifice wider than long with rectangular or anvil-shaped lyrula.

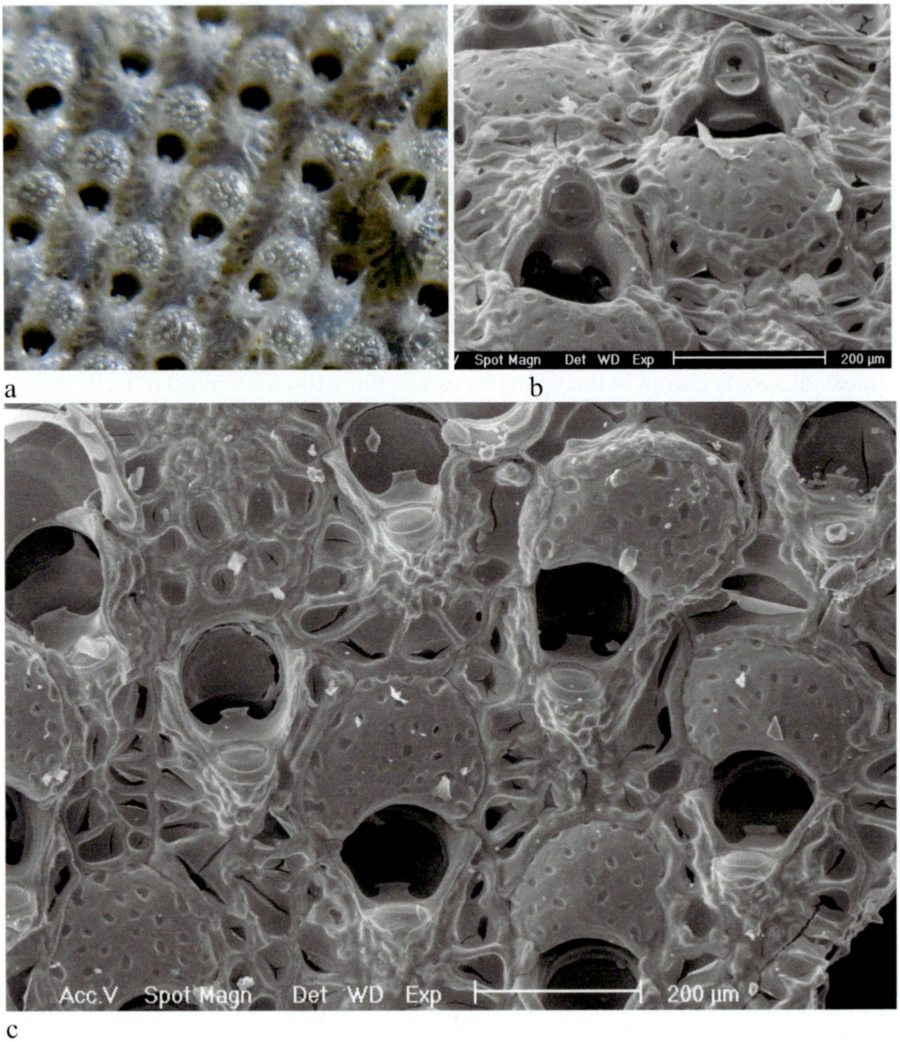

Fig. 4.128 *Smittoidea prolifica.* (**a**) Colony area of ovicellate zooids. (**b**) SEM image showing lyrula and avicularium from a distal position. (**c**) Ovicellate autozooids in frontal view. Goese Meer (the Netherlands)

Condyles sharply pointed and downcurved. Only zooids on the periphery of the colony bear two or three long, distal oral spines. Peristome developed on the lateral borders of the orifice with a deep notch proximally. Polypide with 13 to 15 tentacles. Avicularium immediately proximal to the peristomial notch, parallel to the plane of the frontal shield and proximally directed. Mandible broad triangular with rounded tip. Ovicell prominent, not closed by the zooidal operculum, globular ooecium formed by the distal autozooid, broader than long, ectooecium with numerous

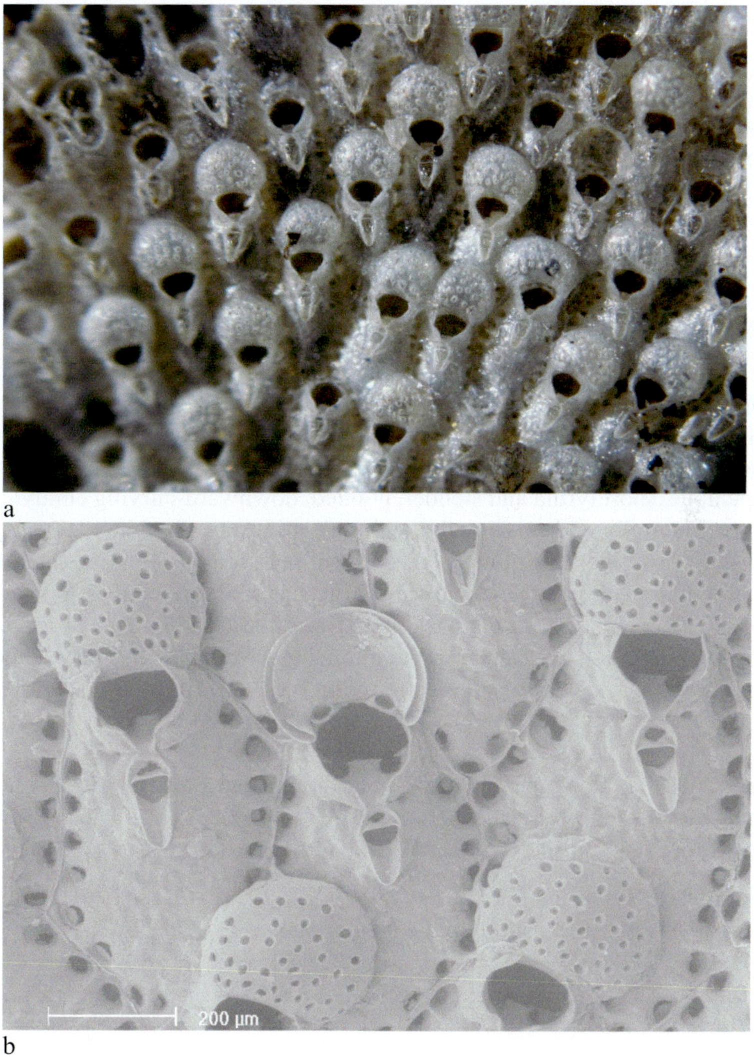

Fig. 4.129 *Smittoidea reticulata.* (**a**) Part of the colony growing on beached plastic, Oostende (Belgium). (**b**) Ovicellate zooids with fully formed and developing ooecia, condyles, lyrula, and avicularia

pseudopores. Embryos orange or yellow-orange. Ancestrula tatiform, surrounded by nine spines (Fig. 4.129).

Occurrence

Belgium Only known from washed-up plastic and a gas bottle (De Blauwe 2000b).

The Netherlands Some old mentions are those of van der Sleen (1920) at Den Helder and of Maitland (1897) without location. Reports from Zeeland most likely refer to *S. prolifica* (De Blauwe and Faasse 2004). A colony washed up on a lobster trap in Westkapelle in the Netherlands in 2004. On washed-up plastic (de Ruijter 2015d, 2019a).

Germany Collected in the Tiefe Rinne (Kuhlenkamp and Kind 2012; Kind and Kuhlenkamp 2016).

Genus *Parasmittina*

Parasmittina trispinosa (Johnston, 1838)

Description Colony encrusting, unilaminar, yellowish-white, often extensive. Zooids 0.4–1.0 mm long, rather flat. Frontal shield with areolae only, smooth and translucent at first, becoming rugose in later ontogeny. Orifice round with a rather inconspicuous, short lyrula and slender, pointed, downward-curving condyles. Peristome erect laterally and proximally, with a deep fissure proximally. Two or three short distal oral spines are quickly lost. Avicularia typically proximolateral to the orifice and directed away from or toward to it, mandible triangular. Avicularia not present on all zooids. Very small additional avicularia may occur at the areolae. Ovicells hyperstomial, ectooecium with a few large, irregular frontal pseudopores, surrounded by the secondary calcification. Embryos red (Fig. 4.130).

Occurrence

Belgium In 1902, a colony was dredged at the Stroom Bank (RBINS collection). Six times on washed-up plastic between 1999 and 2005. In 2005, two deceased colonies were found on shells at the Westhinder Bank.

The Netherlands De Ruijter (2013c, 2015a, d, 2017c, 2018c, 2019a) reports a colony on plastic washed ashore in the Netherlands.

Genus *Phylactella*

Phylactella labrosa (Busk, 1854)

Description The zooids often grow in one or more linear rows before the colony forms a patch. Zooids 0.4–0.6 mm long, oval, and strongly convex. Frontal shield translucent at first, later opaque white, dense, and regularly pseudoporous by round, deeply recessed pores. Orifice round, with sharp condyles and a tapered, truncated, or pointed lyrula. A very well-developed erect peristome forms a flaring cup around the proximal and lateral margin of the orifice. Ovicells prominent, ooecium formed by the distal zooid, ectooecium smooth, with numerous pseudopores. Ancestrula tatiform, membrane surrounded by eight spines and underlain by a broad cryptocyst. Polypide with 11 tentacles in Belgian material (Fig. 4.131)

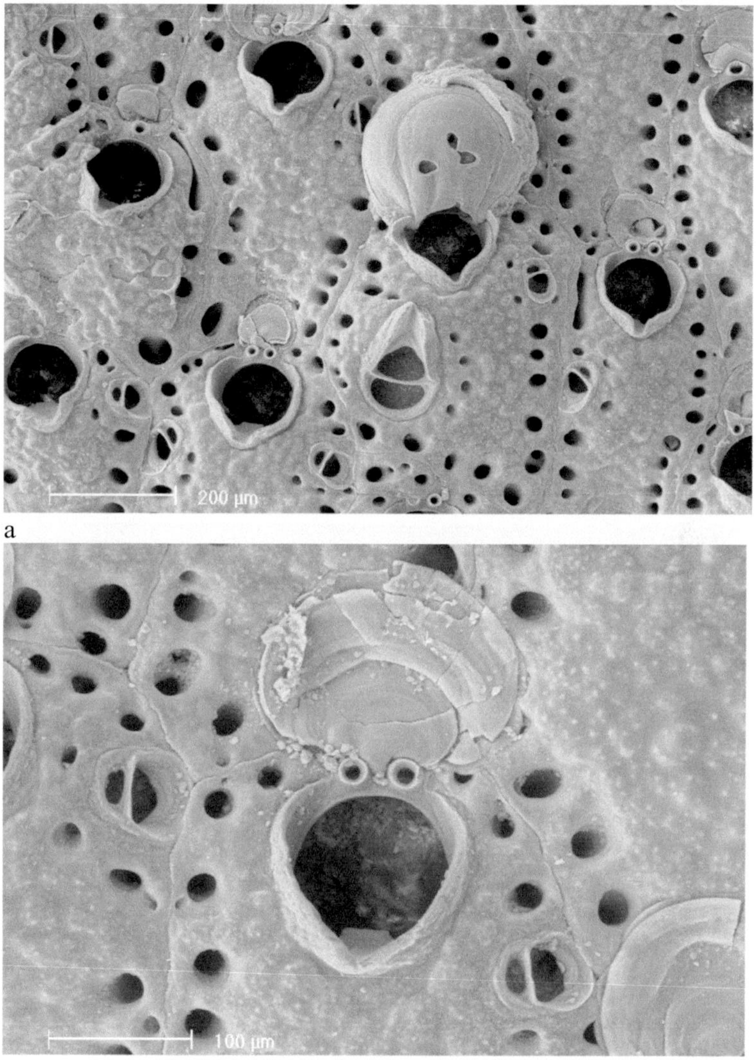

a

b

Fig. 4.130 *Parasmittina trispinosa.* (**a**, **b**) Colony growing on plastic beached in Belgium. (**a**) Autozooids with ovicells and avicularia of two sizes. (**b**) Distal part of autozooid showing developing ooecium; orifice with lyrula and two avicularia

Occurrence

Belgium From 1904 to 1906, several colonies were collected at the Flemish Banks (RBINS collection). In 2004–2005, living colonies were found on shells from the Kwinte Bank and the Westhinder Bank. On washed-up plastic in Zeebrugge in 1999.

The Netherlands Colony in a shell valve on the beach of Nieuwvliet in 2001.

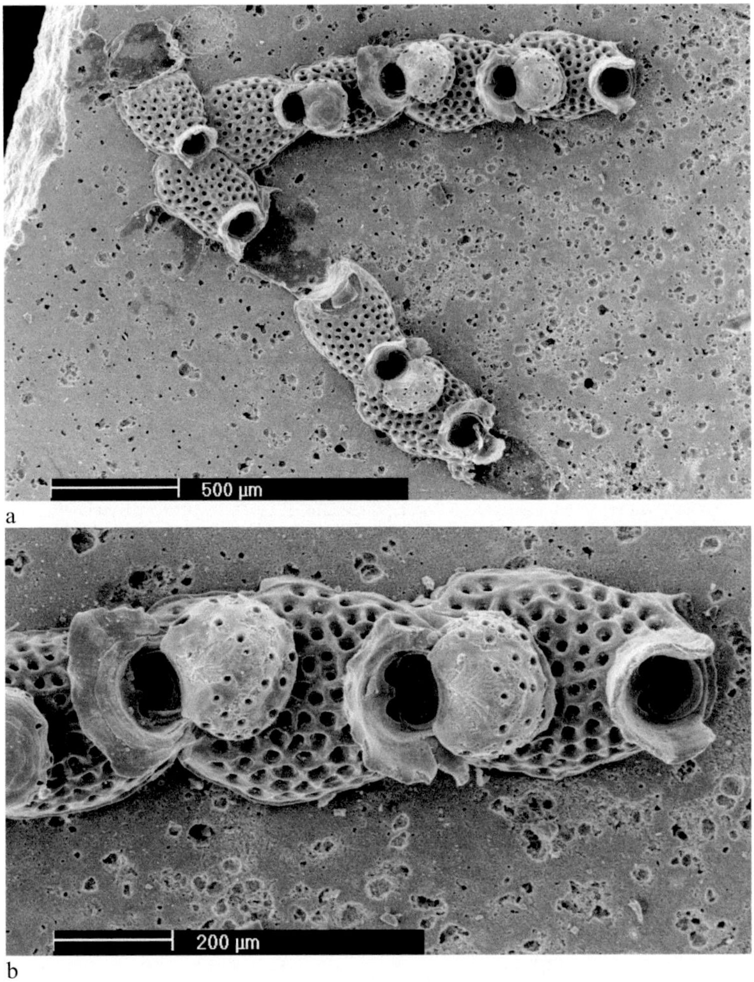

Fig. 4.131 *Phylactella labrosa*. (**a**) Uniserial colony growing in a dead shell from the Kwinte Bank (Belgium). (**b**) Terminal part of the branch showing two ocicellate and a non-ovicellate zooids

Family Bitectiporidae

Genus *Hippoporina*

Hippoporina pertusa (Esper, 1796)

Description Colony encrusting, unilaminar, developing a reddish-orange (in live) spreading sheet. Zooids 0.6–0.7 mm long. Frontal shield rather flat, initially translucent, becoming opaque, evenly pseudoporous by round pores. Proximal to the orifice develops a cruciform median umbo. Orifice wider than long, with concave proximal border, condyles rounded. No spines. No avicularia. Ovicell prominent, not closed by the zooidal operculum, ooecium formed by the distal autozooid, ectooecium regularly pseudoporous. Ancestrula tatiform, membrane underlain by an extensive cryptocyst and surrounded by about seven spines, proximal to the opesia is a smooth gymnocyst (Fig. 4.132)

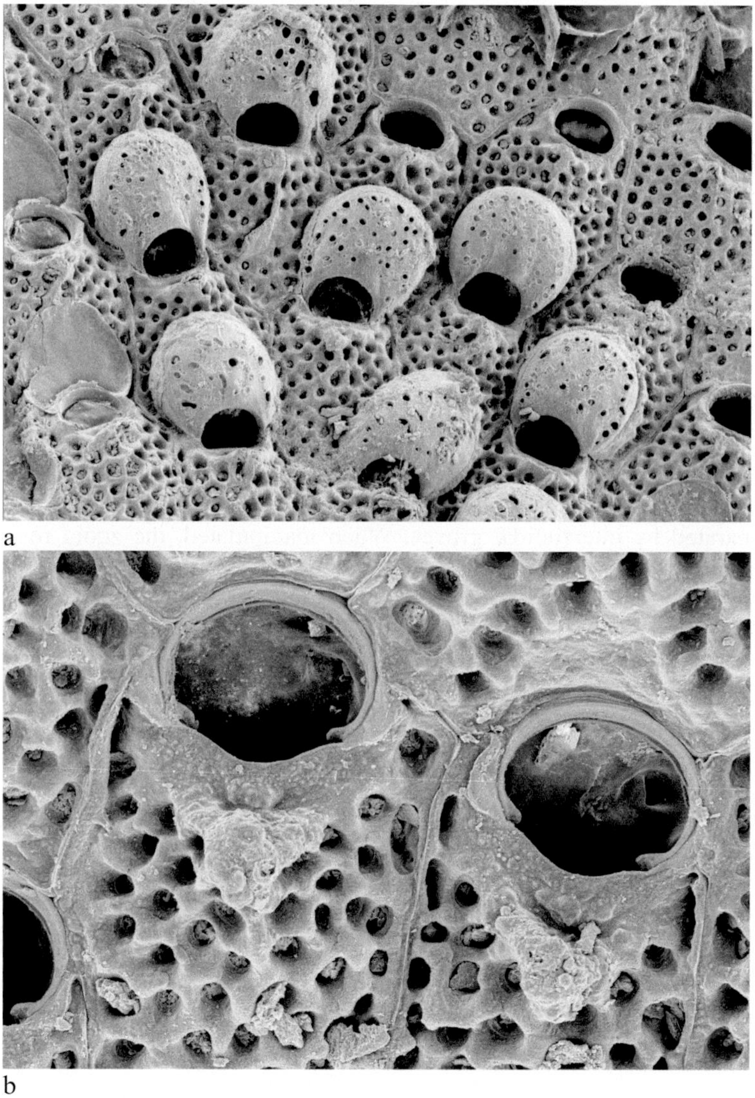

Fig. 4.132 *Hippoporina pertusa*. (**a**) Colony area with ovicellate zooids. (**b**) Two non-ovicellate autozooids showing orifice and cruciform median umbo. Colony from the Flemish Banks (Belgium)

Occurrence

Belgium In 1904 and 1906, a colony was gathered on the Flemish Banks (collection RBINS). In 2004–2005, four colonies were collected at the Westhinder Bank. All colonies had died before the discovery. A colony washed up on an object in September 2001 (De Blauwe 2003a). Small stones were collected by Francis Kerckhof (RBINS) in the extreme north-western part of the BCS, on the border with the United Kingdom on July 12, 2018, during a haul with a Gilson dredge with starting coordinates 51°35′26″ N, 2°15′14″ E and ending point 51°35′36″ N, 2°15′28″ E, 100 stones were examined on epibionts and 9 of them contained living colonies of *H. pertusa*.

The Netherlands On Schouwen (Lacourt 1949), without mentioning whether it concerns washed-up material. De Ruijter (2014b) reports the washing-up in 2013 of a colony on plastic near Egmond aan Zee.

Genus *Metroperiella*

Metroperiella cf. *biformis* (Zhang and Liu, 1995)

Description Colony, encrusting, unilaminar, or forming erect, unilaminar, folded plates. Living orange. Zooids rectangular, arranged in quincuncial series, distinct and separated by interzooidal groves. When manipulated, the zooid rows easily break apart along the vertical walls. The basal surface of zooids from erect plates shows often a pore giving rise to a chitinous rootlet. Orifice oval, broader than long, condyles blunt but serrated, projecting to each other. Peristome thin and somewhat raised. No oral spines. Operculum yellowish, with a chitinous sclerite around its periphery. Frontal wall flat or somewhat raised proximal to the orifice, and uniformly pseudoporous. Suboral avicularia directed proximally, in the plane of the frontal shield or slightly oblique, with a semielliptical or slightly spatulate mandible. Crossbar narrow and complete. Avicularia may be absent in parts of the colony. Ovicells prominent, ooecium usually longer than wide, globose, ectooecium with numerous pseudopores. Ancestrula not observed in European material (Fig. 4.133).

Occurrence In Europe known since 2018 from Stavanger (coll. Marco Faasse) and Huldra oil field (2019, pers. comm. Mali Ramsfjell) (Norway) and since 2020 from one wind farm in the German North Sea.

Belgium Not known.

The Netherlands Not known.

Germany Sampled in September 2020 and October 2021 in one wind farm in the German Bight located at 53.98 N–6.84 E by Lisa Schüller (IfAÖ).

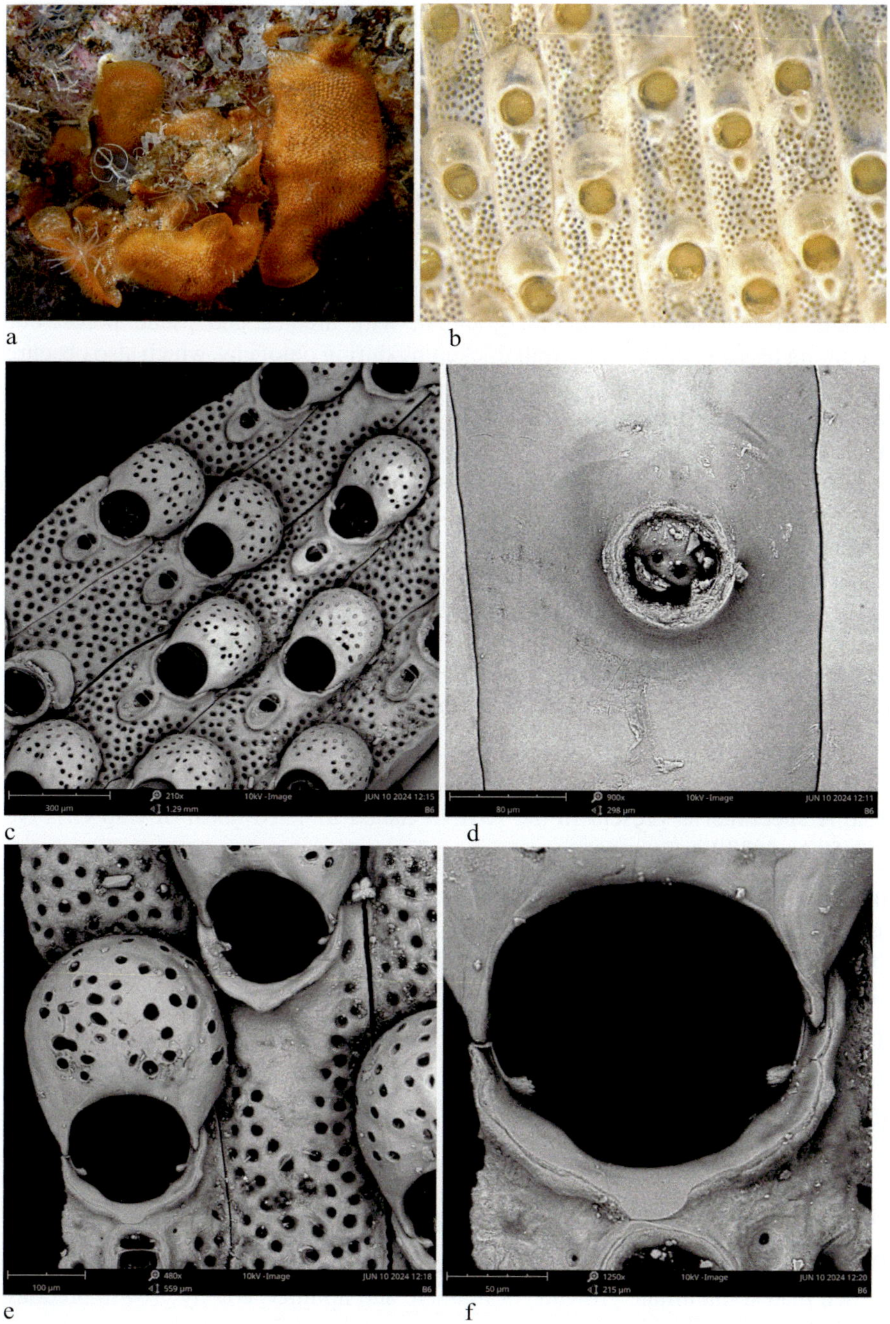

Fig. 4.133 *Metroperiella* cf. *biformis*. (**a**) Living colony in situ at Stavanger (Norway, photo by M. Faasse). (**b**) Colony area with ovicellate zooids, Stavanger. (**c–f**) SEM images of a colony from the German Bight showing. (**c**) Ovicellate zooids with avicularia. (**d**) Basal surface of unilaminar erect colony with pore plate. (**e**) Ovicellate zooids. (**f**) Orifice of ovicellate zooid with serrated condyles

Genus *Pentapora*

Pentapora foliacea (Ellis and Solander, 1786)

Description Colony at first encrusting, unilaminar, later developing erect, stiff, folding, anastomosing, bilaminar plates forming a large, three-dimensional form reaching a diameter of 20 cm. Living orange, dead yellowish. Zooids 0.7–0.9 mm long, arranged in straight rows, alternating with the zooids in adjacent rows. Orifice longer than wide, oval or angular, with a pair of condyles and a convex proximal margin. Frontal shield lepralioid, granular, with areolae and pseudopores, both becoming less distinct through wall thickening during ontogeny, which also obscures zooidal boundaries. Primary orifice with a pair of downturned condyles between anter and poster; secondary orifice slightly oval to trifoliate because of the development of lappets. Operculum brown, lustrous. No oral spines. Ovicell prominent, ooecium elliptical, with a few scattered pseudopores arranged in a band proximally above the orifice, becoming totally overgrown by secondary calcification. Avicularia adventitious, suboral, usually placed on an umbo, inconspicuous, small, longer than wide, rostrum semielliptical. Avicularia monomorphic and small, giant avicularia not observed and presumed lacking (Lombardi et al. 2010), but here reported for the first time as dimorphic, see Fig. 4.134c. Avicularia absent on many zooids. Confusion of young encrusting colonies with *Cryptosula pallasiana* is possible, the shape of the proximal edge of the orifice is an important distinction (Fig. 4.134).

Occurrence Forms large three-dimensional colonies on stones in the English Channel. The area of distribution reaches just into the southern bend of the North Sea. A specimen is present in the RBINS fished in 1910 north of Cap Gris-Nez (50°59′N, 1°33′E).

Belgium A small encrusting, unilaminar colony washed up in Wenduine on September 15, 2001, attached to an object made of plastic and wood (De Blauwe 2003a). Two colonies washed up on a lobster pot at De Panne in January 2024 (pers. comm. Wilfried Laforce).

The Netherlands De Ruijter (2014b, c, 2015b, 2017c, 2021a, c) reports washing-up on plastic.

Germany Not reported.

Genus *Schizomavella* (adapted to Reverter-Gil et al. 2015)

Diagnosis Colony encrusting, unilaminar to multilaminar, sheet-like or developing folded plates, mamillate growths, or massive and nodular; colony growth predominantly by zooidal budding, occasionally by frontal budding. Autozooids with cryptocystidean frontal shield, with a variable number of small pseudopores, plus a row of peripheral areolae. Communication via small uniparous septula. Primary orifice

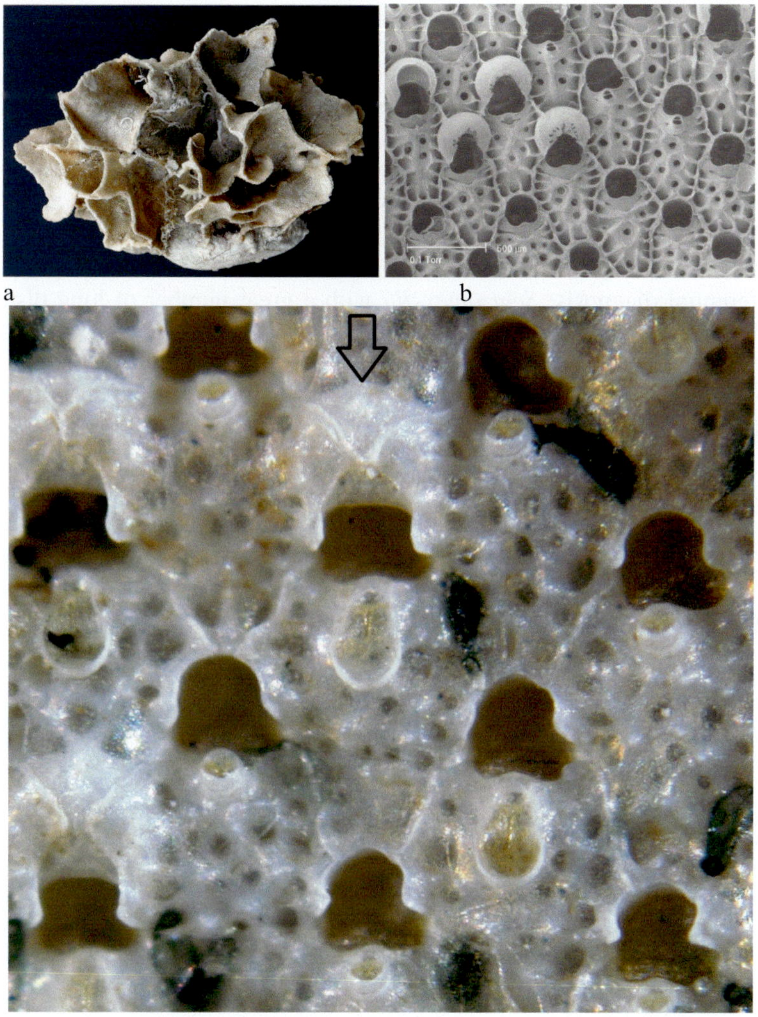

a b c

Fig. 4.134 *Pentapora foliacea.* (**a–c**) Colonies from Brittany (France). (**a**) A small colony on pebble. (**b**) Young autozooids with developing ovicells. (**c**) Colony area showing ovicellate (*arrow*) and non-ovicellate zooids with normal and enlarged avicularia

sinuate, condyles present, often pronounced. Oral spines present though often lost during ontogeny or absent from adult zooids in some species. Avicularia adventitious, monomorphic or polymorphic, suboral or lateral to orifice. Ovicell prominent, often becoming subimmersed by secondary calcification during ontogeny, not closed by autozooidal operculum, ectooecium with conspicuous frontal pseudopores. Ancestrula tatiform or with a well-developed proximal cryptocyst forming a pseudosinus.

Subgenus *Schizomavella*

Differential diagnosis: *Schizomavella* with few oral spines (with some exceptions), usually present only in early ontogeny, absent in ovicellate zooids. Periorbital region not raised. Avicularia monomorphic, single, suboral, rarely double, lateral; rostrum usually oval, rarely pointed. Ooecium marginally covered by a thick layer of secondary calcification, with a proximal area of exposed pseudoporous ectooecium; or later almost completely covered by a thick, nodular secondary calcification that does not occlude the pseudopores.

Subgenus *Calvetomavella*

Differential diagnosis: *Schizomavella* with a great number of long distolateral oral spines forming a semicircle, present in almost all zooids throughout ontogeny; one or two pairs in ovicellate zooids, or replaced by a peristome during ontogeny. Periorbital region usually imperforated and raised. Reduced secondary calcification of frontal shield in comparison with shallow-water *Schizomavella* s.s. Avicularia paired, distolateral, often enlarged and parallel to the distal zooidal margin, frequently asymmetrical developed, small if suboral; rostrum generally pointed, small avicularia occasionally oval. Ovicells prominent. Exposed ectooecial area large, evenly distributed circular or irregular shaped pseudopores, never thickly covered or immersed by secondary calcification during ontogeny; if present, the secondary calcification forms a narrow periooecial rim.

1 (a) Autozooids with six to seven oral spines. .... *Schizomavella (Calvetomavella) biancae* (p. 327)
(b) Less than six oral spines. 2

2 (a) Orifice with a very high peristome proximally in which a pointed avicularium occupies a perpendicular position to the frontal surface. .... *Schizomavella cristata* (p. 330)
(b) Orifice with a very wide and shallow sinus. Avicularium peripendicular to the zooidal plane on the side of a high spiny suboral umbo (see also *Schizomavella (Schizomavella) linearis*). .... *Schizomavella (Schizomavella) hastata* (p. 333)
(c) Not so.. 3

3 (a) Avicularium pointed. One or two avicularia proximolateral to the sinus and directed toward the center of the zooid or one on the side of a suboral umbo and perpendicular to the zooidal surface. Sometimes a large avicularium occurs between two successive zooids, as wide as the zooids, and looks like a slit divided transversely in two. .... *Schizomavella* (*Schizomavella*) *linearis* (p. 333)
(b) Avicularium rounded. Avicularia perpendicular to the frontal surface, in the proximal edge of the orifice or more parallel to the zooidal surface but directed proximally. 4

4 (a) Orifice tear-shaped, widest distally. Condyles with a distally directed tip......................... *Schizomavella (Schizomavella) auriculata* (p. 328)
(b) Orifice horseshoe-shaped, with parallel sides. Never a distal point on the condyles. 5

5 (a) Avicularia short or long, with Y-shaped opening. .... *Schizomavella (Schizomavella) cornuta* (p. 330)
(b) Avicularia short with a round opening. .... *Schizomavella (Schizomavella) teresae* (p. 336)

Schizomavella (Calvetomavella) biancae Pica and Berning, 2022

Description Colony encrusting, unilaminar, multiserial. Autozooids rectangular to irregular polygonal, slightly convex, and separated by a fine suture. Autozooidal frontal shield with a coarsely granular to nodular surface and few small pseudopores around a central area. Orifice suborbicular in shape and with a narrow and deeply U-shaped sinus and broad condyles sloping toward the sinus. Six or seven oral spines. Avicularia dimorphic: large avicularia, single or paired, situated along distolateral margins; oblong in outline, i.e., rostrum distally rounded with slightly raised and serrated edges, palate mostly calcified by a pair of cryptocystal shelves forming central ridges and an elongate foramen at about mid-distance, crossbar with tiny columella. Small avicularia single and usually suboral or less frequent distolateral, oval to oblong in outline, slightly raised distal part of rostrum serrated, distal foramen Y-shaped, crossbar with tiny columella. Ovicells prominent, with large globular ooecium, often occupying over half the frontal area of the distal zooid, ectooecium almost entirely exposed apart from a narrow peripheral rim and with c. 25 rimmed tubaeform pseudopores. Ovicellate zooids form a peristome with a large U-shaped suboral notch. Ancestrula oval, longer than wide, opesia mushroom-shaped surrounded by nine or ten mural spines (Fig. 4.135).

Occurrence

Belgium A deceased colony was found on the concave side of a shell valve on the Westhinder Bank in 2005, reported in De Blauwe (2009a) as *Schizomavella discoidea* (Pica et al. 2022).

The Netherlands A small colony on beached plastic in 2023 collected by Rien de Ruijter at Egmond aan Zee, unpublished.

Germany Twelve colonies reported from the Tiefe Rinne (Kuhlenkamp and Kind 2012), probably misidentified (pers. comm. B. Kind).

France On *Chlamys distorta,* with ovicells and peristome, at the quai of Le Conquet (July 2002), and on a lobster pot at St-Malo (2005), coll. H. De Blauwe.

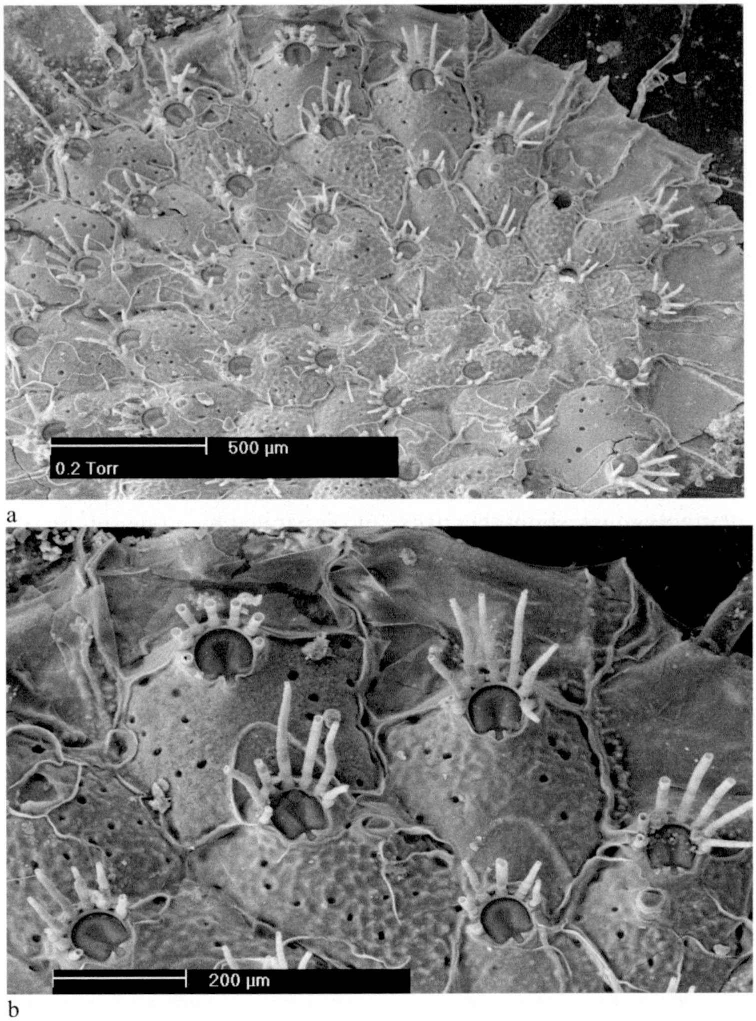

Fig. 4.135 *Schizomavella (Calvetomavella) biancae.* (**a**, **b**) Non-ovicellate zooids with long oral spines, lobster pot, St-Malo (France)

Schizomavella (Schizomavella) auriculata (Hassall, 1842)

Description Colony encrusting, unilaminar to multilaminar, sheet-like. Autozooids in radiating linear series, rectangular or polygonal, 0.4–0.7 mm long. Frontal shield cryptocystidean, evenly pseudoporous by small pseudopores plus a row of peripheral areolae. Frontal wall becoming nodular, frequently with a small umbo proximal to the orifice. Orifice drop-shaped, widest distally, with a shallow sinus in the proximal edge. The condyles have a sharp triangular tip on their distal side.

Some zooids can bear one to three oral spines. Avicularia flat or oblique on the frontal surface, directed proximally, just proximal to the sinus. Ovicell not closed by autozooidal operculum, initially prominent, further becoming covered by a thick layer of secondary calcification, resulting in a reduced proximal area of exposed pseudoporous ectooecium (Fig. 4.136).

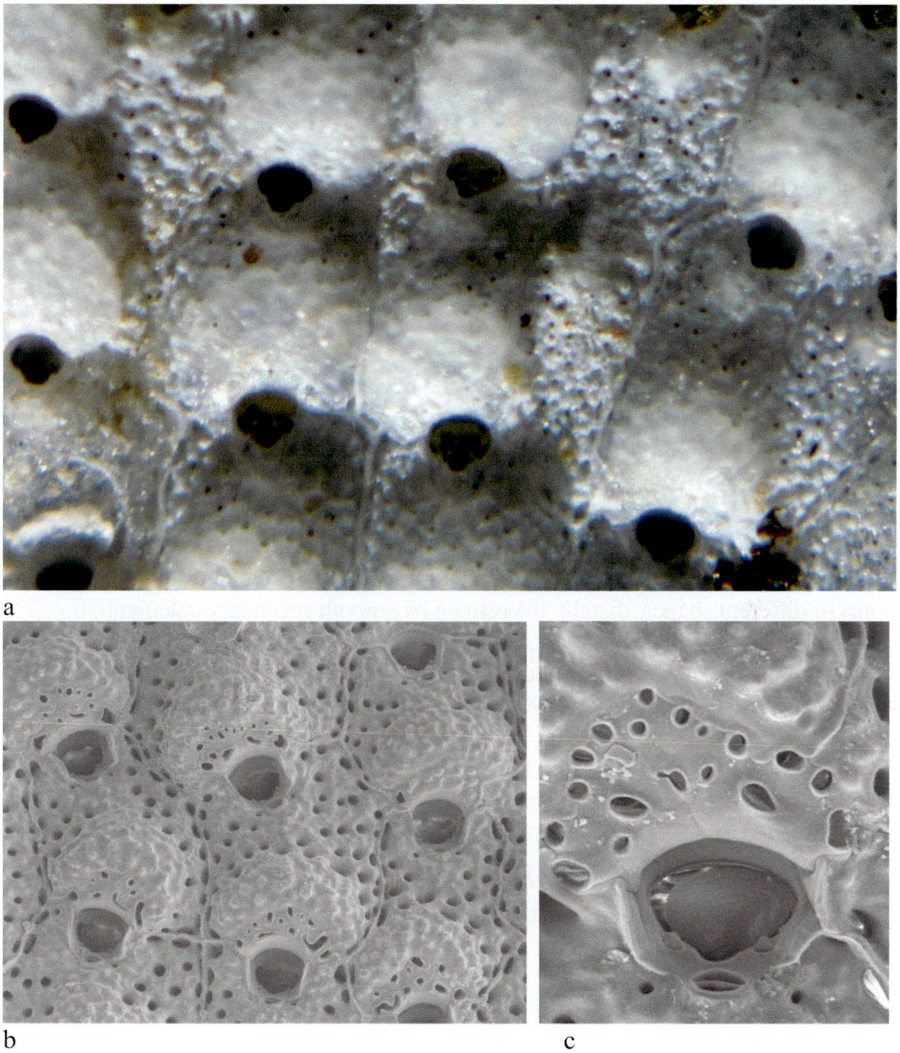

Fig. 4.136 *Schizomavella (Schizomavella) auriculata.* (**a**) Part of the colony of ovicellate zooids, on beached plastic, Wales (United Kingdom). (**b**, **c**) Ovicellate autozooids, on beached plastic, Koksijde (Belgium). (**c**) Detail of ovicellate zooid showing pseudoporous part of ectooecium, and zooidal orifice with pointed condyles

Occurrence

Belgium Loppens (1906) reports this species as rare on shells offshore, but he was then unable to distinguish it from *Schizomavella (Schizomavella) teresae*, which was only described in 1996.

Lives on stones at the Westhinder Bank. Rare on washed-up plastic (Blankenberge January 15, 2001). Small stones were collected by Francis Kerckhof (RBINS) in the extreme north-western part of the Belgian continental shelf, on the border with the United Kingdom on July 12th 2018 during a haul with a Gilson dredge with starting coordinates 51°35′26″ N, 2°15′14″ E and ending point 51°35′36″ N, 2°15′28″ E, 100 stones were examined on epibionts and 5 of them contained colonies of *S. auriculata.*

The Netherlands De Ruijter (2014b, 2015a, d, 2017b, 2019c) reports a colony washing-up on plastic.

Schizomavella cristata (Hincks, 1879)

Description Colony encrusting bivalve shells, unilaminar, forming a small patch. Autozooids oval to hexagonal, up to 0.4 mm long. Frontal shield cryptocystidean, pseudoporous, with scattered pseudopores. Orifice as wide as long with shallow sinus. The condyles give the sinus an angular impression. Four or five slender oral spines in zooids without ovicell. Peristome present proximal to the spines, laterally thin and somewhat flared, proximally highest with a perpendicularly placed avicularium mid-proximal to the opening. Rostrum and mandible sharply triangular and directed upward. Ovicell prominent, ooecium somewhat wider than long, evenly pseudoporous with pseudopores of different shapes, distally overgrown by smooth secondary calcification.

Hincks' type material, which has been lost, is described as having more and smaller pores on zooids and pseudopores on ovicelles and an "angular" sinus. It is assumed here to be the same species that shows some variability in pores (Fig. 4.137).

Occurrence

Belgium Of seven colonies on shell valves from the Westhinder Bank in 2005, five were collected alive.

The Netherlands A *Schizomavella* cf. *cristata* washed up on plastic near Petten in November 2016 (de Ruijter 2017b).

Schizomavella (Schizomavella) cornuta (Heller, 1867)

Note Reported as *Schizomavella cuspidata* (Hincks, 1880) in De Blauwe (2009a), this name is now unvalid.

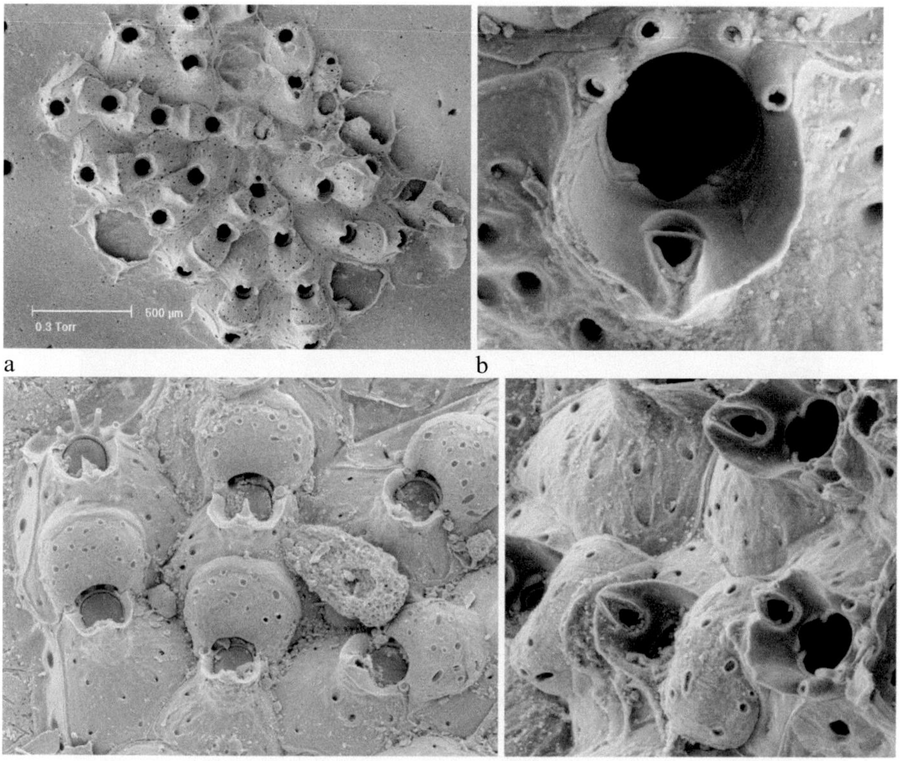

a b c d

Fig. 4.137 *Schizomavella cristata.* (**a–d**) Colony growing in a dead shell from the Westhinder Bank (Belgium). (**a**) Entire colony. (**b**) Orifice showing condyles and suboral avicularium. (**c**) Ovicellate zooids. (**d**) Oblique distolateral view on the zooidal orifices and peristomes

Description Colony encrusting, forming an extensive unilaminar or multilaminar sheet. Autozooids in radiating linear series, angular, 0.4–0.6 mm long. Orifice as long as wide with angular transition from the lateral sides to the proximal side. The sinus is rather small U-shaped but apparently enlarged by the condyles that protrude beyond the sides of the sinus and somewhat constrict it. Mostly three but up to five oral spines, lost in later ontogeny. Autozooids with cryptocystidean frontal shield, with a variable number of frontal pseudopores, a row of peripheral areolae, and often a suboral umbo. Avicularia always just proximal to the sinus. Two shapes are possible: either small and oblique or perpendicular to the zooidal plane on the distal side of the suboral umbo, or larger and parallel to the frontal shield. Large avicularia have a Y-shaped opening.

Ovicell prominent, ooecium wider than long, frontally somewhat flattened, with large scattered pseudopores, in later ontogeny covered by the secondary calcification with nodular surface (Fig. 4.138).

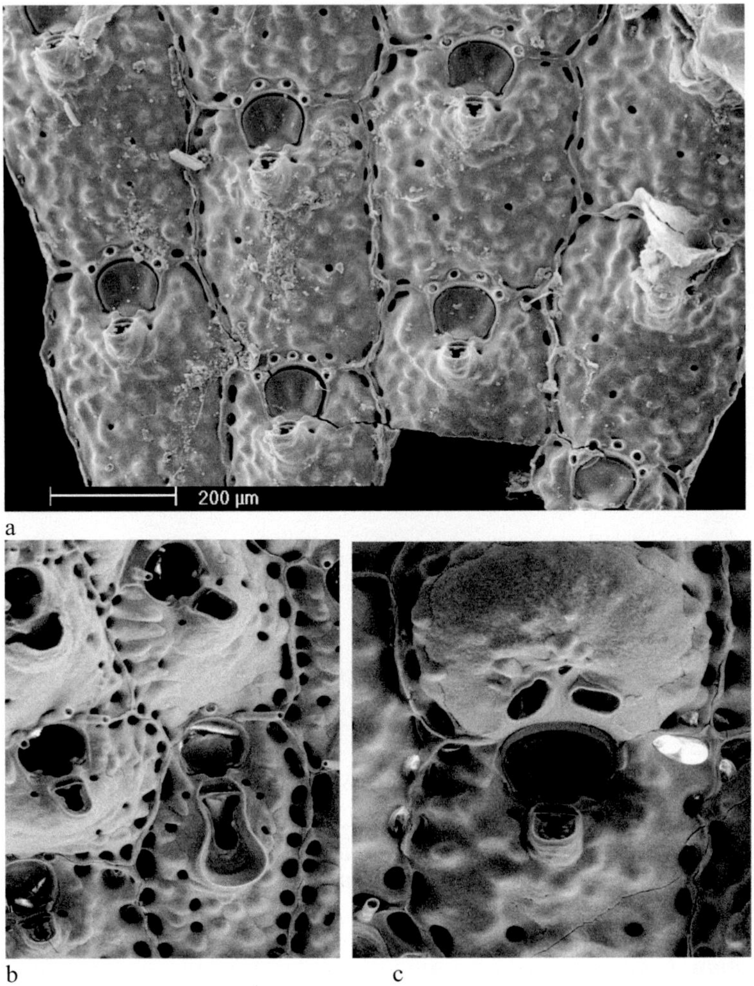

Fig. 4.138 *Schizomavella (Schizomavella) cornuta.* (**a**–**c**) colony growing on a lobster pot at St-Malo (France). (**a**) Part of the colony of non-ovicellate autozooids. (**b**) Autozooids with small and enlarged avicularium. (**c**) Ovicellate autozooid, ectooecium overgrown by the secondary calcification and two large pseudopores

Occurrence

Belgium Only known from washed-up plastic, in Blankenberge in 1999.

The Netherlands Beached near Egmond aan Zee and Petten (de Ruijter 2012a, 2013c, 2014b, 2018a, 2023).

Schizomavella (Schizomavella) hastata (Hincks, 1862)

Description Colony forming an extensive unilaminar or multilaminar sheet, encrusting but sometimes erecting from the substrate. Zooids angular, 0.4–0.7 mm long. Orifice wider than long, between the condyles the proximal border is shallow concave, so that there is hardly a sinus. This is the main distinction from *Schizomavella (Schizomavella) linearis* which can develop a similar suboral umbo with avicularium. Young zooids may bear two to four oral spines. Autozooids with cryptocystidean frontal shield, at first smooth, later coarse with large round pseudopores. A large, spiky umbo can develop very high. An avicularium with an elongated mandible is present on the left or right side of the spiky umbo and is upward directed. The polypide has 13 tentacles. Ovicell prominent, not closed by autozooidal operculum; ooecium frontally flattened, ectooecium with about 20 irregular pseudopores becoming overgrown by secondary calcification and a tall conical umbo during ontogeny. Embryos yellow or orange (Fig. 4.139).

Occurrence

Belgium In Zeebrugge on plastic washed up on November 28, 1999, and July 18, 2000, and in October 2004 on an empty oyster doublet between Oostduinkerke and De Panne (De Blauwe 2005a).

The Netherlands On plastic (de Ruijter 2017b, c, 2023).

Schizomavella (Schizomavella) linearis (Hassall, 1841)

Description Colony forming an extensive unilaminar sheet, encrusting but sometimes partly erect. Colonies are light pink to red, and washed-up colonies are white. Autozooids 0.4–0.7 mm long, angular, in radiating linear series. Orifice with a U-shaped sinus, clearly marked by the condyles, half as width as the orifice. Young zooids bear two oral spines. Frontal shield cryptocystidean, pseudoporous by a variable number of pseudopores and a row of peripheral areolae. This species shows a wide variety of avicularia. Usually, one or two avicularia lie proximolateral to the orifice and are directed obliquely toward the midline of the zooid. A third avicularium may exist elsewhere on the frontal surface. Another possibility (especially in colonies from Zeeland) is that one avicularium occurs on the side of a suboral umbo, almost perpendicular to the zooid surface. In this case, confusion with *Schizomavella (Schizomavella) hastata* is possible and the shape of the sinus is decisive for identification. In some zooids, but not in all colonies, there is a large slit-like avicularium between two successive zooids, which is divided transversely. The presence of such avicularium is of great help in identification. Ovicell prominent, not closed by autozooidal operculum; ectooecium with irregular frontal pseudopores, becoming enclosed by secondary calcification. Embryos orange-red (Fig. 4.140).

Occurrence Often present on washed-up plastic in Belgium and the Netherlands.

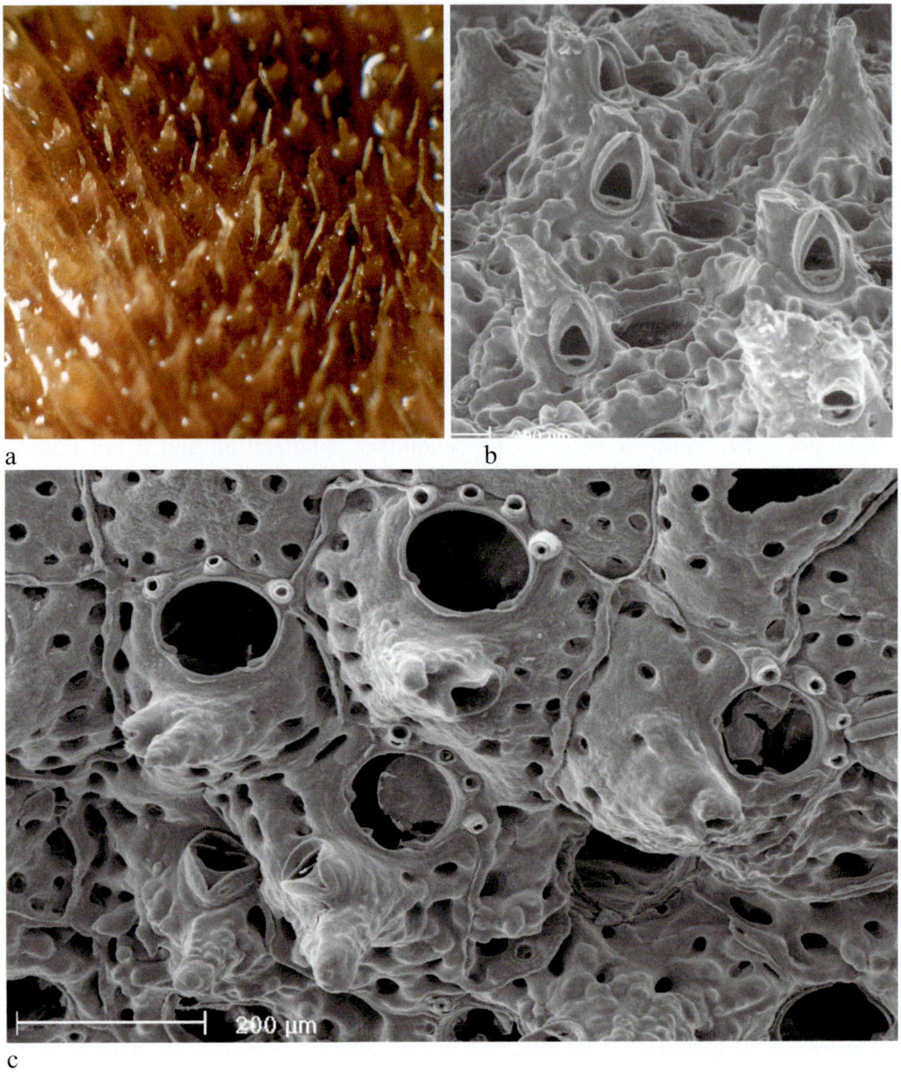

Fig. 4.139 *Schizomavella (Schizomavella) hastata*. (**a**) Living colony from Brittany (France). (**b**) Side view on the colony surface showing suboral avicularia on high proximal spiky umbo. (**c**) Umbo, oral spines, orifices with broad, shallow sinus and suboral avicularia

Belgium In 1902, a colony was collected at "de Spleet," another at the Stroom Bank, and in 1904, a colony at 51°24′ N, 2°28′ E (RBINS collection). At least 23 colonies were collected at the Westhinder Bank in 2005, of which two were alive. A living colony was collected in the Baai van Heist in March 2016. Small stones were collected by Francis Kerckhof (RBINS) in the extreme north-western part of the Belgian continental shelf, on the border with the United Kingdom on July 12, 2018,

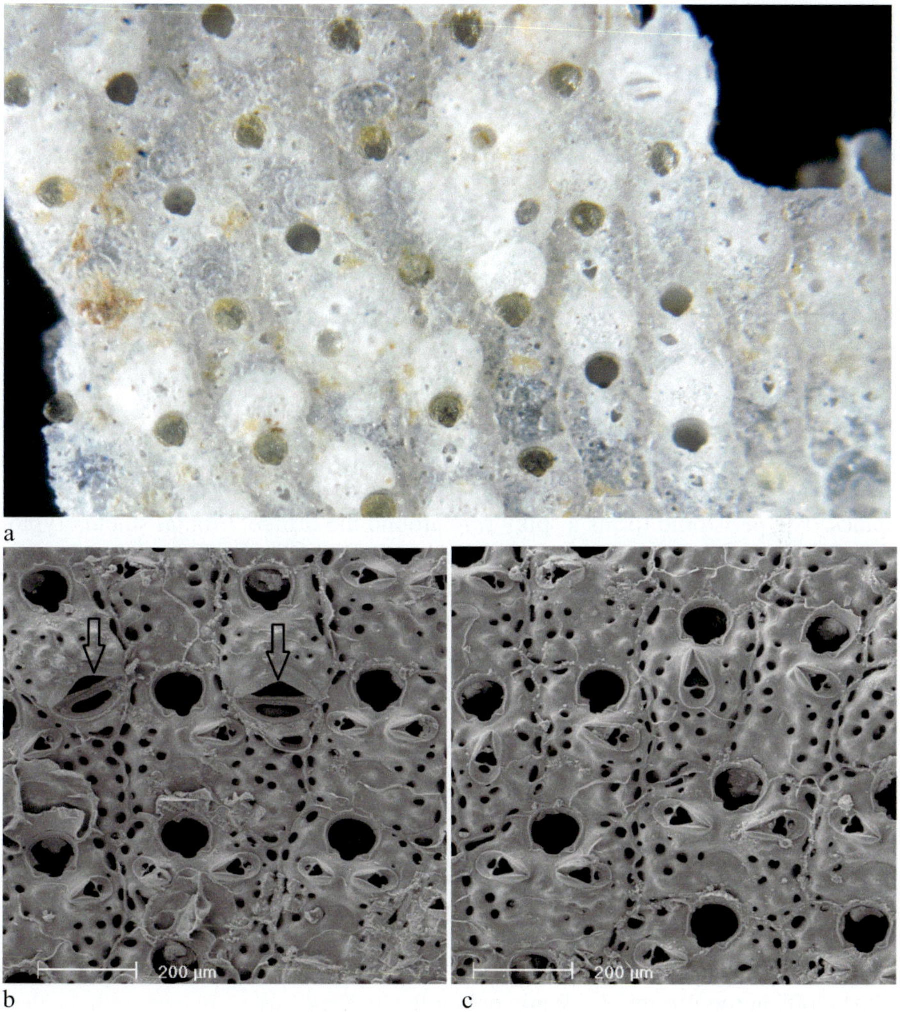

Fig. 4.140 *Schizomavella (Schizomavella) linearis.* (**a**) Part of colony growing on beached plastic, Belgium. (**b–c**) Zooids with normal and large (*arrows*) as well as single and paired avicularia, pointing in different directions

during a haul with a Gilson dredge with starting coordinates 51°35′26″ N, 2°15′14″ E and ending point 51°35′36″ N, 2°15′28″ E, 100 stones were examined on epibionts and 2 of them contained colonies of *S. linearis*.

The Netherlands Lacourt (1949) reports this species from Katwijk (washed up?).

Reported from the Eastern Scheldt (Lacourt 1949; de Kluijver 1997), in the Westerschelde estuary between Westkapelle and the port area "Vlissingen-Oost" (Faasse and De Blauwe 2004), on the Cleaver Bank (pers. comm. G. van Moorsel), at the Dutch Borkum Reef Grounds (Coolen et al. 2015). On shipwrecks in Dutch waters, namely on the HMS Hogue, HMS Scott, and SS Nautilus (Lengkeek et al. 2013). Washed up in July 2020 on Texel on a *Crepidula* hanging on *Sargussum* (Rob Dekker, NIOZ, det. HDB).

Germany Collected in the Tiefe Rinne (Kuhlenkamp and Kind 2012). Collected from the wreck of the SMS Mainz (54°00.143 N, 06°43.847 E) in 2021 (van Leeuwen 2023).

Schizomavella (Schizomavella) teresae Reverter-Gil and Fernandez Pulpeiro, 1996

Description Colony an encrusting, unilaminar, small sheet. Autozooids 0.4–0.5 mm long, angular, in radiating series. Orifice more or less round with a small sinus and broad condyles that project over the sinus. Young zooids can bear two or three oral spines. The peristome forms a spiky ring around the orifice. Frontal shield cryptocystidean, nodular to spiky, pseudoporous by relatively large and few pseudopores and a row of peripheral areolae. Avicularia perpendicular to the zooidal plane, just proximal to the sinus in the interior of the peristome. Distinguishable from *S. cornuta* by the more rounded opening, the small sinus, and the shape of the condyles. Ovicell prominent, ectooecium with few large frontal pseudopores. Embryos pink (Fig. 4.141).

Occurrence

Belgium Probably very common on stones near the Westhinder Bank. It is advisable to bleach the colonies and to take pictures with an electron microscope in order to arrive at a reliable identification. Confusion with *S. cornuta* and *S. auriculata* is not imaginary. The collection of the RBINS may contain many colonies of this species, collected between 1899 and 1911 on the Flemish Banks. These colonies were not identified at the species level during the revision. Of the material collected in 2004–2005 at the Westhinder Bank, only a few colonies were imaged with an electron microscope. Small stones were collected by Francis Kerckhof (RBINS) in the extreme north-western part of the Belgian continental shelf, on the border with the United Kingdom on July 12, 2018, during a haul with a Gilson dredge with starting coordinates 51°35′26″ N, 2°15′14″ E and ending point 51°35′36″ N, 2°15′28″ E; 100 stones were examined on epibionts, and seven of them contained living colonies of *S. teresae*.

The Netherlands A colony (cf) washed up near Bergen aan Zee on nylon rope (de Ruijter 2011a) and on plastic (de Ruijter 2014b).

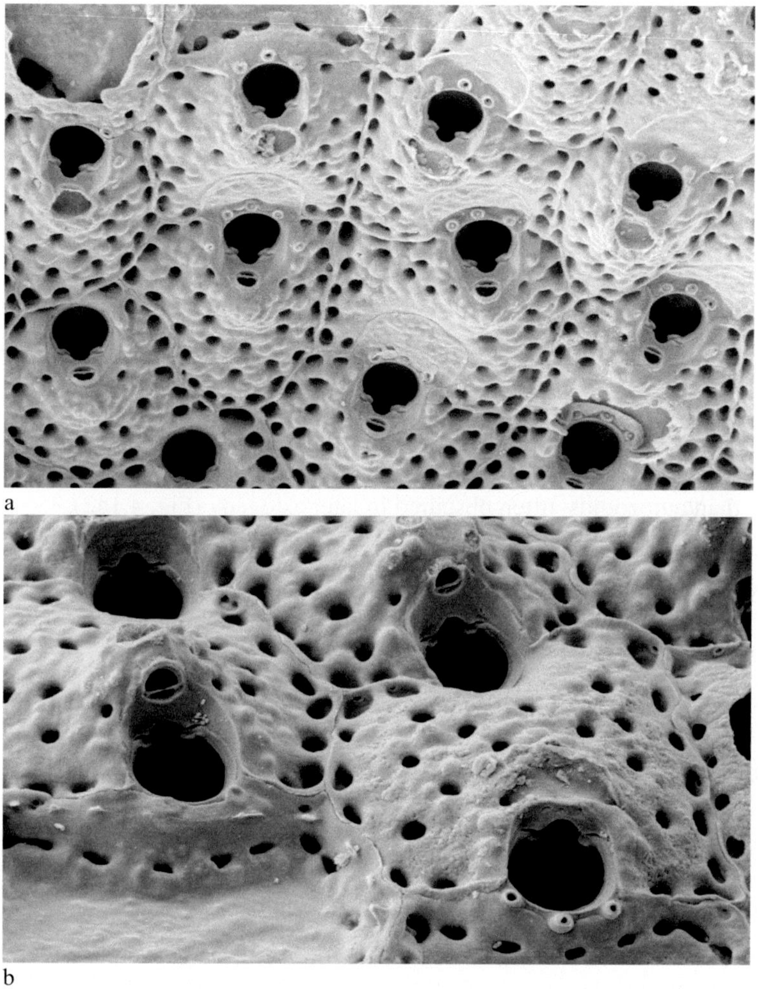

Fig. 4.141 *Schizomavella (schizomavella) teresae*. (**a**) Portion of colony growing on a dead shell, Westhinder Bank (Belgium). (**b**) Peripheral zooids of the same colony showing zooidal orifices with oral spines and suboral avicularium

Family Watersiporidae

Genus *Watersipora*

More invasive *Watersipora* species might arrive in the area covered by this book; for identification, see Vieira et al. 2014a. At this moment, only one species is recorded here.

Watersipora subatra (Ortmann, 1890)

Description Colonies encrusting, multiserial, uni- to multilamellar; sometimes erect, foliaceous, and bilamellar; color from orange to brownish-purple or grayish to black. Zooids subrectangular to hexagonal, about twice as long as wide, separated by slightly raised lateral walls. Cryptocystidean frontal shield slightly convex, with numerous round pseudopores; two large latero-oral intrazooidal septula near lateral zooidal margin, proximolateral to the orifice, each with three to eight (often five) small pores. Frontal shield obscured by translucent cuticle. Orifice subcircular to oval, slightly wider than long, with broad, well-defined proximal, U-shaped sinus; orificial rim thin, sometimes slightly raised; narrow bar-shaped condyles, sometimes inconspicuous. Operculum reddish-brown to black in color with broad, biconcave proximal band, without lucidae. Spines absent. Multiparous mural pore plates in distolateral and transverse distal walls. Avicularia absent. Ovicells absent, embryos incubated in the internal brood sac. Ancestrula schizoporelloid, single, smaller than autozooids, often obscured in later astogeny (Fig. 4.142).

Occurrence Japan, Australia, New Zealand, California, and on the Atlantic coast of Europe (Ryland et al. 2009; Kelso and Jackson 2012; Vieira et al. 2014a; Aleman et al. 2018). In Brittany (France) in shaded places from the low water mark sublittoraly since 1999 (De Blauwe 2000d, 2003b, 2005b).

Belgium Washes ashore on a *Himanthalia* button and plastic since October 2011 originating from the English Channel (De Blauwe 2011).

The Netherlands Washes ashore since 2009 (Faasse and de Ruijter 2010).

Established in the Grevelingen: On wood in the northern marina of the island of Ossenhoek in the Grevelingen, September 2018 (Aleman et al. 2018) and October 2020 (de Ruijter 2021a); in 2023, on stone on a breakwater of the Springersdiephaven, in 2024, at the Brouwersdam (pers. comm. Ad Aleman).

Germany In 2012 on floating or stranded holdfast of the non-indigenous *Himanthalia elongata* on the island of Heligoland (Kuhlenkamp and Kind 2013). Few colonies have established on mussels on pontoons in the south harbor of Heligoland in 2022, already common in the harbor in October 2023 (pers. comm. F. Hofmann).

Genus *Terwasipora*

Terwasipora complanata (Norman, 1864)

Description (after Reverter-Gil and Souto 2019). Colony encrusting, unilaminar to multilaminar by overgrowing, forming broad, irregular, whitish to cream-colored incrustations, mainly on shells and rocks. Autozooids subrectangular to hexagonal or rhomboidal, 0.6–0.85 mm long, in alternating series. Frontal shield flat, crypto-

a

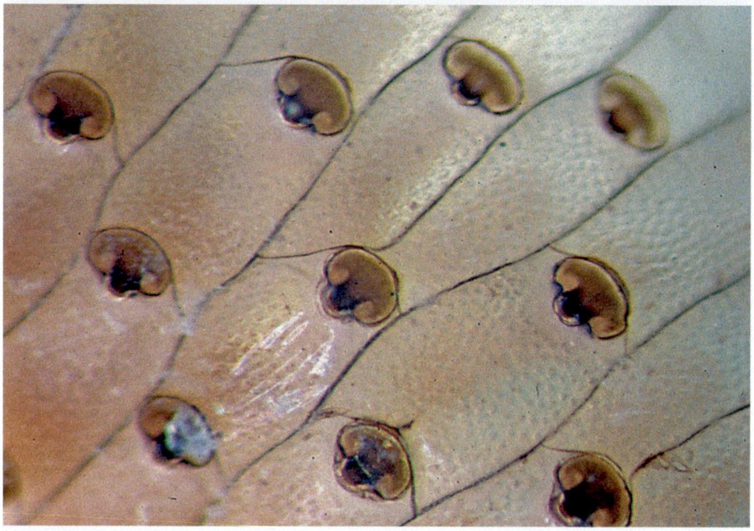

b

Fig. 4.142 *Watersipora subatra.* (**a**) Living colony in Brittany (France, photo by M. Faasse). (**b**) Autozooids showing typical dark operculum (photo by M. Faasse)

cystidean, closely punctured by small rounded pseudopores, and covered by a thin, transparent cuticle. Frontal shield and orifice surrounded by a continuous strip of smooth, gymnocystidean calcification. Primary orifice campanulate, wider than long; anterior margin D-shaped, with edge slightly raised; proximal border straight or slightly convex. Two small latero-proximal condyles. Operculum pale brown, with a distal sclerite and two proximal lucidae adjacent to the condyles. Successive gymnocystal calcification of the orifice rim produces a raised, recurved ridge which, in the corners between it and the lateral walls, may overarch to give an impression of shallow opesiular indentations, which actually correspond to latero-oral intrazooidal septula, visible internally; in later ontogeny, a pair of knob-like umbones proximolateral to orifice may appear, or a single, medioproximal triangular umbo; and, especially in older zooids, a broadened triangular plateau mid-distally, which has been confused in the past with an ovicell. Multiparous mural pore plates in distolateral and transverse distal walls. Avicularia and spines absent. Ovicells absent, embryos incubated in the internal brood sac. Ancestrula tatiform, oval, about half the size of autozooids, with lateral walls calcified, very thin, and frontal surface completely membranous. Distal wall forming an arch coextensive with the operculum. Outline asymmetrical, with the lateral wall of the side of the first bud more curved (Fig. 4.143).

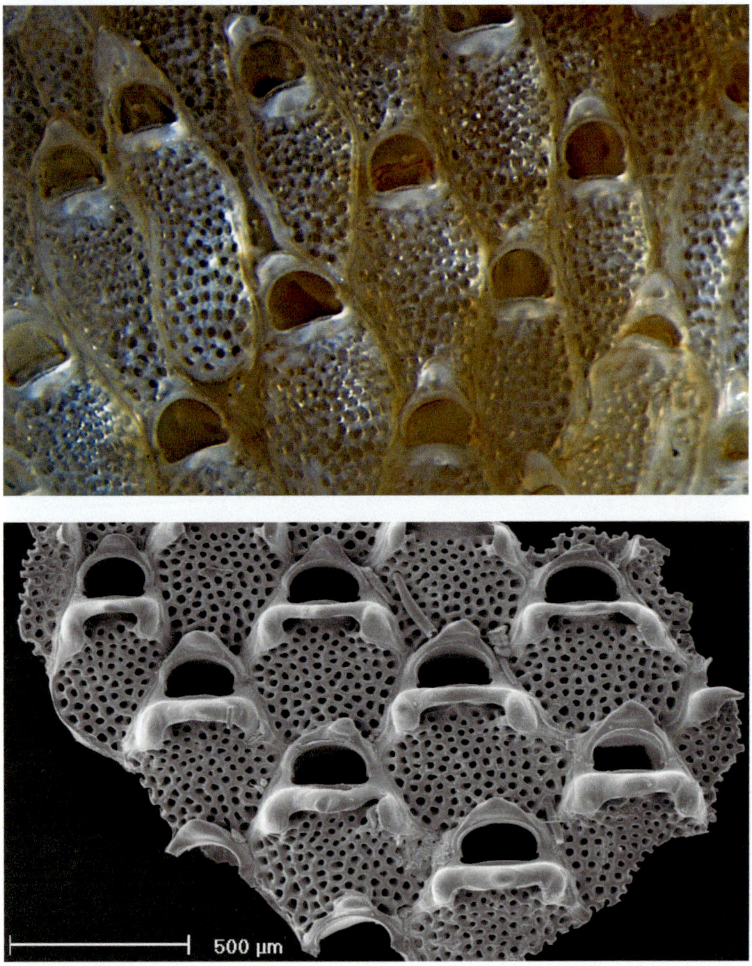

b

Fig. 4.143 *Terwasipora complanata.* (**a**) Part of colony growing on beached *Himanthalia* (Belgium). (**b**) SEM image of colony fragment

Occurrence Southern species that reach the British Isles, at Scilly (Hayward 1976) and France (Molène) on a stone, in July 2018 by F.-X. Decaris, confirmed by photo.

Belgium Rarely washes up on *Himanthalia* buttons (De Blauwe 2003a; Vanhaelen et al. 2006), November 2023 at Zeebrugge (own obs., unpublished); Loppens (1906) mentions this species as *Micropora complanata* Norman as rare on shells, presumably those shells were supplied on *Himanthalia* buttons.

The Netherlands Rarely washes up on *Himanthalia* buttons (de Ruijter 2017c, 2024a). Later on, a *Himanthalia* button beached on Texel (Rob Dekker, October 2023).

Germany Doubtful record from Tiefe Rinne (Kuhlenkamp and Kind 2012).

Family Schizoporellidae

Genus *Schizoporella*

1 (a) Sinus shallow U-shaped, wider than deep 2
(b) Sinus U-shaped, deeper than wide. ....
.... *Schizoporella hesperia/magnifica* (p. 341)

2 (a) Rostrum triangular. 3
(b) Rostrum sharper, with inward bended sides. Orifice often not centered. ....
.... 4

3 (a) Prominent suboral umbo. One or two avicularia. Ovicell with few or no pseudopores. Sinus shallow U-shaped. ....
.... *Schizoporella unicornis* (p. 345)
(b) Incipient umbo. Number of avicularia variable. Ovicell with pseudopores. Sinus with wider proximal border. ....
.... *Schizoporella japonica* (p. 342)

4 (a) Frontal shield with sunken pseudopores. ....
.... *Schizoporella* cf. *variabilis* (p. 345)
(b) Pseudopores not sunken.................................. *Schizoporella spec.* (p. 347)

Schizoporella hesperia Hayward and Ryland, 1995 *or S. magnifica* Hincks, 1886

Description Colony encrusting, unilaminar. Autozooids, oval to polygonal, 0.6–0.7 mm long, 0.43–0.56 mm wide. Orifice longer than wide, sinus one-third of the length of the orifice and apparently even longer due to the presence of two broad rounded condyles, sinus almost parallel-sided. No oral spines. Frontal shield cryptocystidean, densely pseudoporous by small round pseudopores and a row of peripheral areolae. Avicularia single or paired, distal to sinus (in favor of *S. magnifica*), proximal portion level with the sinus, tip of rostrum level with distal rim of zooid, crossbar slender lacking columella. Ovicell prominent, ooecium produced by the distal autozooid, membranous ectooecium lost in dry colonies, calcified endooecium with numerous marginal pits, frontal pits absent (argument in favor of *S. hesperia* (Fig. 4.144)).

Occurrence Rare on beached plastic in Belgium and the Netherlands.

Belgium In November 2000, in Oostende (De Blauwe 2009a) and in January 2024, in De Panne (W. Laforce, unpublished).

The Netherlands In 2013, near Egmond aan Zee, reported in de Ruijter (2014b) as *S. hesperia.*

Schizoporella japonica Ortmann, 1890

Description Colonies unilaminar but frequently with elevated lobes, rapidly becoming extensive, pale whitish-pink to vivid orange-red. Zooids in linear series, quincuncial; rectangular; conspicuously longer (often twice as long) than broad (0.5–0.7 mm long); the dividing line between series distinct, slightly depressed; frontal shield with sunken pseudopores (c. 600 mm^{-2}, range 400–800 mm^{-2} (Ryland et al. 2014); but 1250 mm^{-2} in Fig. 4.145c) and marginal areolae, the distolateral pair of areolae somewhat larger. Sometimes an incipient suboral umbo. Orifice wider than long, more oval than in *S. unicornis*; the sinus shallow and broad (80–120 μm between condyles), with sinuous margins (more rounded quadrate than in the other species), obtusely pointed condyles. A single avicularium lateral to the orifice but frequently none; distolaterally directed, inner end of the hinge-line level with the condyles; mandibles triangular, their height scarcely exceeding the hinge width. Frontal enlarged avicularia may occur, their elevated chamber having small pores, the rostrum nearly perpendicular to the frontal plane (Fig. 4.145c). Ovicell prominent, subglobular, ooecium produced by the distal autozooid, membranous ectooecium lost in dry colonies, calcified endooecium with numerous pits except near the mid-proximal margin; with slender sinuous ridges ascending from the distal zooid and converging in a mid-proximal direction. Polypide with c. 19 tentacles. Embryos reddish. Ancestrula with D-shaped orifice and eight marginal spines; a central, patterned circular area on the frontal calcification.

Occurrence First recorded in European waters in 2010 and later reported from further locations in Great Britain and Norway (Ryland et al. 2014; Loxton et al. 2017).

Belgium *Schizoporella* cf. *japonica* on an oyster, imported from Norway as a baby oyster and deposited on Belwind windmill park in June 2021 and brought up in September 2021 (collected by F. Kerckhof, identified by H. De Blauwe, unpublished).

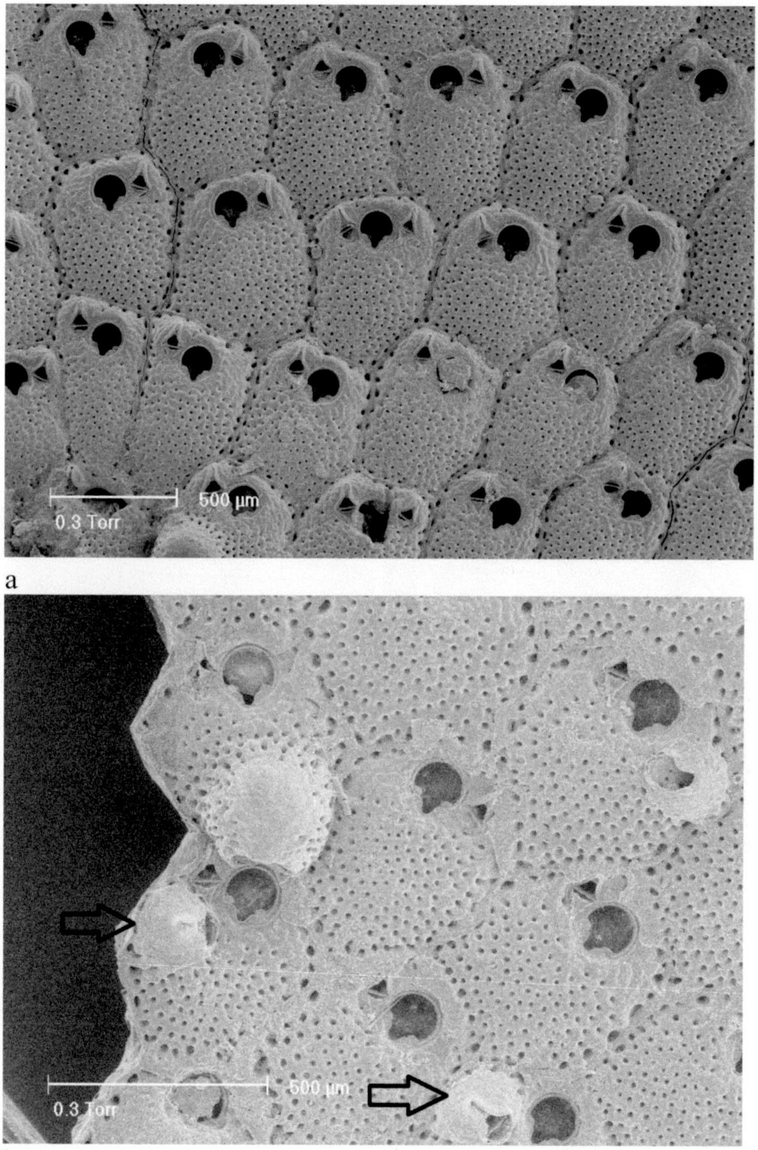

Fig. 4.144 *Schizoporella hesperia* or *magnifica*. (**a**) Zooids with avicularia in a colony on beached plastic, Oostende (Belgium). (**b**) Zooids with normal and enlarged (*arrows*) avicularia, and ovicell without frontal pits

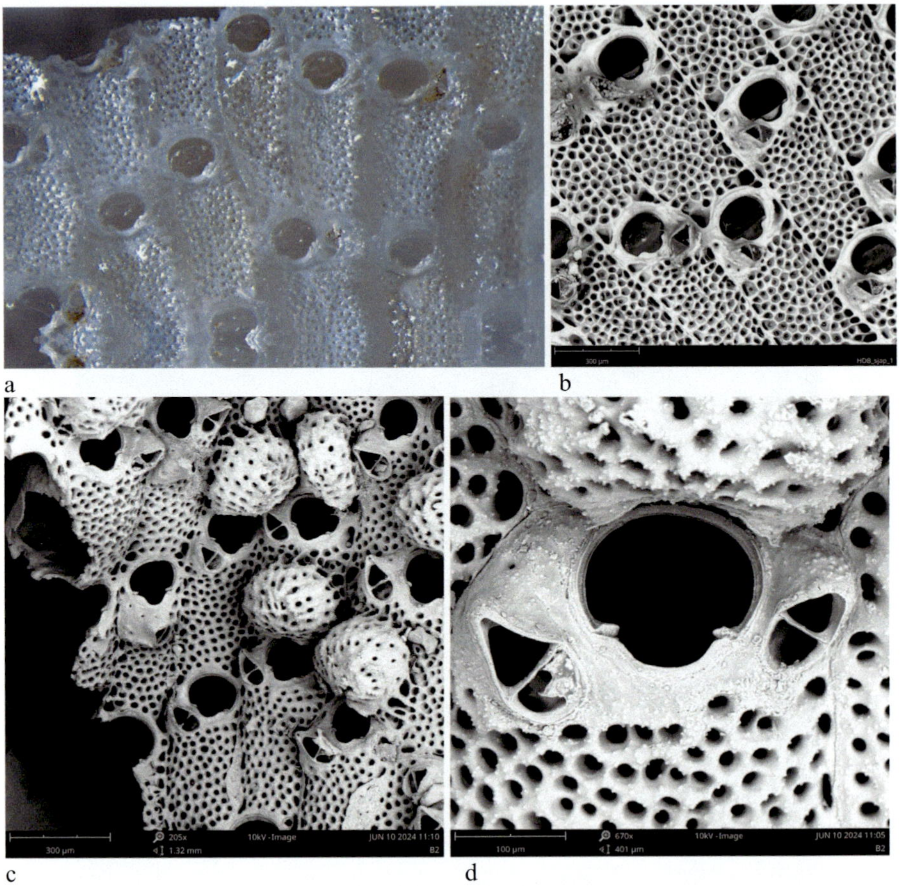

Fig. 4.145 *Schizoporella japonica.* (**a**) Part of colony growing on pontoon in Heligoland (Germany). (**b**) Same colony, autozooids with pseudopore density 800 mm^{-2}. (**c**) Ovicellate autozooids in colony collected near the windmill farm in the German Bight (pseudopore density of 1250 mm^{-2}). (**d**) Distal part of the autozooid showing the shape of the orifice and avicularia

The Netherlands *Schizoporella* cf. *japonica* washed ashore on plastic, associated with *Cribrilina mutabilis* (de Ruijter 2017b).

Germany A colony was sampled in October 2021 on a windmill farm in the German Bight (53.99 N-6.87E) by L. Schüller (IfAÖ). Several colonies on pontoons extracted from the northwest harbor of Heligoland in October 2023.

Schizoporella unicornis (Johnston in Wood, 1844)

Description Colonies encrusting, unilaminar, whitish-gray or pinkish-red. Autozooids angular, on average about 1.5 times as long as broad, 0.4–0.7 mm long. Frontal shield covered with rather small pseudopores and often a conspicuous umbo proximal to the orifice. Frontal pseudopores numerous (c. 525 mm^{-2}, range 350–780 mm^{-2}). Orifice D-shaped, a little wider than long with a shallow U-shaped sinus; inter-condyle distance 75–80 µm. Usually 1–2 avicularia per autozooid, proximolateral to orifice, hinge-line level or proximal to condyles; mandible length 65–85 µm, orientation variable but ~45° from the medial axis. Ovicell prominent, globular, ooecium produced by the distal autozooid, membranous ectooecium lost in dry colonies, calcified endooecium with radial ridges, pits absent or few (Fig. 4.146).

Occurrence

Belgium A colony washed up on a *Himanthalia elongata* button in July 2000 between Zeebrugge and Blankenberge (De Blauwe 2000c).

The Netherlands Lacourt (1949) reports a colony on a *Patella vulgata* near Scheveningen (washed ashore on a *Himanthalia elongata* button?). Reported from 2009 on as washed up on plastic, on *Himanthalia* buttons and styrofoam (de Ruijter 2011a, 2012a, 2015a, 2017a, 2018c, d). Found by Marco Faasse in the Eastern Scheldt in October 2016 between Zeelandbrug and Kulkenol.

Germany Reported from the Tiefe Rinne (Kuhlenkamp and Kind 2012).

Schizoporella cf. *variabilis* (Leidy, 1855)

Description Colony encrusting in one to many layers, sometimes with knobs or tubular extensions, up to 1 cm thick; growing on rigid or flexible surfaces such as worm tubes and algae. Color of living colonies varies from white to pink, orange, and brick red. Zooids hexagonal, quadrangular, or irregularly polygonal. Frontal shield calcification granular in texture, with many large, sunken pseudopores (400–600 mm^{-2}) and slightly larger areolae. The orifice is roughly semi-circular, with a U-shaped sinus on the proximal, roughly straight, edge. Condyles on a distance from the sinus, distance between condyles 67–92 µm; orifice width 118–153 µm, orifice height 111–147 µm. Orifice often not centered, closer to one side of the zooid. Sinus shape somewhat variable in width and depth, and condyles may vary in length. An avicularium below and on the right or left side of the sinus and oriented distolaterally or laterally, or absent in large parts of the colony. Avicularia have a more incurved rostrum and a broader palate than the similar warm-water species *S. errata*. Larger interzooidal avicularia, similar in shape, may occur in frontally budded areas of the colony.

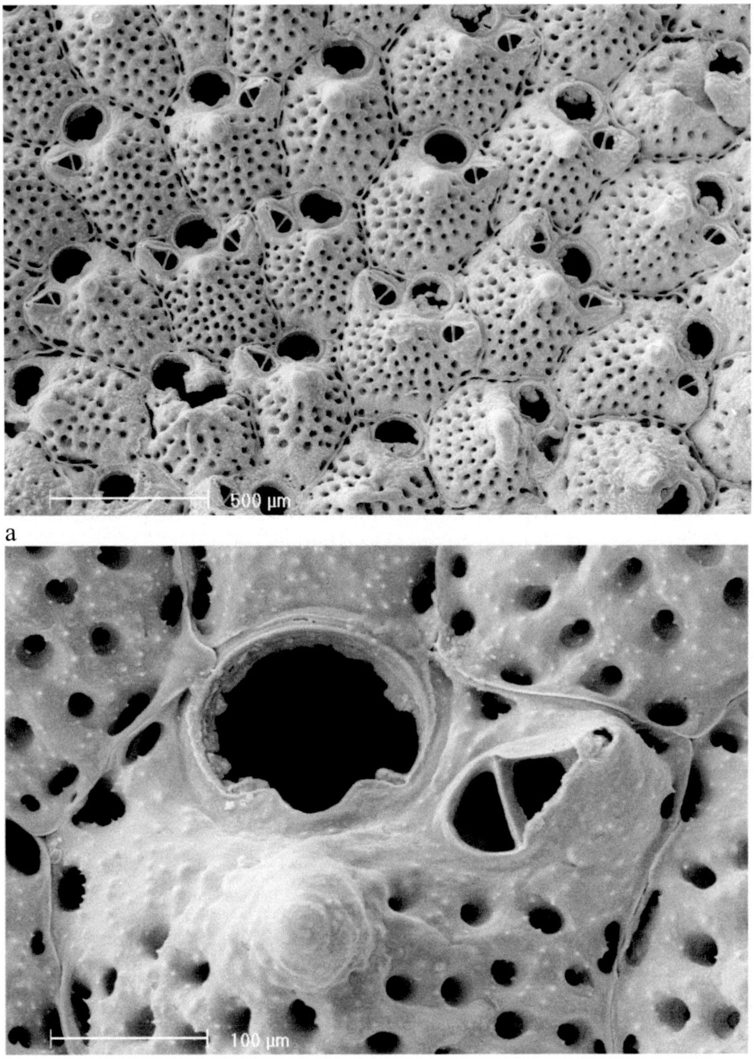

Fig. 4.146 *Schizoporella unicornis*. (**a**) Part of a colony from Presqu'île de Quiberon (France). (**b**) Distal half of an autozooid with umbo and avicularium

Adventitious, large, frontal avicularia, may be rare. Spines absent. Ovicells prominent, globular, ooecium produced by the distal autozooid, membranous ectooecium lost in dry colonies, calcified endooecium with a radiating pattern of pseudopores and a rough texture. Its calcification thickens and the pits deepen with later development. Ancestrula oval, without pores, and with a large, semicircular orifice, surrounded by four pairs of spines (Fig. 4.147).

Occurrence West Atlantic from the Gulf of Maine to Cape Hatteras. This is the first report for the East Atlantic.

Belgium No records.

The Netherlands Present author collected a colony without avicularia in April 2007 at the Emelissedijk (west of Colijnsplaat) on an oyster. Later, a colony was found on an oyster shell within an alien-species-focused survey of the Wadden Sea in 2018 (Gittenberger et al. 2019). Subsequently, several colonies were found in 2020 in the Eastern Scheldt, just outside of the sluices to the Veerse Meer (Gittenberger et al. 2021). Within the Marine Alien Species Detection Network in 2022 at several sites east of Texel in the Wadden Sea (Gittenberger et al. 2023c) the species was again collected. One of these colonies (Fig. 4.147a) had almost no avicularia and a distance between condyles of 67 µm (AG 7448). The detection network in 2023 revealed that the species was already widely distributed in the Eastern Scheldt, spanning from the central Zeeland bridge inland toward the northeast and southeast regions (Gittenberger et al. 2024). DNA analyses with colonies from all locations confirm that they concern the same species, here named *S.* cf. *variabilis*, but previously misidentified in the Netherlands as either *S.* cf. *japonica* or *S.* cf. *errata*. A colony (Fig. 4.147b, c) collected in June 2023 by D. Oude Luttikhuis in the Wadden Sea, distance between condyles 75–82 µm (WM 2023, MPA 7, PosKey 21,082, N 53°7.03′ E 4°58.24′).

Germany Collected on a windmill farm in the German Bight on October 2021 by L. Schüller (IfAÖ) on two locations: 54.02° N-6.52° E (Fig. 4.147d) and 53.99° N-6.87° E (Fig. 4.147e, f).

Schizoporella sp.

Description Colony encrusting in one or few layers. Color of living colonies not known. Zooids square or irregularly polygonal. Frontal shield calcification granular in texture, with many small round pseudopores (400–500 mm^{-2}), with relatively large spacing, not sunken in pits and larger areolae. The orifice is roughly semicircular, with a U-shaped sinus on the proximal, roughly straight, edge. Orifice often not centered, closer to one side of the zooid. Orifice semicircular distally, distance between condyles 60–74 µm. Avicularia single or absent, besides or below the sinus and on the right or left side of the orifice, oriented distolaterally, laterally, or slightly proximolaterally. Avicularia have also an incurved rostrum. Larger avicularia may be present on some zooids. Spines absent. Ovicells and ancestrula not observed (Fig. 4.148).

Remarks
Quite similar to *Schizoporella* cf. *variabilis* but the orifice is smaller (orifice width 119–125 µm, orifice height 111–125 µm), the rostrum relatively longer, and the frontal pseudopores not sunken and smaller.

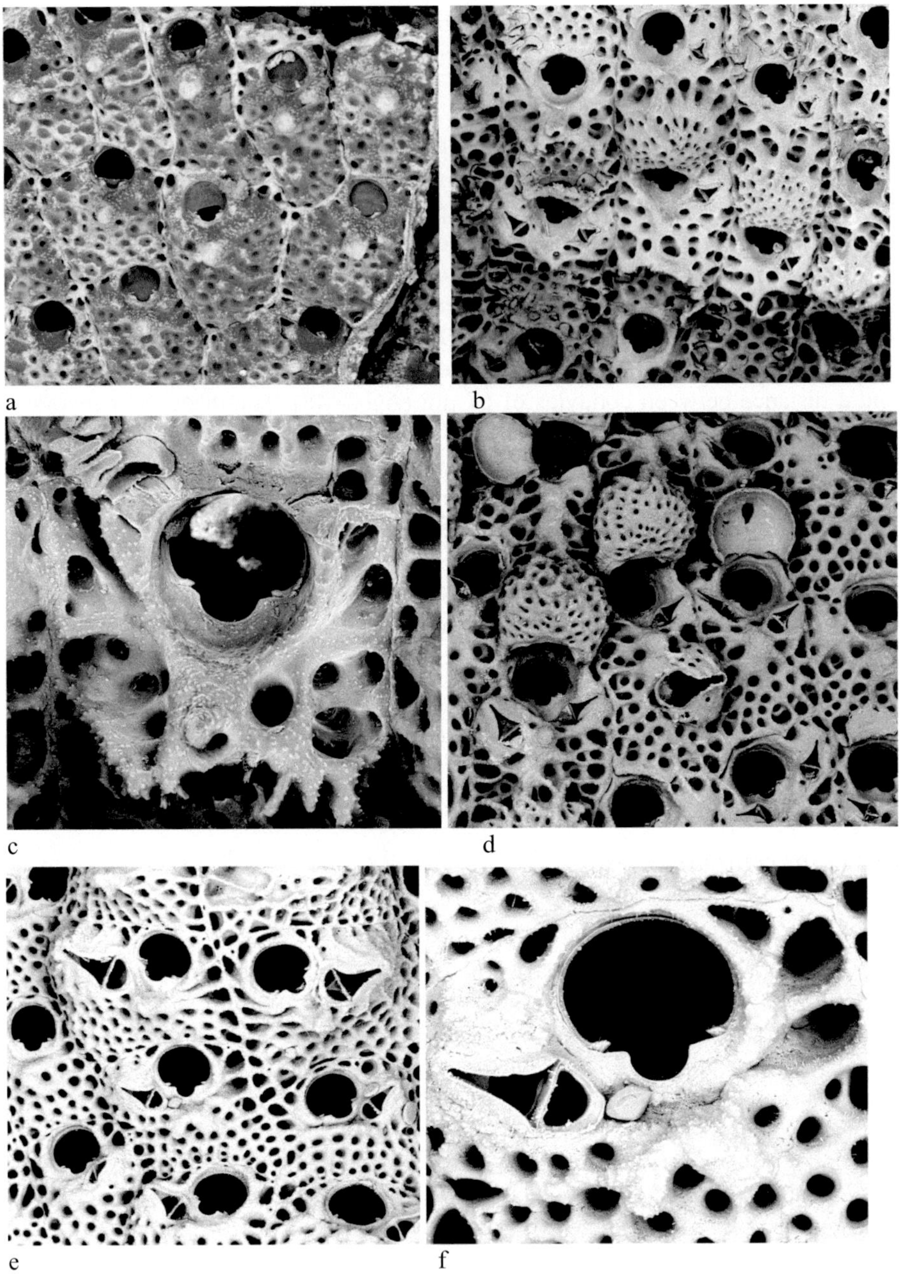

Fig. 4.147 *Schizoporella variabilis*. (**a**) An unbleached colony area of non-ovicellate zooids, sample AG7448. (**b**) Ovicellate zooids in the colony from the Dutch Wadden Sea. (**c**) Detail of zooidal orifice, also showing rough calcification and umbo in the same colony. (**d**) Ovicellate and non-ovicellate zooids and enlarged avicularium in the colony from the German Bight. (**e**, **f**) Details of non-ovicellate zooids from colony collected in the German Bight

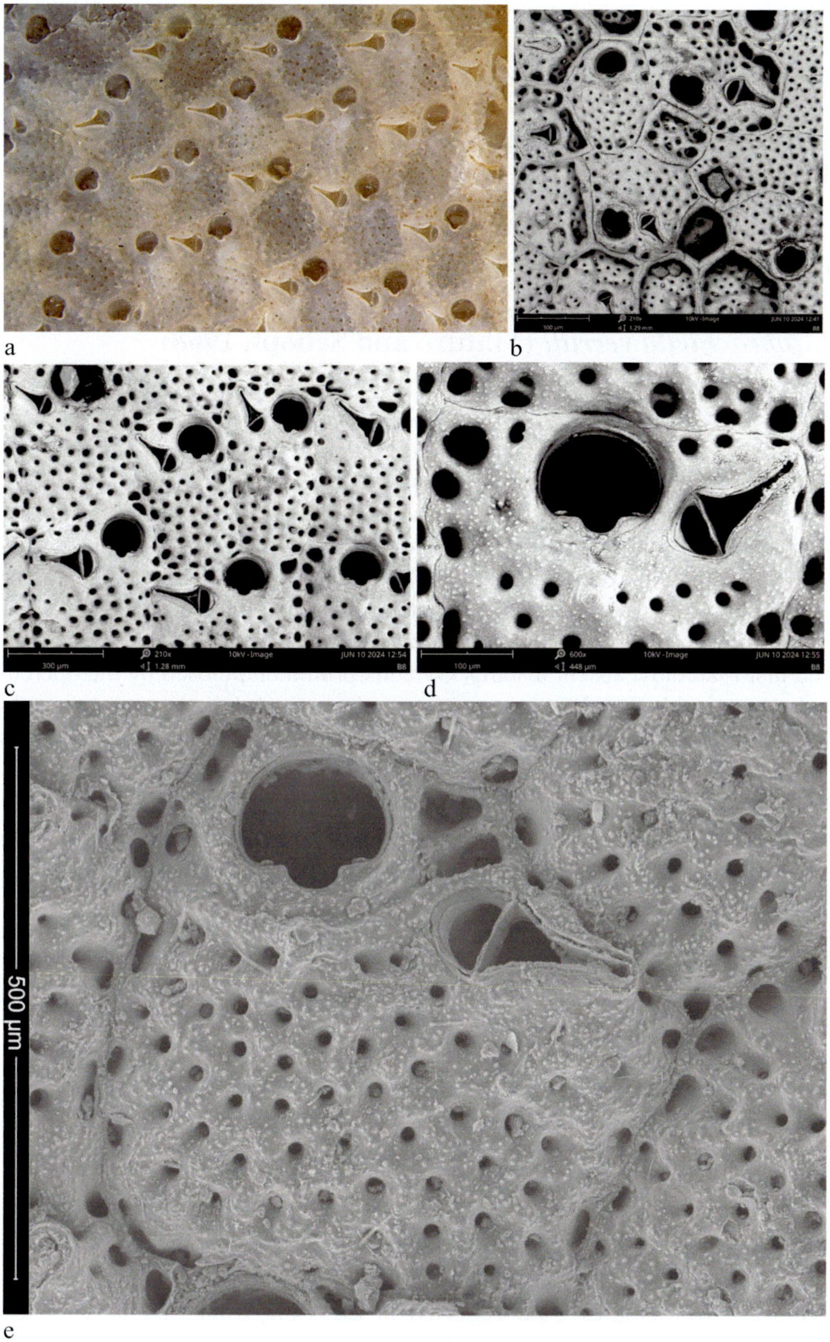

Fig. 4.148 *Schizoporella* sp. (**a**) Part of living colony, German Bight. (**b**) Part of the colony showing autozooids with smaller and larger avicularia and kenozooids. (**c**, **d**) Details of non-ovicellate autozooids and their avicularia. (**e**) Autozooid from colony collected at Kats (the Netherlands)

Occurrence

The Netherlands One colony collected at Kats in August 2022 by the present author.

Germany One colony on a crab shell in the German Bight in November 2020 at 54.08 N-6.61E collected by Lisa Schüller (IfAÖ).

Genus *Schizobrachiella*

Schizobrachiella verrilli (Maturo and Schopf, 1968)

Description Colonies form small rounded sheets, usually encrusting empty bivalve shells. Living colonies beige, changing to brown or white after death. Autozooids 0.3–0.5 mm long, broadly rectangular to polygonal, slightly convex, separated by a slightly raised ridge. Orifice slightly broader than long, with a broad, very shallow U-shaped sinus between small rounded condyles. Frontal shield evenly pseudoporous by large round pores, except proximal to the orifice, becoming deeply immersed in the surface and enlarged laterally in later ontogeny. Peristome thick, incorporating a prominent proximal umbo and flaring lappets laterally, abruptly ending, often tipped disto-laterally, absent distally. Ovicells prominent, closed by the operculum of the parent zooid, globular, ectooecium membranous, lost in dry material, endooecium calcified, with rugose surface and evenly distributed pits. Avicularia very rare, mostly absent, small, with a narrow triangular rostrum, pointing proximally outward, situated at the distal end of the autozooid at either side of the orifice. Ancestrula tatiform, surrounded by approximately ten spines, rapidly overgrown by secondary calcification from neighboring autozooids (Fig. 4.149).

Occurrence

Belgium A colony was dredged by Gilson in 1905 at 51°29′ N, 2°32′ E (coll. RBINS). Two colonies were found on bivalve shells extracted from the Kwinte Bank (51°15′ N, 2°40′ E in 2004, coll HDB). Sixteen colonies were dredged by Houziaux on the Westhinder Bank (51°30′ N, 2°30′ E in 2005, coll. RBINS). One colony was extracted from Buiten Ratel (51°15′ N, 2°30′ E in 2012, coll. H. De Blauwe) and another was found on a bivalve shell during sand supply on the beach of Wenduine in 2022 (coll. H. De Blauwe).

The Netherlands In 2005, a living colony was collected by Self Contained Underwater Breathing Apparatus (SCUBA) diving on a mussel shell near the Goese Sas in the Eastern Scheldt (51°32′ N, 3°55′ E).

Germany Fifty colonies were collected in the Tiefe Rinne, 54°11′ N, 7°53′ E (Kind et al. 2015).

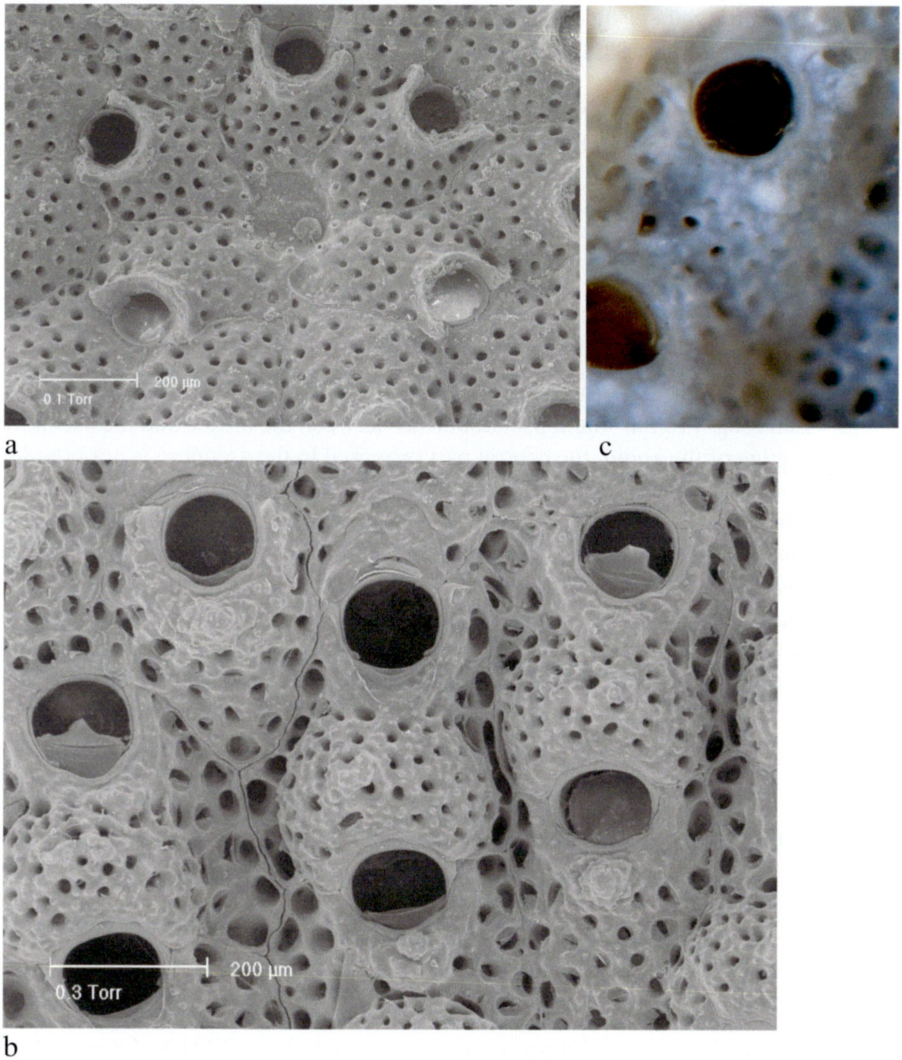

Fig. 4.149 *Schizobrachiella verrilli.* (**a**) Ancestrula and periancestrular zooids growing in a dead shell from the Westhinder Bank (Belgium). (**b**) Ovicellate and non-ovicellate zooids in the colony from the Eastern Scheldt (the Netherlands). (**c**) Autozooid with two avicularia situated at either side of the orifice, colony found in Tiefe Rinne (Germany, photo by B. Kind)

Genus *Fenestrulina*

Following the example of the genus *Microporella*, it is necessary to revise the genus *Fenestrulina* along the west coast of Europe because the population is likely to contain one or more species that are not the same as *Fenestrulina malusii* (Audouin, 1826).

1 (a) Smooth gymnocystal margin of zooids partially encroaching on the lateral pseudopores. One to four oral spines. Endooecium with ribs and irregular transverse ridge. .... *Fenestrulina delicia* (p. 352)

(b) Smooth gymnocystal margin not encroaching on lateral pseudopores. Two to three oral spines. Endooecium without ribs and transverse ridge......... .... *Fenestrulina* sp. (p. 352)

Fenestrulina sp.

Fenestrulina malusii: Hayward and Ryland 1999, pp. 300–301, Fig. 138.

Description Colony encrusting, porcelain at first, later white. Autozooids oval, 0.6–0.7 mm long. Frontal shield pseudoporous between the orifice and the ascopore and at the edge of the zooid, the central part proximal to the ascopore is not pseudoporous. Zooidal periphery is represented by a smooth gymnocystal ribbon. The orifice is D-shaped with minuscule condyles, the light brown operculum has a thickened rim. Young zooids have two or three short distal oral spines. The crescent-shaped ascopore is rimmed by a thick rim and has a serrated inner rim. Polypide with 14 or 15 tentacles. Ovicells prominent, hemispherical, ooecium produced by the distal autozooid with raised margin and large peripheral pores. Membranous ectooecium lost in dry colonies, calcified endooecium smooth. The embryos are yellow-orange. Ancestrula tatiform with about ten marginal spines, with three periancestrular zooids (Fig. 4.150).

Occurrence

Belgium In 1904, a colony was collected near the lightship "Westhinder." In the same area, a second colony was fished up in 1905 (collection RBINS). In June 2005, at least 16 deceased colonies were found on bivalve shells fished from the Westhinder Bank. Washed up on plastic in Blankenberge in 1999, three times in Zeebrugge in 2000, and in the autumn of 2004 between Oostduinkerke and De Panne (De Blauwe 2005a).

The Netherlands On washed-up plastic in Westkapelle in 2004 and reported on washed-up plastic by de Ruijter (2011d, 2013c, 2014b, 2015a, d). De Kluijver (2006) reports the occurrence of *F. malusii* in the Eastern Scheldt near Wemeldinge in 2002, but it probably concerns *F. delicia*. This material could not be traced.

Fenestrulina delicia Winston, Hayward and Craig, 2000

Description Colony encrusting, white. Autozooids oval or irregular, 0.5–0.8 mm long. Smooth gymnocystal margin of zooids partially encroaching on the lateral pseudopores. Large star-shaped pseudopores in two or three rows between the ascopore and the orifice, one or two rows along the lateral margin, and one row along the

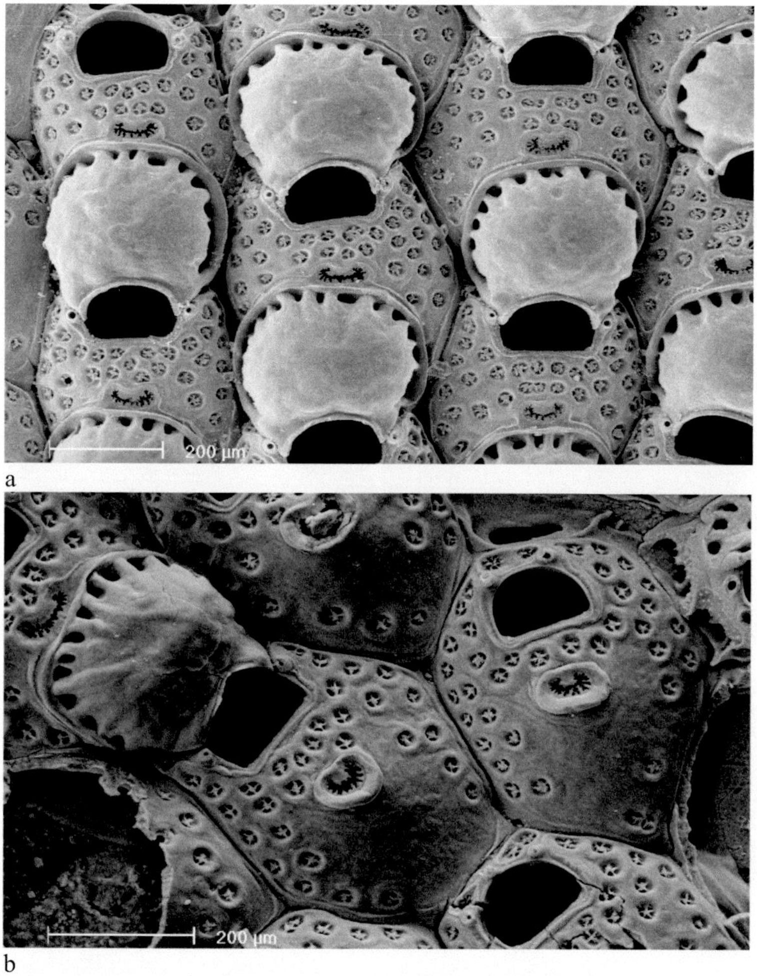

Fig. 4.150 *Fenestrulina* sp. (**a**, **b**) Parts of colony collected in fishing nets, Le Conquet (France), showing ovicellate autozooids, without gymnocystal margin encroaching on lateral pseudopores and ovicells with smooth endooecium without transverse ridge

proximal margin. No pseudopores between the ascopore and the proximal zooid margin. Orifice D-shaped with a small triangular condyle in both proximal angles. Distal and distolateral to the orifice, a row of pores occurs in one or two sharply delineated panels. Ascopore with an erect rim, the pore itself has a serrated inner rim. Non-ovicellate zooids have one to four oral spines, but usually, a remnant of a medial distal spine is found. Ovicellate zooids have a remnant of spine on either side of the orifice. Ovicells prominent, hemispherical, ooecium produced by the distal autozooid with raised margin and large peripheral pores. Membranous ectooecium lost in dry colonies, calcified endooecium often with thick ribs that meet with an irregular transverse ridge. Endooecium in old ovicells may become completely rough with a smooth erect proximal margin. Ancestrula tatiform with ten marginal spines, surrounded by six periancestrular zooids. The polypide has 13 tentacles (obs. M. Faasse) (Fig. 4.151).

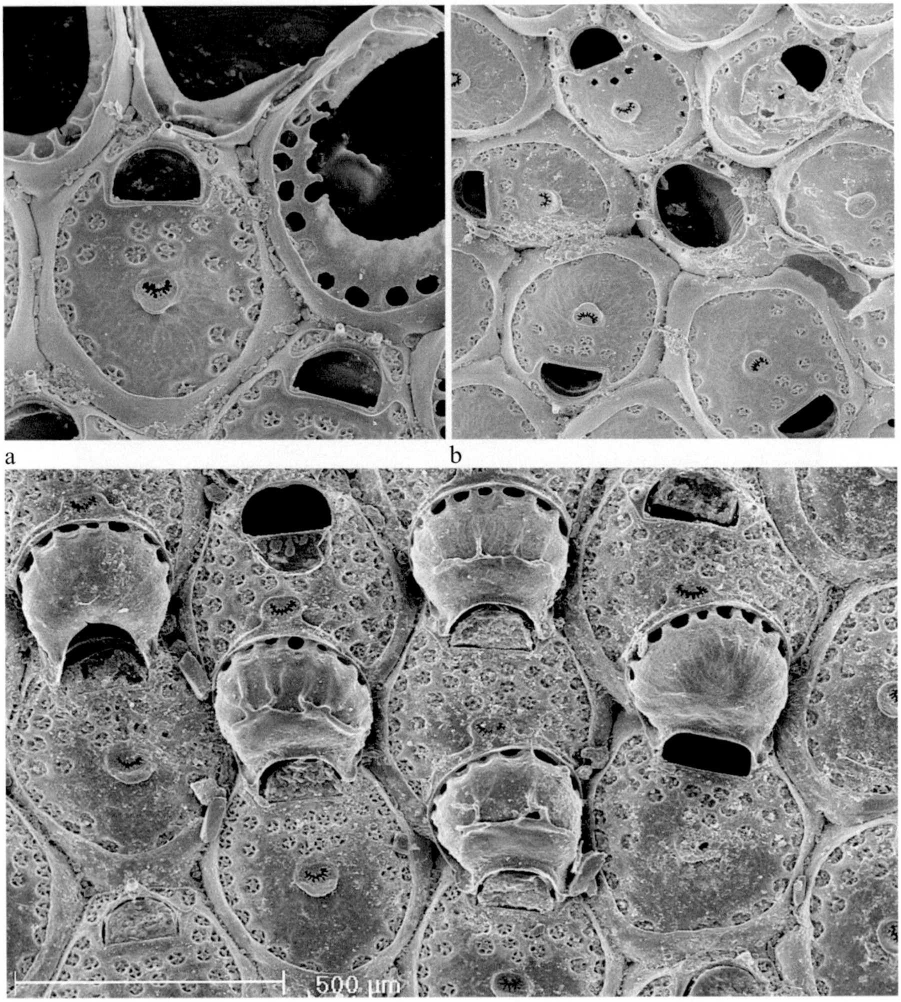

Fig. 4.151 *Fenestrulina delicia*. (**a**) Autozooids near the colony edge, gymnocystal margin of zooids partially encroaching on the forming lateral pseudopores. (**b**) Ancestrular region. (**c**) Part of the colony showing ovicellated zooids having endooecia with ribs and transverse rudge. Colony collected at the Goese Sas (the Netherlands)

Occurrence Recognized in 1994 and described in 2000 from the Gulf of Maine in the Western Atlantic Ocean, discovered very shortly afterward on test panels in Alaska and San Francisco (eastern Pacific). This invasive species has been present in European waters since 2002 or earlier (Wasson and De Blauwe 2014).

Belgium Washed ashore on December 29, 2009, on the plastic entrance of a crab pot (Verhaeghe 2010).

The Netherlands In 2005, some colonies were collected inside empty mussel valves while diving near the Goese Sas in the Eastern Scheldt (De Blauwe 2008). In 2011, on empty bivalve shells near the Anna Frisopolder, more westward in the Eastern Scheldt. In 2012, it was collected near Zierikzee and St. Annaland and therefore is now probably widespread in the entire Eastern Scheldt. In 2011, it has also been collected in the Prinses Amaliawindpark (Vanagt et al. 2013). Since January 1, 2011, the species has been found washed ashore on plastic parts of lobster pots originating from the French Channel coast along the Dutch coast near Goedereede and the Brouwersdam (Faasse et al. 2013). *F. delicia* was absent in the samples from the Dutch gas platforms of Beukhof et al. (2016).

Germany First record from Borkumer Reef Grounds in 2011 (De Blauwe et al. 2014). Collected in the Tiefe Rinne (Kuhlenkamp and Kind 2012, pers. obs. 2024).

Family Escharinidae

Genus *Escharina* and *Herentia*

1 (a) Orifice with high raised rim and without oral spines. Endooecium and frontal surface with one or more umbo. Few avicularia, mandible triangular. .... *Escharina johnstoni* (p. 356)
(b) Orifice without clearly raised edge. Oral spines are present. Umbo can be present. Avicularia with bristle-shaped mandible. 2

2 (a) here is an avicularium on either side of the orifice. The mandible is whip-shaped. .... *Escharina haywardi* (p. 356)
(b) One or two avicularia proximal to the orifice. 3

3 (a) Two avicularia per zooid, rostrum pointed, mandible rarely longer than a zooid. Endooecium evenly granulated. .... *Escharina vulgaris* (p. 358)
(b) One avicularium per zooid, rostrum ovate, mandible often longer than a zooid. Ovicell with membraneous ooecium having conspicuous arch-like frontal calcification. .... *Herentia hyndmanni* (p. 360)

Escharina haywardi Zabala, Maluquer and Harmelin, 1993

Description Colony encrusting, unilaminar. Zooids 0.6–1.0 mm long, broadly oval. Frontal shield with a row of areolae and occasionally an indistinct umbo proximal to the sinus. Orifice D-shaped with a straight proximal border and a notch-like median sinus, occupying about one-sixth of the total width. Condyles inconspicuous. Six short and blunt oral spines, only four remaining in ovicellate zooids. Occasionally kenozooids occur of the same size as an autozooid but without an orifice. Avicularia on both sides of the orifice. The elliptical rostrum is small and has a curved bristle-shaped mandible, half the length of one zooid, directed over the distal zooid. Ovicells terminal (formed by the distal kenozooid not seen from above) or subimmersed, formed by the distal autozooid, sometimes present in the same colony; membranous ectooecium lost in dry colonies, calcified endooecium with granular surface (Fig. 4.152).

Occurrence Offshore on hard substrates.

Belgium Colony on a shell collected (RBINS) in 1905 on the Hinder Banks. In 2004–2005, eight colonies were collected on shell valves and stones on the Hinder Banks.

The Netherlands Twice on beached plastic at Callantsoog (de Ruijter 2020b) and in 2020 at Texel (unpublished).

Escharina johnstoni (Quelch, 1884)

Description Colony encrusting, forming an unilaminar patch on small stones and shells. Autozooids 0.5–0.7 mm long, elongated ovate, and strongly convex. Frontal surface smooth or fine-grained. Only small areolae present, often inconspicuous. Orifice as a semicircle, slightly longer than wide with a short rounded or angular sinus. Broad condyles extend to the edge of the sinus. Proximal to the sinus develops a conical umbo. Oral spines absent. Dwarf zooids and heterozooids with central aperture are uncommon. Avicularia sometimes rare, sometimes common, mid-lateral on the frontal surface or at the base of an ovicell. The triangular mandible is directed laterally. In Belgian material, avicularia are a great rarity, often completely absent in a colony. Ovicells terminal, with membranous ectooecium lost in dry colonies; calcified endooecium convex and frontally somewhat flattened with one or more conical umbones. In ovicellate zooids, the peristome forms a conspicuous lip proximally. Ancestrula round with a small central opening under an oval membrane, surrounded by about nine spines (Fig. 4.153).

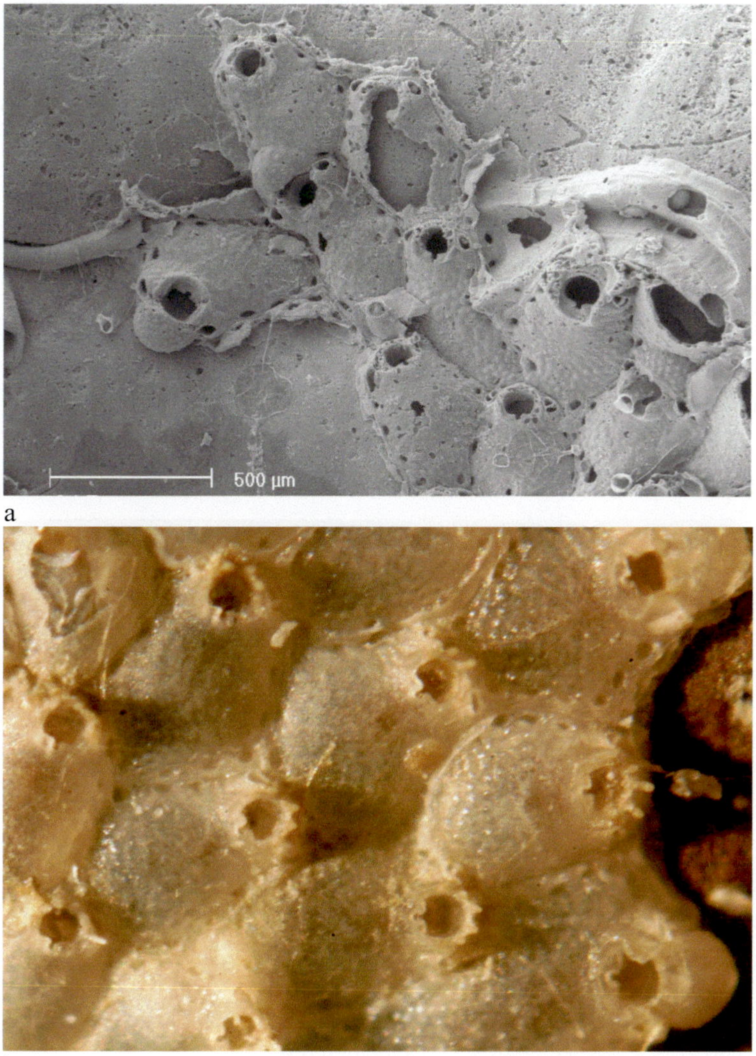

a

b

Fig. 4.152 *Escharina haywardi.* (**a**) Colony area with ovicellate zooids. (**b**) Part of colony with peripheral ovicellate zooid, Westhinder Bank (Belgium)

Occurrence

Belgium Fairly common on shell valves offshore (Kwinte Bank and Westhinder Bank). Often found during sand replenishments on the beach.

The Netherlands De Ruijter (2014b) reports the washing-up in 2013 of a colony on plastic near Egmond aan Zee.

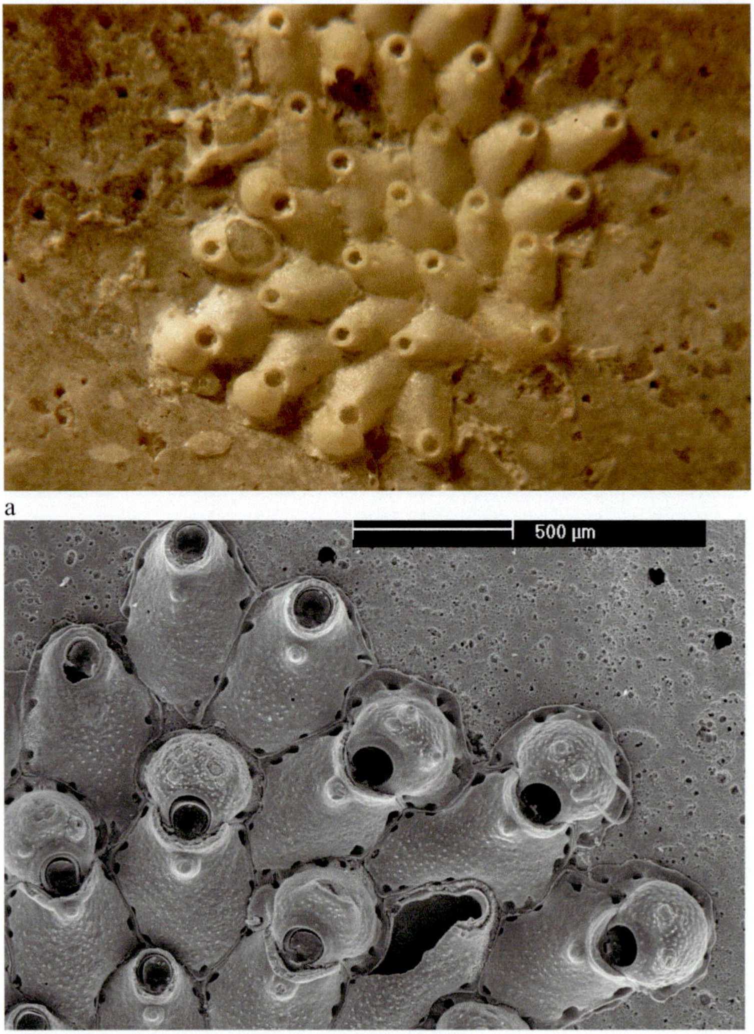

Fig. 4.153 *Escharina johnstoni.* (**a**) Colony growing in a dead shell, Kwinte Bank (Belgium). (**b**) Peripheral colony part with ovicellate autozooids, the same location

Escharina vulgaris (Moll, 1803)

Description Colony encrusting, forming an unilaminar patch on small stones and shell. Autozooids 0.4–0.6 mm long, oval or hexagonal. Frontal shield with few areolae. Orifice slightly longer than wide, with a short, rounded sinus and short condyles. Peripheral zooids bear four slender oral spines. Interstices between the zooids are filled by rounded calcifications that may bear avicularia. Two avicularia per zooid, proximal to the orifice, on the sides of the zooid, budded from areolae. Rostrum slender, acute, distally directed, with setiform mandible. Ovicell subimmersed, formed by the distal autozooid, membranous ectooecium lost in dry colonies, calcified endooecium with finely grained surface. Ancestrula with frontal oval membrane surrounded by spines (Fig. 4.154).

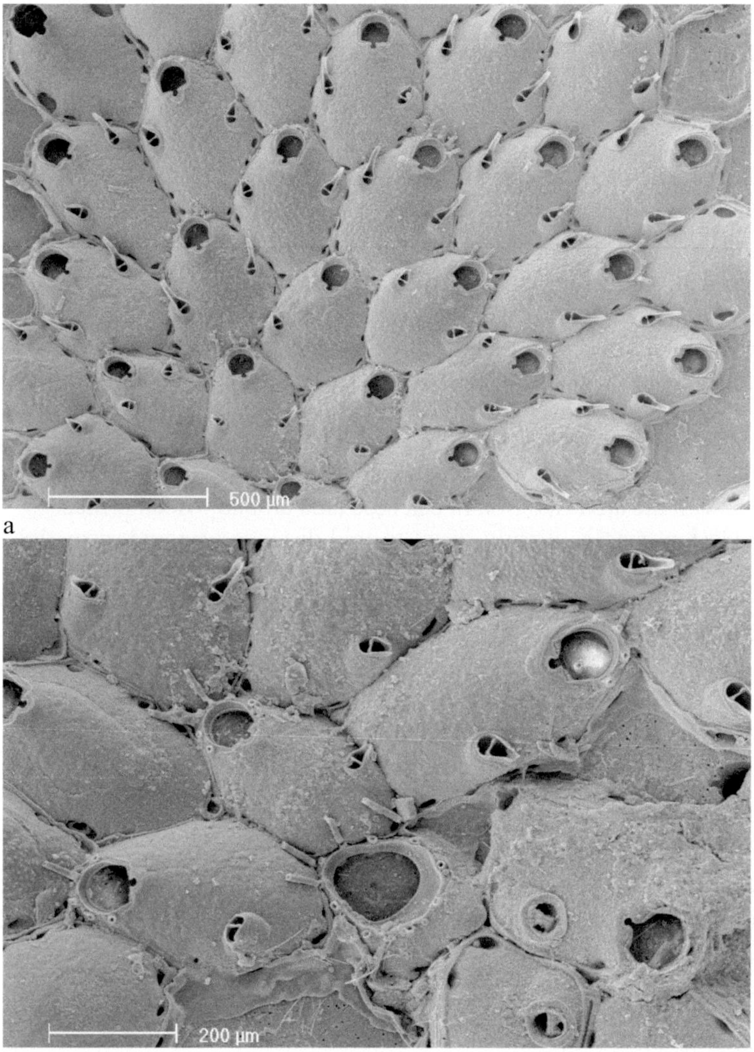

b

Fig. 4.154 *Escharina vulgaris*. (**a**) Part of the colony of non-ovicellate zooids growing in a dead shell. (**b**) Ancestrular region and two autozooids of *Herentia hyndmanni* in the right lower corner, Westhinder Bank (Belgium)

Occurrence Southern species reaching the English Channel and the Southern North Sea.

Belgium In 2004, a small colony was found on a peat block that washed up in Koksijde. Collected in 2005 on the Hinder Banks. Beached on pumice in 2024 in Oostende.

The Netherlands On beached plastic (de Ruijter 2017c, 2019a).

Germany Not reported.

Genus *Herentia*

Herentia hyndmanni (Johnston, 1847)

Description (after Berning et al. 2008): Colony encrusting, forming an unilaminar patch on small stones and shells. Zooids hexagonal to polygonal, separated by deep grooves. Frontal wall rather flat, with 6 to 12 relatively large areolae, five or six in periancestrular zooids. Orifice horseshoe-shaped, proximal margin nearly straight with drop-shaped sinus; aperture in ovicellate zooids more or less semicircular and larger. Four oral spines in the first autozooid, absent in zooids formed during later astogeny. Ovicells terminal, ooecium initially spherical, later for the most part immersed between neighbor zooids; ectooecium mostly membranous, lost in dry colonies, with the arch-like frontal calcified area around ovicell orifice.

Avicularium single, slightly distal to mid-distance of zooid, near right or left zooid margin, proximal to the (or rarely two) distalmost marginal pore; rostrum suborbicular to oval with a broad, smooth, and slightly raised rim. Mandible long and setiform, slightly sigmoidal, and up to twice the length of an autozooid.

Ancestrula kenozooidal, about as long as wide, with an irregular lobate outline, gymnocystal frontal wall moderately convex, rather flat centrally, smooth, lightly but entirely calcified apart from a small central hole, some six articulated spines forming a circle at about mid-distance between margin and ancestrula center (Fig. 4.155).

Occurrence

Belgium In 1906, a colony was collected at the Flemish Banks (RBINS collection). In 2004, a living colony was found on a shell on the Kwinte Bank; in 2005, 5 living and 81 deceased colonies were found at the Westhinder Bank.

The Netherlands Not reported.

Germany Not reported.

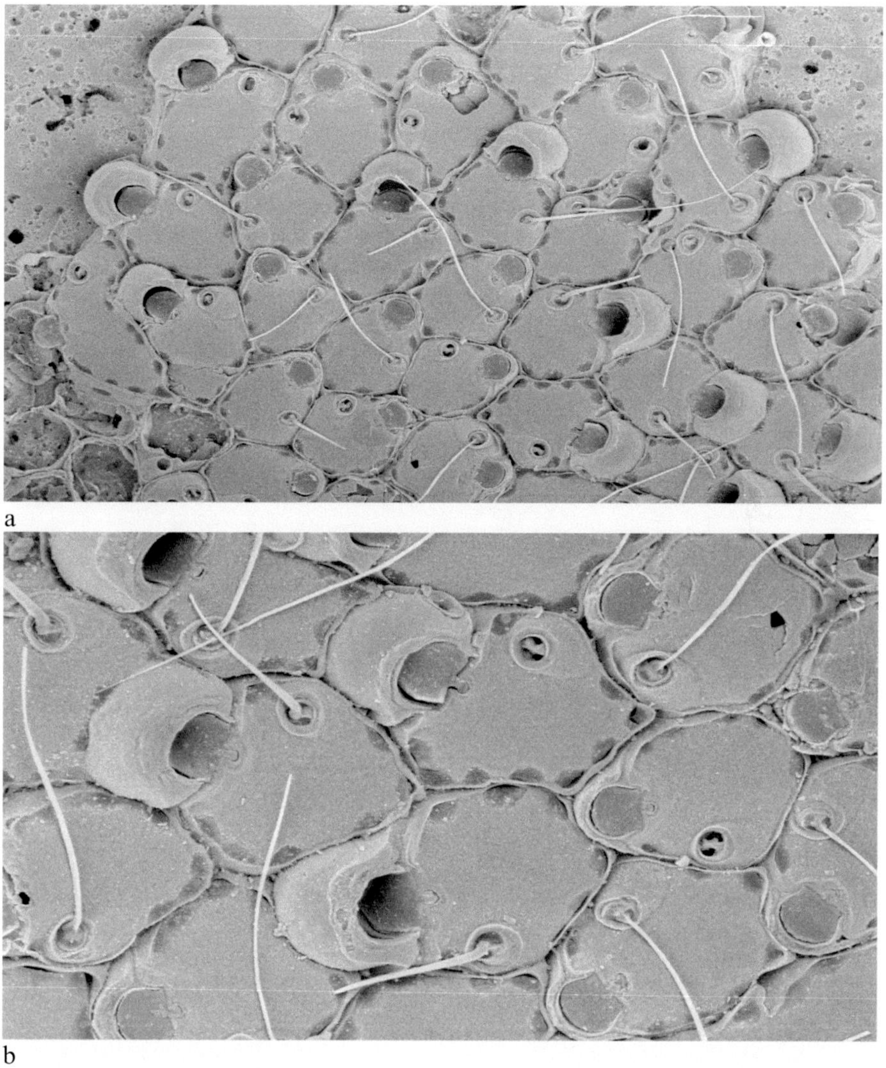

Fig. 4.155 *Herentia hyndmanni*. (**a**) Part of colony growing in an empty shell collected on the Kwinte Bank (Belgium). (**b**) Ovicellate autozooids and avicularia with long setiform mandibles

Genus *Phaeostachys*

Phaeostachys spinifera (Johnston, 1847)

Description Colony encrusting, unilaminar, forming an extensive light brown patch, dried material is white. Zooids 0.5–0.7 mm long, oval to hexagonal, convex. Frontal shield with scattered small pores. Orifice almost as a semicircle with a square small sinus. Five to seven large oral spines, their base is usually chitinous and brown. Operculum with a curved sclerite. Sometimes a low conical umbo proximal to the sinus with a brown chitinous tip, sometimes enlarged as a spine. Avicularia infrequently present, on the frontal surface next to or proximal to the umbo or replaced by a second umbo with spine. The sharply triangular mandible is laterally directed. Ovicells prominent, ooecium formed by the distal autozooid, membranous ectooecium lost in dry colonies, calcified endooecium with conspicuous ribs radiating from the distal base over the surface. Embryos orange-red. Ancestrula tatiform with about 12 spines, or spineless and frontally calcified (Fig. 4.156b), giving rise to 6 periancestrular zooids (Fig. 4.156).

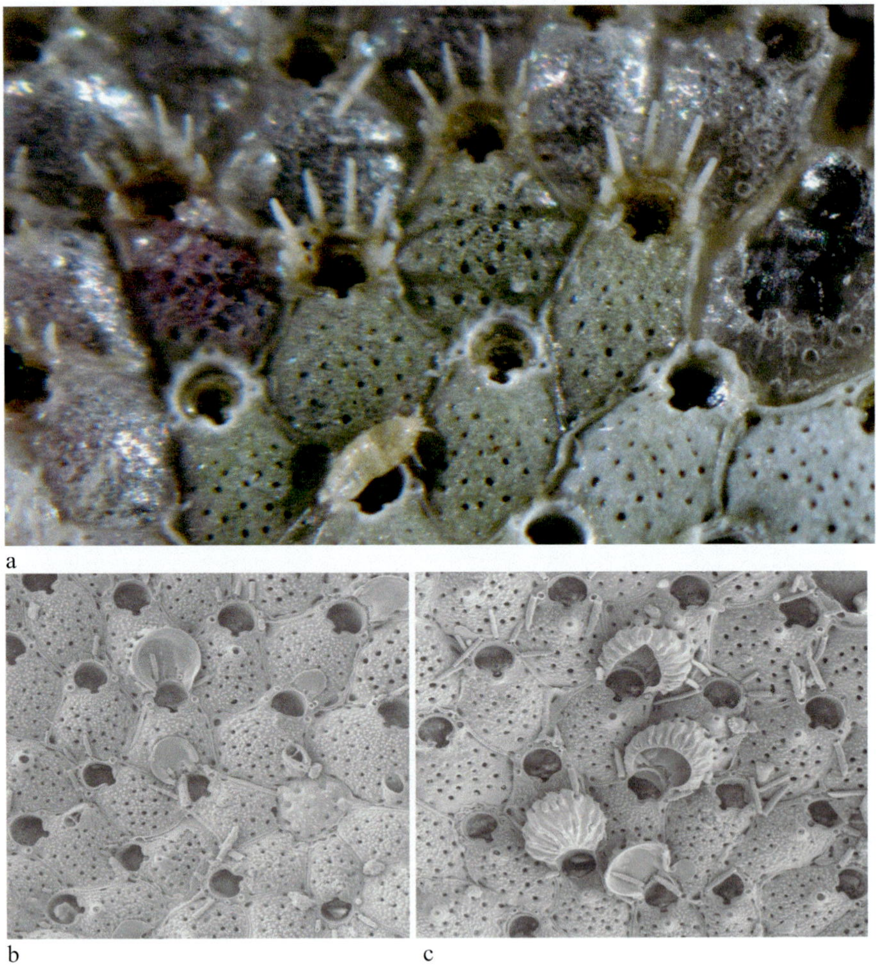

Fig. 4.156 *Phaeostachys spinifera.* (**a**) Part of living colony showing four autozooids with undamaged spines. (**b**) Ancestrular region, two zooids bear avicularia. (**c**) Colony area with forming and fully formed ooecia. Île Chausey (France)

Occurrence

Belgium Washed up in Zeebrugge on plastic and rubber once in 1999 and twice in 2000.

The Netherlands Van der Sleen (1920) reported this species on stones on the Noorderhoofd in Den Helder. On washed-up plastic (de Ruijter 2012a, b, 2018a).

Family Pacificincolidae

Colony encrusting or encrusting-erect, sometimes rising into a frilled massive colony. Zooids irregularly rectangular to subhexagonal. Orifice subcircular, with wide shallow sinus. Peristome distinct or confined to a raised rim, with proximal raised umbo and a small opening of a heterozooid (avicularium or glandular kenozooid) between the umbo and the orifice. Frontal shield lepralioid, uniformly pseudoporous. Spines and avicularia lacking. Ovicell prominent, ectooecium membranous, endooecium calcified and sometimes forming extreme spiky processes; orifice of ovicellate zooid sometimes larger than autozooidal orifice. The frontal shield of the ancestrula is covered by two lateral daughter zooids (*Primavelans,* Fig. 4.157b) or remains free frontally (*Pacificincola,* Fig. 4.157a) (De Blauwe 2006a).

1 (a) Ancestrula without longitudinal ridge, ooecium with large pores distally, without smooth proximal border and proximolateral flanges, orifice with raised peristomal lateral lappets. .... *Pacificincola perforata* (p. 363)
- *idem, ooecium in some places with three to five spiky processes.* .... *Pacificincola hozawai* (p. 366)
(b) Ancestrula with longitudinal ridge, ooecium with small pits, mostly lateral, with a smooth proximal border if mature, and with proximolateral flanges in early development. Orifice without lateral peristomal lappets. .... *Primavelans glabricollaris* (p. 369)

Remarks

Preliminary and limited genetic research suggests that *Pacificincola perforata* is a younger synonym for *P. hozawai.*

Genus *Pacificincola*

Pacificincola perforata (Okada and Mawatari, 1937)

Description Colony encrusting, unilaminar, light yellow when alive, becoming grayish white. In its native habitat (South China Sea), the colony is capable of producing erect growths. Zooids subhexagonal, elongate ovoid, or irregularly rectangular, 0.4–0.7 mm long. Frontal shield slightly convex, granulated, evenly pseu-

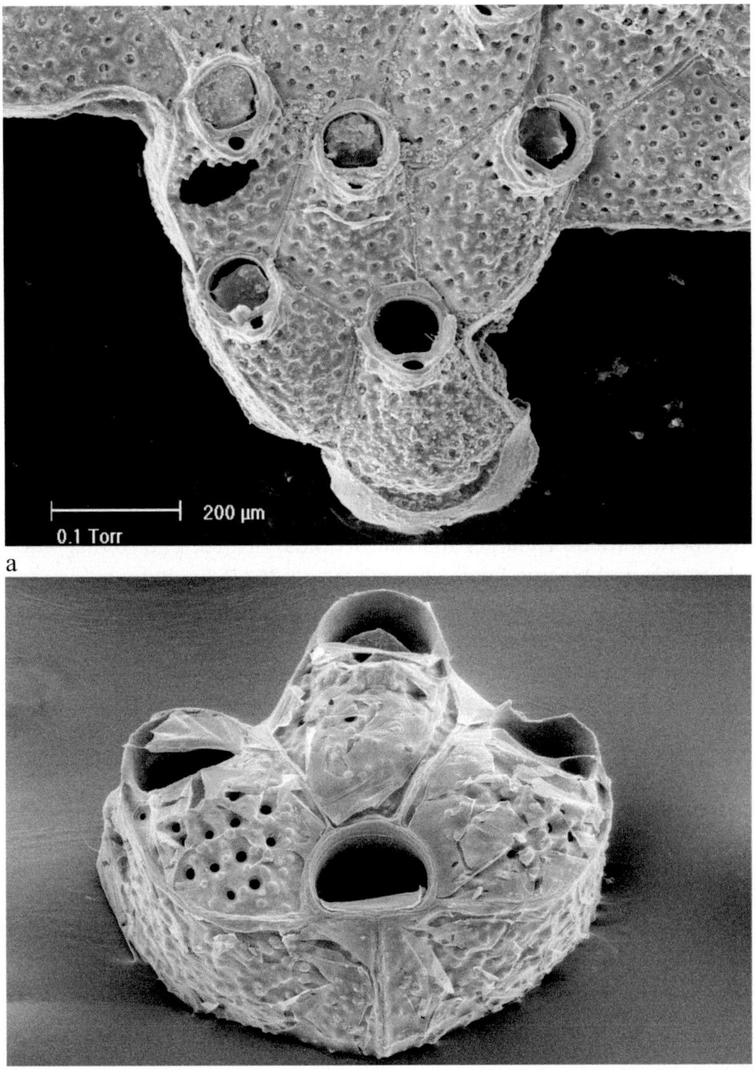

a

b

Fig. 4.157 (**a**) *Pacificincola perforata*, part of the colony showing ancestrula and first zooids, Goese Sas (the Netherlands). (**b**) *Primavelans glabricollaris,* ancestrula and three early zooids, Dutch North Sea

doporous with circular pores. Orifice subcircular, longer than wide, the wide shallow sinus laterally confined by triangular condyles. The proximal border of the sinus is quite straight. Operculum brownish with lateral sclerite. Peristome distinct lateral lappets, not developed distally. Prominent suboral umbo encorporating a heterozooid, with a foramen facing distally, nearly perpendicular to the frontal plane. No oral spines and avicularia. Ovicell prominent, globose, closed by an operculum, ooecium formed by the distal zooid and partially overhangs zooidal orifice; ectooecium membraneous, endooecium calcified, with irregular nodular, sometimes with finely tuberculate radiating ribs, surface surrounded by a series of pores. Pink embryos were observed in late summer. Interzooidal communication via multiparous septula. Ancestrula similar to autozooids, frontal wall finely granulated with evenly distributed numerous pseudopores. Ancestrular orifice roughly circular. Ancestrula similar to autozooids, surrounded by a proximolateral smooth gymnocystal border, budding two distolateral and one lateral zooids that do not cover the ancestrula (Figs. 4.157a, 4.158).

a b c

Fig. 4.158 *Pacificincola perforata*. (**a**) Living colony encrusting Pacific oyster shell. (**b**) Ovicellate and non-ovicellate zooids with distinct lateral lappets of the peristome. (**c**) Autozooids with kenozooidal opening proximal to orifice, and two ovicellate zooids. Goese Sas (the Netherlands)

Occurrence *Pacificincola perforata* was first discovered in Japan (Okada and Mawatari 1937), later it was reported from Hong Kong and its neighboring waters and from China. It is one of the commonest bryozoan foulers in the coastal waters of the southern China seas (Liu and Liu 1999). Recently introduced to the Atlantic coast of Europe. Discovered in 2001 on oysters in Arcachon, France (De Blauwe 2006a).

Belgium Since 2016 common in the Spuikom of Oostende (pers. comm. M. Faasse).

The Netherlands *Pacificincola perforata* was first collected on an empty mussel shell at a depth between 5 and 10 m near Goese Sas, Eastern Scheldt on August 26, 2004 (De Blauwe 2006a). In 2006, it was already the most common Cheilostomate on a beach with boulders just west of the harbor entrance of Yerseke, although not found there in 2005. In March 2008, several colonies on a roof tile in the tide pool in Kattendijke. Occurs en masse in the eastern half of the Eastern Scheldt since the beginning of 2008, since 2007 in the Veerse Meer (Faasse et al. 2013), and since 2015 in the Grevelingen (Gittenberger et al. 2015b). Beached at Langevelderslag in January 2016 on *Sargassum muticum* (pers. comm. Kees Peeters).

Pacificincola hozawai (Okada, 1929)

Escharella hozawai (Okada, 1929) in McCuller and Carlton (2018).
Mucronella hozawai Okada, 1929.

Description Colony uni- or multilaminar. Zooids distinct, typically hexagonal, 0.5–0.6 mm long. Frontal wall granulated, regularly pseudoporous except for the suboral area. Suboral area with heterozooid chamber, raised to form a high arch, smooth at its peak, with median cavity. Orifice higher than wide, separated by condyles. Peristome distinct lateral lappets, not developed distally. Prominent suboral umbo encorporating a heterozooid, with a foramen facing distally, nearly perpendicular to the frontal plane. Ovicell prominent, ooecium formed by the distal zooid, ectooecium membraneous, endooecium calcified finely granular, often with a central umbo and a pair of proximolateral spines that curve inward. Four large multiparous septula on lateral walls. Ancestrula as in other *Pacificincola*.

Tatiform ancestrula described in McCuller and Carlton (2018) must be of another species (Fig. 4.159).

Remarks *Pacificincola hozawai* differs from *Pacifincola perforata* in the presence of paired curved spines on either side of the lower margin of the ovicell. Japanese tsunami marine debris (JTMD) and Dutch material showed considerable variation in spine presence on the ovicell. Paired spines were absent in some parts of the colony, but long and distinct in other parts. JTMD material consists of fouling colonies on fiberglass and plastic, Dutch material consists of one colony on *Sargassum muticum*.

Fig. 4.159 *Pacificincola hozawai*. (**a**) Part of colony growing on *Sargassum muticum*. (**b**) Distal view of autozooids showing one to three spiky processes on ooecia, Kats (the Netherlands)

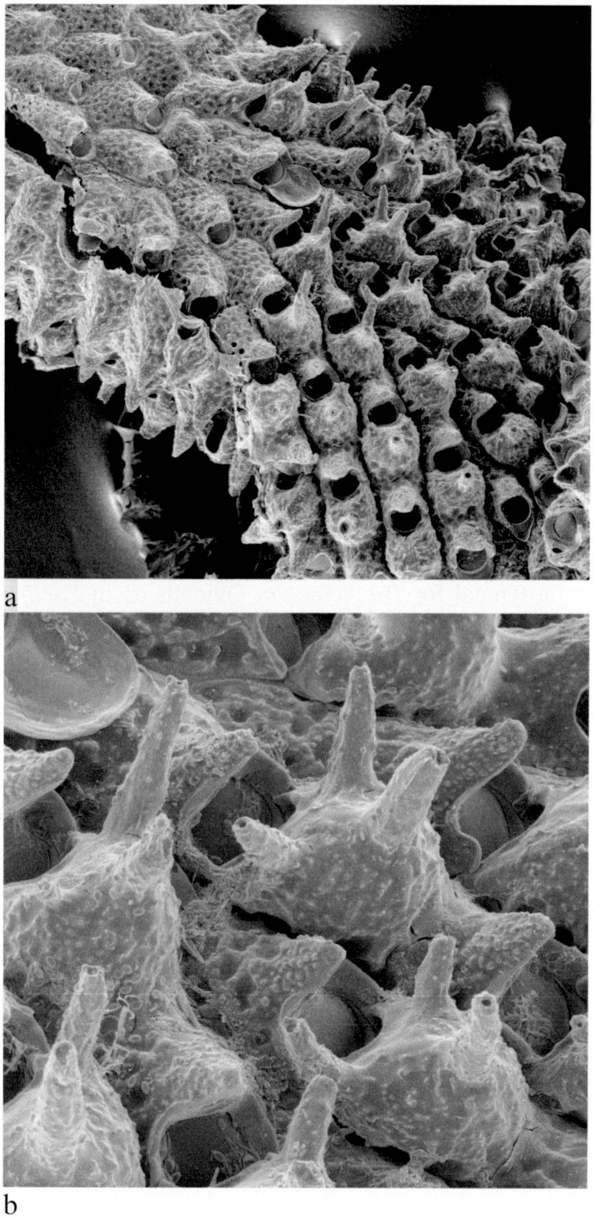

Distribution Japan: Mutsu Bay, Honshu (Okada 1929) on seagrasses such as *Zostera marina* and appears to have not been recorded since (McCuller and Carlton 2018, pers. comm. Bumsik Min, Chihiro Kubo, Masato Hirose). Found on Japanese tsunami debris at the west coast of the USA (McCuller and Carlton 2018).

Occurrence

The Netherlands One colony on *Sargassum muticum* at Kats in 2022.

Remarks For the genetic marker 16S, Gimaris became a sequence that 100% matches with the sequence of *Pacificincola perforata* from South Korea identified by J.E. Seo (De Blauwe et al. 2024) indicating that both species are conspecific. Nevertheless, recent bryozoologists never saw *P. perforata* with spines on the ovicell, nor in Japan nor in South Korea.

Pacificincola cf. *hozawai* with up to Five Long Spines on the Ovicell

Description Colony encrusting on *Sargassum muticum*, unilaminar, soon producing bilaminar folded, erect growths, light yellow when alive. Autozooids appearing identical to the description of *Pacificincola perforata*. Interzooidal communication via multiparous septula, two in the distal wall and two in each lateral wall. Ancestrula typical for *Pacificincola*. Ovicells as in *Pacificincola perforata* but in areas where the colony is folded so that the zooids are tightly appressed, the large, hollow, spinous processes are formed. A median spine is much thicker than the others, often broken, standing up, sometimes with a horizontal side branch pointing proximally. Two long, curved spinous processes in the proximal corners of ooecia and one or two extra spinous processes proximal to the big medial one. Some of the spinous processes are bifid at their tips or have an irregular side branch. They have sometimes the appearance of the frozen arms of an octopus. Why they only occur on folded areas is not known, it seems they keep space for their own tentacle crown (Fig. 4.160).

Occurrence

Belgium In June 2021, collected in the Spuikom in Oostende, a center of oyster culture, by F. Kerckhof.

Remarks

For the genetic marker 12S, the sequence is 100% identical for the three growth forms, which are described as *Pacificincola perforata*, *Pacificincola hozawai,* and the form with locally five long spines on the ovicell (De Blauwe et al. 2024, analysis by Gimaris). If further research confirms the conspecificity of the three growth forms (no spines, 3, or up to 5), we should name them *Pacificincola hozawai* (Okada 1929), as this is the oldest description.

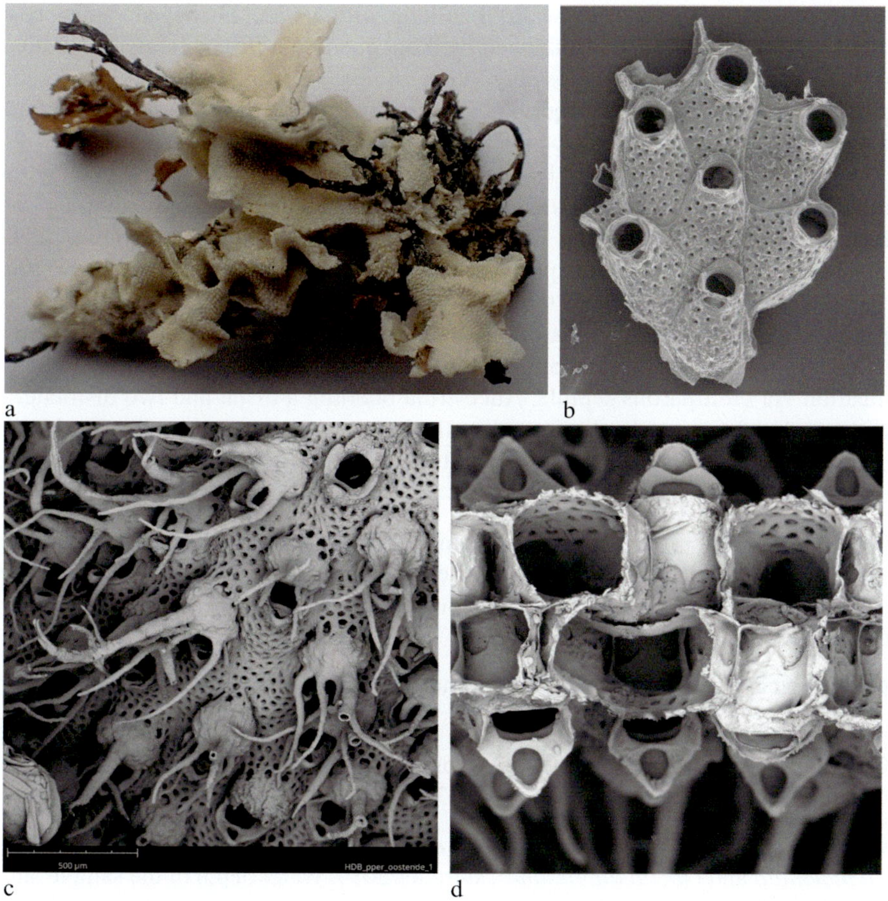

Fig. 4.160 *Pacificincola hozawai*. (**a**) Colony growth on *Sargassum* (photo by F. Kerckhof). (**b**) Ancestrula and periancestrular autozooids. (**c**) Ovicellated zooids with ooecia having spinous processes. (**d**) Bilaminar plate of the colony in cross-section with two large multiparous septula in the distal wall. Oostende spuikom (Belgium)

Genus *Primavelans*

Primavelans glabricollaris Min, Chae, Yang, Noh, Lee and Seo, 2021

Description Colony unilaminar, encrusting when young; growing into bilaminar erect growth. The size of the colonies found is very small, and the maximal size is unknown. Color of preserved material yellowish-white. Zooids irregularly rectangular to hexagonal, 0.5–0.8 mm long. Frontal wall flat to slightly convex, evenly pseudoporous. Orifice mostly wider than long with a wide, D-shaped, sinus broad with a straight proximal edge, flanked by blunt condyles. Orifice rarely longer than wide, especially in smaller (possibly male) zooids. Peristome not developed laterally nor distally. Prominent suboral umbo encorporating a heterozooid, with a foramen facing distally, nearly perpendicular to the frontal plane.

Ovicells prominent, closed by the zooidal operculum. Ooecium formed by the distal autozooid. Ectooecium membraneous, with calcified proximolateral lappets lateral to the ovicell opening. Endooecium completely calcified; young ovicells show tubercles and pits and depressions in radiating series from the proximal border to the contact zone with the distal zooid. Mature ovicells have only some proximolateral pits left. In mature ovicells, endooecium has a smooth proximal edge, typical for this species. Spines and avicularia lacking. Interzooidal communication via multiparous septula. Ancestrula budding a distal and two distolateral primary daughter zooids; the frontal wall of ancestrula covered by the two lateral daughter zooids, having a median longitudinal ridge, only the ancestrular orifice remaining free (Fig. 4.161).

Distribution The type locality is Munseom Island, Korea. The species occurs widespread along the southern and eastern coast of Korea (Min et al. 2021). Its occurrence in Europe is clearly due to its introduction. Found in Japan (pers. comm. Chihiro Kubo).

Occurrence Schüler et al. (2022), see map Fig. 4.161a.

The Netherlands A few loose colonies were collected by Hamon Happer on the Cleaver Bank on March 19, 2019, in a mixed sample of 35-L sediment, consisting of sand, gravel, and pebbles (coll. RWS, 3.01175927, 53.88163941, EPSG:4258).

Germany Two of the colonies grew epizoically on campanulariid hydroids (*Obelia bidentata* and *Obelia longissima*), summer 2019. Water depth of the sample sites varied between 19 and 21 m.

Family Cryptosulidae

Genus *Cryptosula*

Cryptosula pallasiana (Moll, 1803)

Description Colony encrusting, unilaminar, sometimes with pleated or curled edges, orange when alive, becoming white. Zooids 0.6–1.0 mm long. Frontal shield cryptocystidean, uniformly pseudoporous. Pseudopores are each in a deep pit which gives the surface a rough appearance. Orifice longer than wide, bell-shaped, and large (one-third to one-half of the frontal zooid length) surrounded by an erect flared peristome. Proximal margin of the orifice between the condyles is concave. Operculum brown. Spines absent. An umbo can develop proximal to the orifice. Avicularia absent or distal in the suboral umbo. Embryos orange, incubated in the internal brood sac (Fig. 4.162).

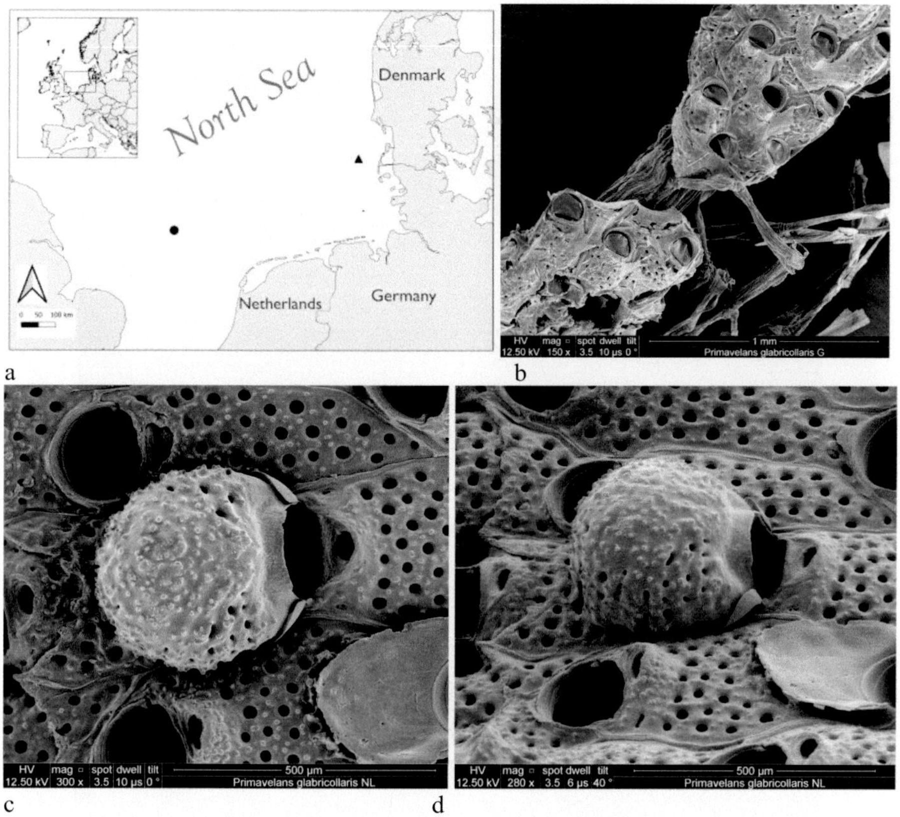

Fig. 4.161 *Primavelans glabricollaris*. (**a**) Map showing two locations where this species has been collected (Schüler et al. 2022). (**b**) Portions of two colonies on a hydroid stem, the upper one with the ancestrular region (Germany). (**c**) Frontal and (**d**) lateral view of the mature ooecia with proximolateral lappets and smooth proximal edge (the Netherlands)

Occurrence

Belgium Lives in the Mercator marina in Ostend (Kerckhof 2000) and since 2001 annually in the marina of Zeebrugge on floats and pontoons. Sometimes found on washed-up plastic.

The Netherlands In the Delta area, this species is common on hard substrates and sometimes on seaweed. In addition, there are records from Texel and Den Helder. Also in brackish and stagnant water, *C. pallasiana* is very common in the Goese Meer. In Egmond aan Zee, about 20 colonies washed up on a fish crate in November 2007 and again in December 2011 (de Ruijter 2008, 2012a). In 2018, Sytske Dijksen found a colony on a silex stone on Texel.

Germany Collected at Heligoland and in the Tiefe Rinne (Kuhlenkamp and Kind 2012).

Fig. 4.162 *Cryptosula pallasiana.* (**a**) Living colony. (**b**) SEM image of a bleached colony, Brittany (France)

Family Cheiloporinidae

Genus *Hagiosynodos*

Hagiosynodos latus (Busk, 1856)

Description Colony unilaminar, encrusting empty shells, small stones, and living gastropods. Autozooids oval, angular or irregular, rather flat. Frontal surface with densely and evenly distributed small round pseudopores. Orifice longer than broad, the distal part horseshoe-shaped, separated from the broad and shallow proximal part by blunt condyles. Operculum brown and conspicuous with marginal sclerites. Oral spines absent. Avicularia not observed. Ovicells subimmersed, ooecium formed by either distal zooid or distal kenozooid, longer than wide; ectooecium membraneous, endooecium calcified, proximally reaching the condyles and surrounded by large round or slit-like pores separated from each other by ribs radiating to the center of endooecium. About a dozen small basal pore chambers are visible laterally and distally at peripheral autozooids (Fig. 4.163).

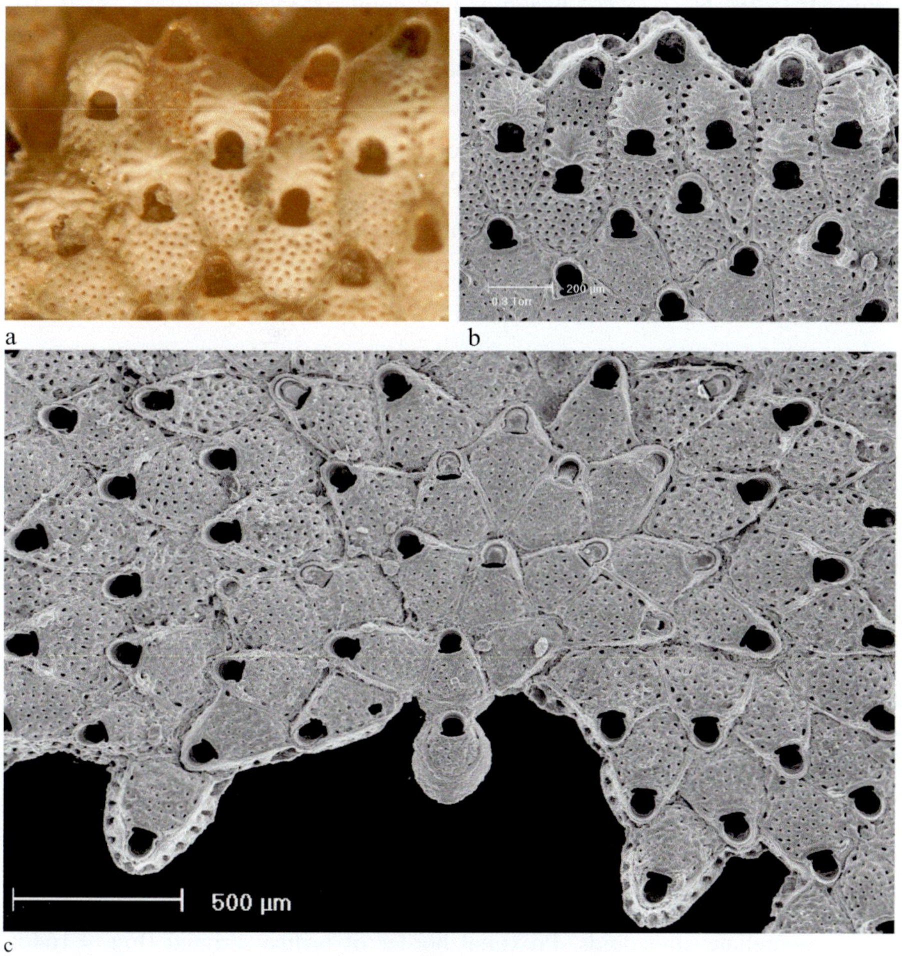

Fig. 4.163 *Hagiosynodos latus.* (**a**) Part of colony growing in a dead shell, Westhinder Bank (Belgium). (**b**) Colony periphery with ovicellate and non-ovicellate zooids. (**c**) Part of the same colony showing ancestrula and periancestrular zooids. On dead shells collected in fishing nets at Le Conquet (France)

Occurrence

Belgium Twenty-two colonies were collected in a very small area on the Flemish Banks on shells and stones in 1903–1907 (RBINS collection). This area is bounded by the trapezoid with the following angles: 51°19′ N, 2°24′ E; 51°23′ N, 2°24′ E; 51°25′ N, 2°34′ E; and 51°30′ N, 2°34′ E. Despite intensive dredging at other locations on the Belgian continental shelf, no other colonies have been collected.

In 2005, 14 deceased colonies were fished up on shell valves on the Westhinder Bank. Small stones were collected by Francis Kerckhof (RBINS) in the extreme north-western part of the Belgian continental shelf, on the border with the United Kingdom on July 12, 2018, during a haul with a Gilson dredge with starting coordinates 51°35′26″ N, 2°15′14″ E and ending point 51°35′36″ N, 2°15′28″ E, 100 stones were examined on epibionts and 37 of them contained living colonies of *H. latus*! A colony was found in December 2020 in a whelk *Buccinum undatum* on the beach of Zeebrugge by A. Beidts.

The Netherlands A colony washed up on plastic (de Ruijter 2012b, 2014a). Rien de Ruijter found another one in January 2022.

France Near Cap Gris-Nez two colonies were found in June 1992 (Davoult et al. 1999).

Family Microporellidae

Genus *Microporella*

Awaiting a general revision of specimens reported from all over the world, focusing on the Mediterranean Sea, the occurrence of ***Microporella ciliata* (Pallas, 1766)** to date is only confirmed in the Gulf of Naples (SE Tyrrhenian Sea) and the Gulf of Noto (W Ionian Sea) (Di Martino and Rosso 2021).

Microporella cf. *ciliata* has been reported from several locations, in situ as beached. SEM photography revealed different forms that are described as five new species hereafter.

1 (a) Ooecium associated with peristome surrounding zooidal orifice (personate zooid), and incorporating the ascopore (Fig. 4.164a). .... *Microporella dirupoi* nov. sp. (p. 378)
(b) Ovicell without peristome, at most two lateral lappets next to the orifice, ascopore never hidden. 2

2 (a) Ooecium forming lateral lappets (Fig. 4.164b), no spines remaining in ovicellate autozooids. Proximal border of orifice serrated (Fig. 4.164c). .... *Microporella kindi* nov. sp. (p. 379)
(b) No lateral lappets, zero to two spines remaining in ovicellate zooids, proximal border of orifice not serrated between the condyles. 3

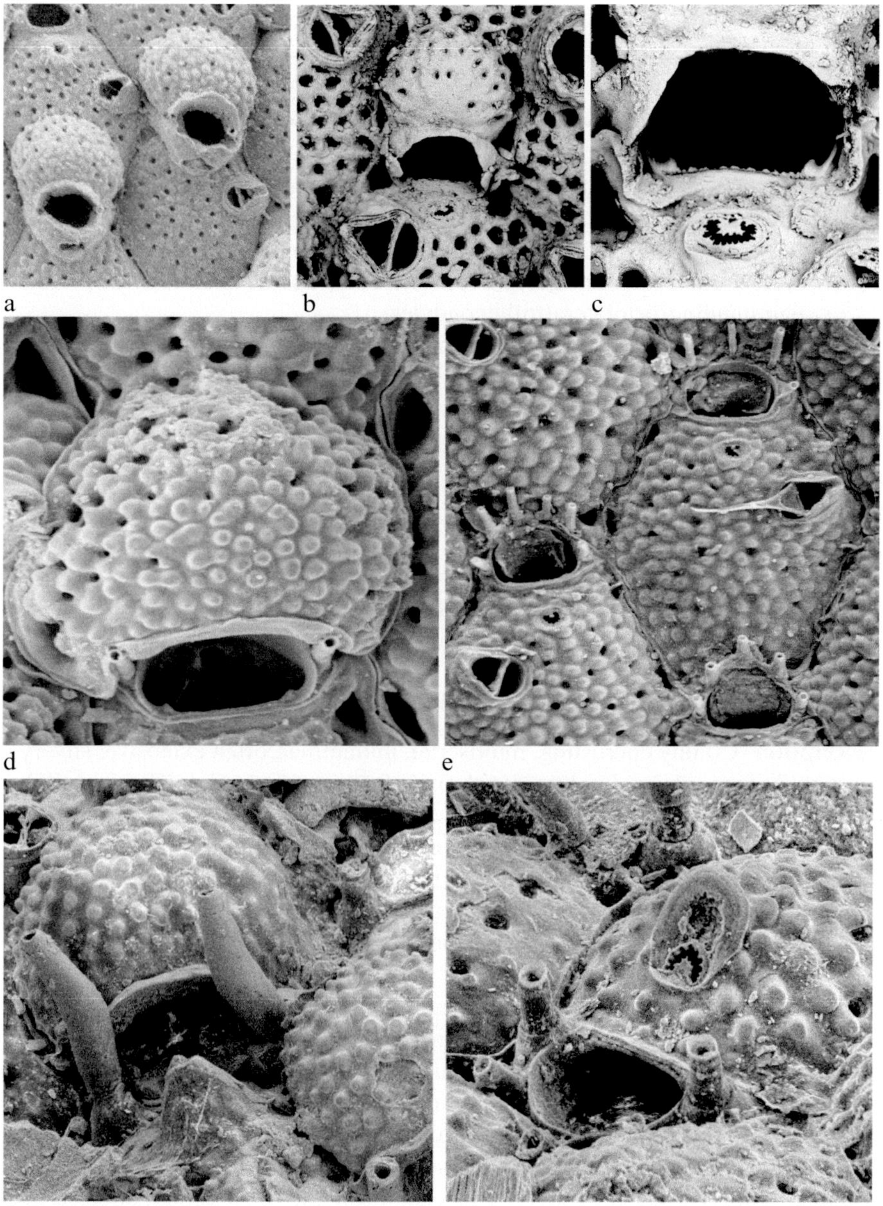

a b c d e f g

Fig. 4.164 Details of ovicellate and non-ovicellate zooids. (**a**) *Microporella dirupoi* nov. sp. with peristome hiding orifice and ascopore. (**b**, **c**) *M. kindi* nov. sp. with lateral lappets and serrated proximal edge of orifice. (**d**, **e**) *M. vanoutryvei* nov. sp. with non-personate ovicell, five to six oral spines seen in non-ovicellate zooids. (**f**, **g**) *M. annae* nov. sp. zooid with non-personate ovicell and two oral spines, and non-ovicellate zooid showing ascopore on the distal side of an umbo

3 (a) Autozooids with five or six oral spines, exceptionally four or seven ascopore not associated to an umbo. .... *Microporella vanoutryvei* nov. sp. (p. 376)
(b) Ascopore distal on a prominent spiky umbo. 4

4 (a) Autozooids with four to five oral spines, the proximal pair thicker than the others and standing at the height of the condyles. .... *Microporella annae* nov. sp. (p. 380)
(b) Autozooids with three to eight spines, all of the same thickness, the proximal pair standing distolateral to the condyles. .... *Microporella helgolandensis* nov. sp. (p. 382)

Microporella vanoutryvei nov. sp.

Microporella ciliata (Pallas) in Hayward and Ryland 1999 p. 296, Fig. 136 A–C.

Type Material

I.G. 34,864 / BRYT.225 holotype on plastic, Marloes Sands, Wales, UK, collected by Hans De Blauwe in April 2001.

I.G. 34,864 / BRYT.222/1b paratype on shell, gold coated for SEM, sampled by J.S. Houziaux (JSH 36) on the Flemish Banks with Gilson dredge in 2005.

Description Colony encrusting, multiserial, unilaminar, often extensive, silvery or vitreous. Autozooids oval to hexagonal, 0.41–0.54 mm long. Frontal shield slightly convex, densely granular and vitreous, pseudoporous by numerous (20 to 40) minute pseudopores. Few marginal areolae. Interzooidal communications through pore chamber windows along lateral walls. Orifice transversely D-shaped, outlined by a thin, raised rim; hinge-line straight, smooth, with a pair of condyles at corners. Oral spines five or six (rarely four or seven); the proximal pair standing distolateral to the condyles; base of proximalmost pair of spines mostly remain visible in ovicellate zooids.

Ascopore field a very narrow, subcircular area of gymnocystal calcification, located proximal to the orifice, slightly raised relative to the adjacent frontal shield; opening transversely C-shaped, with a small, subcircular tongue projecting from distal edge and tiny radial denticles. Avicularium single, relatively large, acute to the frontal plane, located laterally, on either side, at about half zooidal length; crossbar complete, thin; rostrum long lanceolate, channeled and open-ended, directed laterally or distolaterally. Mandible lanceolate, much longer than the rostrum, about 2.28 times longer than rostrum. Ovicellate zooids non-personate. Ovicell prominent, semispherical, ooecium produced by and continuous with frontal shield of distal zooid; ectooecium membranous, calcification of endooecium with smaller and closer-spaced granules than in the frontal shield, and with lateral pits, absent frontally. Ancestrula tatiform with about nine spines (Fig. 4.165).

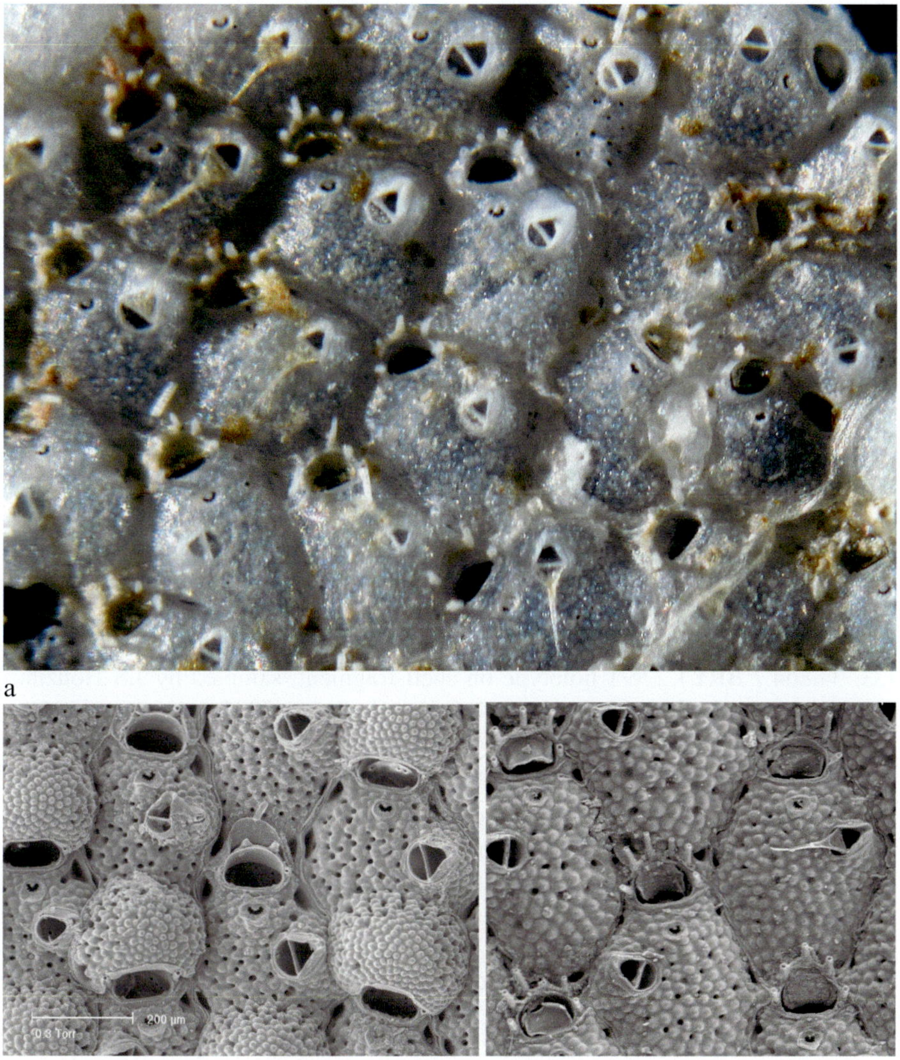

a

b c

Fig. 4.165 *Microporella vanoutryvei* nov. sp. (**a**) Part of colony growing on plastic beached in Wales (the United Kingdom). (**b**, **c**) SEM images of a colony growing on a lobster pot beached St-Malo (France). (**b**) Colony area showing ovicellate zooids with two oral spines. (**c**) Non-ovicellate zooids with five to six oral spines

Etymology

Named after René van Outryve (1937–2017) who always supported me in my research and photography of bryozoans.

Occurrence

Belgium On shell valves dredged from the Hinder Banks in 1905 and 2005 (RBINS collection). Common on beached *Flustra foliacea* (Loppens 1906). On an erosion stone of the Belwind wind farm, in October 2012 and January 2021 at a depth of 15 m. Eleven colonies were found on 100 stones collected by F. Kerckhof (RBINS) in the extreme north-western part of the Belgian continental shelf, on the border with the United Kingdom in July 2018 during a haul with a Gilson dredge between 51°35′26″ N, 2°15′14″ E and 51°35′36″ N, 2°15′28″ E. Often on beached plastic and *Himanthalia* buttons.

The Netherlands Lacourt (1949) reports the species from several locations and substrates. Collected in 2002 on the Cleaver Bank (Van Moorsel 2003). In 2011, in the Prinses Amalia windpark (Vanagt et al. 2013). In 2016, on worm tubes on nets on the wreck Niponia nr 2305 (52°44.520′N–3°58.702′E) (van Leeuwen et al. 2017).

Germany At Heligoland and in the Tiefe Rinne (Kuhlenkamp and Kind 2012).

Microporella dirupoi nov. sp.

Microporella ciliata personate form in Hayward and Ryland 1999 p. 296, Fig. 136 D.

Type Material

I.G. 34,864 / BRYT.226/1 holotype on shell fragment, sampled by J.S Houziaux (JSH 36) on the Flemish Banks with Gilson dredge in 2005.
I.G. 34,864 / BRYT.226/2 paratype on shell fragment gold coated for SEM, sampled by J.S Houziaux (JSH 45) on the Flemish Banks with Gilson dredge in 2005.

Description Colony encrusting, multiserial, unilaminar. Autozooids oval to hexagonal, 0.37–0.43 mm long. Frontal shield slightly convex, densely granular, pseudoporous by numerous (20 to 44) minute pseudopores. Few small marginal areolae. Interzooidal communications through pore chamber windows along lateral walls. Orifice transversely D-shaped, outlined by a thin, raised rim; hinge-line straight, smooth, with a pair of condyles at corners. Oral spines five or six in autozooids.

Ascopore field a very narrow, subcircular area of gymnocystal calcification, located proximal to the orifice, raised to the adjacent frontal shield; opening transversely C-shaped, with a small, subcircular tongue projecting from distal edge and tiny radial denticles. Avicularium single, relatively large, acute to the frontal plane, located laterally, on either side, at about half zooidal length; cross-bar complete or not, thin; rostrum long lanceolate, channeled and open-ended, directed laterally or distolaterally. Mandible not observed. Ovicellate zooids personate. Ovicell prominent, semispherical, ooecium with a proximal visor, produced by and continuous with frontal shield of distal zooid; ectooecium membranous, calcification of endooecium with more pronounced granules than in the frontal shield and pits over most of its surface, less common centrally and proximally. A personate peristomial collar with a pair of oral spines at the junc-

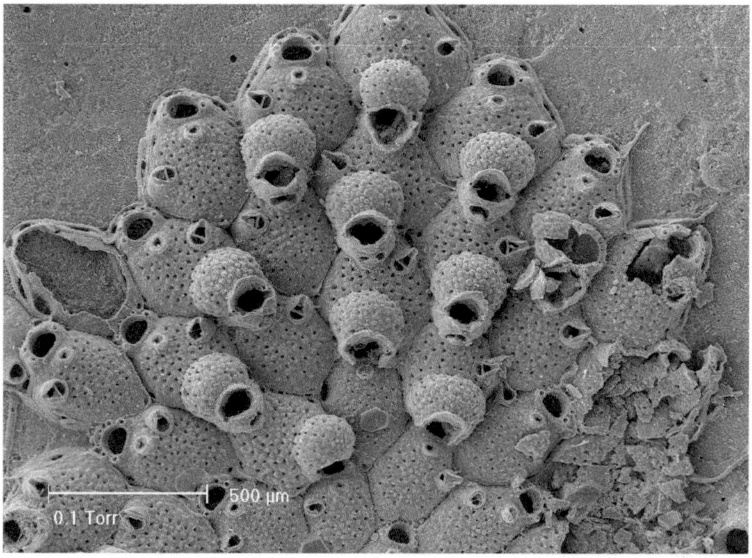

Fig. 4.166 *Microporella dirupoi* nov. sp., SEM image of a colony growing on a dead shell, Westhinder Bank (Belgium)

tion with the vizor of the ooecium. The peristome is formed by the lateral lappets of endooecium, encloses the ascopore, and may form a complete tube or may have a large or two smaller proximal foramen above the ascopore. Ancestrula not observed (Fig. 4.166).

Etymology

Named after the Belgian politician Elio Di Rupo, who characteristically always wears a bow tie, masking his neck, like the peristomal collar in ovicellate zooids.

Occurrence

Belgium Few colonies on shell valves dredged at the Hinder Banks in 1905 and June 2005 (collection RBINS).

Microporella kindi nov. sp.

Type Material

NHMUK 2021.10.15.4, AW214 Holotype: Vatlestraumen, Norway. Scale bar = 100 µm. The complete mitochondrial genome of this specimen is sequenced (Andrea Waeschenbach) (https://www.ncbi.nlm.nih.gov/nuccore/OR460702).

RBINS I.G. 34,864 / BRYT.224 Paratype: on shellfragment, Tiefe Rinne, Germany, 2023.

Description Colony encrusting, multiserial, unilaminar, frontal shield thick, white. Autozooids oval to hexagonal, 0.40–0.58 mm long. Frontal shield slightly convex, a bit granular, with numerous (24 to 37) relatively large pseudopores. Few marginal areolae. Orifice transversely D-shaped, outlined by a thin, raised rim; hinge-line straight, roughly serrated, with a pair of prominent, out-turning condyles at corners. Oral spines five; none remaining in ovicellate zooids.

Ascopore field an oval area of gymnocystal calcification, located proximal to the orifice, slightly raised relative to the adjacent frontal shield, a bit higher proximally; opening transversely C-shaped, with a small, subcircular tongue projecting from the distal edge and tiny radial denticles. Avicularium single, relatively large, acute to the frontal plane, located laterally, close to the ascopore; crossbar complete, thin; rostrum long lanceolate, channeled and open-ended, directed laterally or distolaterally. Mandible lanceolate, about 1.5 to 2 times as long as the rostrum. Ovicellate zooids semi-personate (lateral flaps not enclosing the ascopore). Ovicell prominent, ooecium relatively small (about 190 µm long and wide); ectooecium membranous, calcified endooecium produced by and continuous with the frontal shield of the distal zooid, with fewer pits centrally and proximally. Ancestrula not observed (Fig. 4.167).

Etymology

Named after Britta Kind, who first observed the aberrant morphology from other *Microporella* species in the area.

Occurrence

Belgium Not observed but *Microporella* material should be re-examined.

The Netherlands One small colony on an oyster shell on August 17, 2023, in the Eastern Scheldt collected by GIMARIS, det. H. De Blauwe.

Germany Tiefe Rinne, not uncommon (obs. Britta Kind, present author).

Microporella annae nov. sp.

Type Material

I.G. 34,864 / BRYT.223/1: holotype on shell fragment, Pleneuf-Val-André, France, collected by Hans De Blauwe in March 2014.

I.G. 34,864 / BRYT.223/2: paratype on shell fragment, Isle Chausey, France, collected by Hans De Blauwe on 31 March 2002.

I.G. 34,864 / BRYT.223/3: 2 paratypes on shell fragment, Pleneuf-Val-André, France, collected by Hans De Blauwe in March 2014.

Description Colony forming a small encrusting patch on disarticulated shells or plastic. Autozooids oval to hexagonal, 0.31–0.39 mm long. Frontal shield convex, densely granular, pseudoporous by less numerous (6 to 25) minute pseudopores. Few large marginal areolae. Interzooidal communications through pore chamber windows along lateral walls. Orifice transversely D-shaped, outlined by a thin,

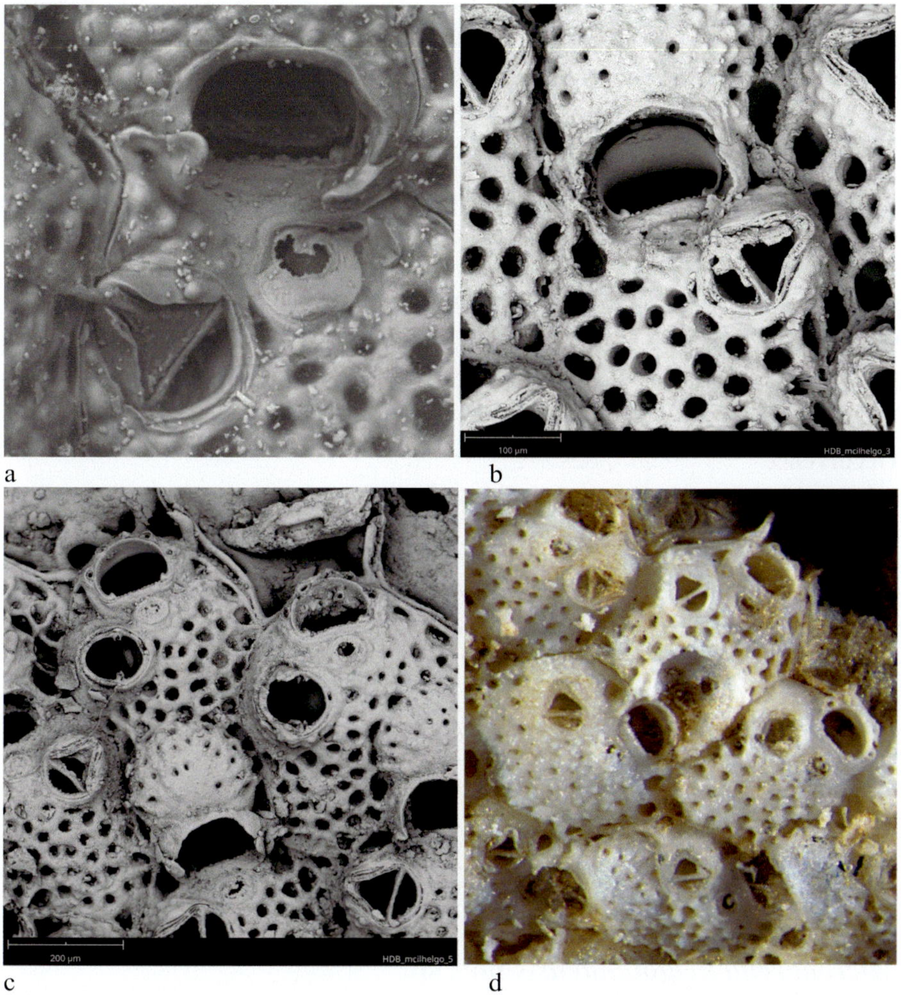

Fig. 4.167 *Microporella kindi* nov. sp. (**a**) SEM image of the holotype NHMUK 2021.10.15.4, AW214, Norway. (**b–d**) Paratypes, Tiefe Rinne (Germany). (**a**, **b**) Ovicellate semi-personate zooids showing condyles and serrated proximal edge of zooidal orifice. (**c**) Ovicellate and non-ovicellate zooids. (**d**) Colony area showing rugose, heavily calcified appearance

raised rim; hinge-line straight, smooth, with a pair of inconspicuous condyles at corners. Oral spines four or five; the proximal pair of spines is thicker than the others, the proximal pair thicker than the others and standing at the height of the condyles; and persists in ovicellate zooids. Spines have a brown chitinous base. Ascopore field acute to the frontal plane, elongated proximally and becoming rounded spiky, located proximal to the orifice; opening transversely C-shaped, with a small, subcircular tongue projecting from the distal edge and tiny radial denticles.

Avicularia are sparse, single, elevated above the frontal surface, located laterally, on either side, at about half zooidal length; crossbar complete; rostrum long lanceolate, channeled and open-ended, directed proximolaterally, laterally or distolaterally. Mandible not observed. Ovicells are produced in very early astogeny compared to other species of *Microporella*.

Ovicellate zooids non-personate. Ovicell prominent, semispherical, ooecium with a narrow proximal visor, produced by and continuous with a frontal shield of distal zooid; ectooecium membranous, calcification of endooecium with equally sized granules as in the frontal shield and with lateral pits, absent frontally. Ancestrula tatiform with ten spines ($n = 3$) (Fig. 4.168).

Remarks This species is very similar to *Microporella pachyspina* Di Martino and Rosso 2021, especially in having a more robust proximal pair of oral spines, but the new species does not have such ring of large pores around the ovicell.

Occurrence

UK Marloes Sands (Wales) one colony on beached plastic in April 2001.

France One colony on *Cerastoderma edule*, Isle Chausey, March 2002. Many colonies in empty shells at Pléneuf-Val-André (De Blauwe 2019). Several colonies of *Microporella* cf. *annae* were collected by F. Gully in September 2023 on *Saccharina latissima* at Erquy. These zooids have three oral spines and occasionally four spines.

Belgium Material of *Microporella* cf. *ciliata* in collections should be re-examined, this species is expected to be found.

The Netherlands On beached plastic sampled by Rien de Ruijter at Egmond aan Zee in 2013, re-examined and confirmed in 2024. On the wrek of Rode Poon (SS Trevier) 52°26.201′ 03°43.970′ in June 2021 (van Leeuwen 2023).

Microporella helgolandensis nov. sp.

Type Material

I.G. 34,864 / BRYT.227/1: holotype, on SEM-stub; on *Laminaria* holdfast on Heligoland, collected by Hans De Blauwe in 2024.
I.G. 34,864 / BRYT.227/2: paratype, on SEM-stub; on *Laminaria* holdfast on Heligoland, collected by Hans De Blauwe in 2024.

Description Colony forming a small encrusting patch on *Laminaria* holdfasts. Autozooids oval to hexagonal, 0.34–0.48 mm long. Frontal shield convex, densely granular, pseudoporous by minute pseudopores. Few large marginal areolae. Interzooidal communications through pore chamber windows along lateral walls. Orifice transversely D-shaped, outlined by a thin, raised rim; hinge-line straight, smooth, with a pair of condyles at corners. Oral spines four or five, but sometimes three to

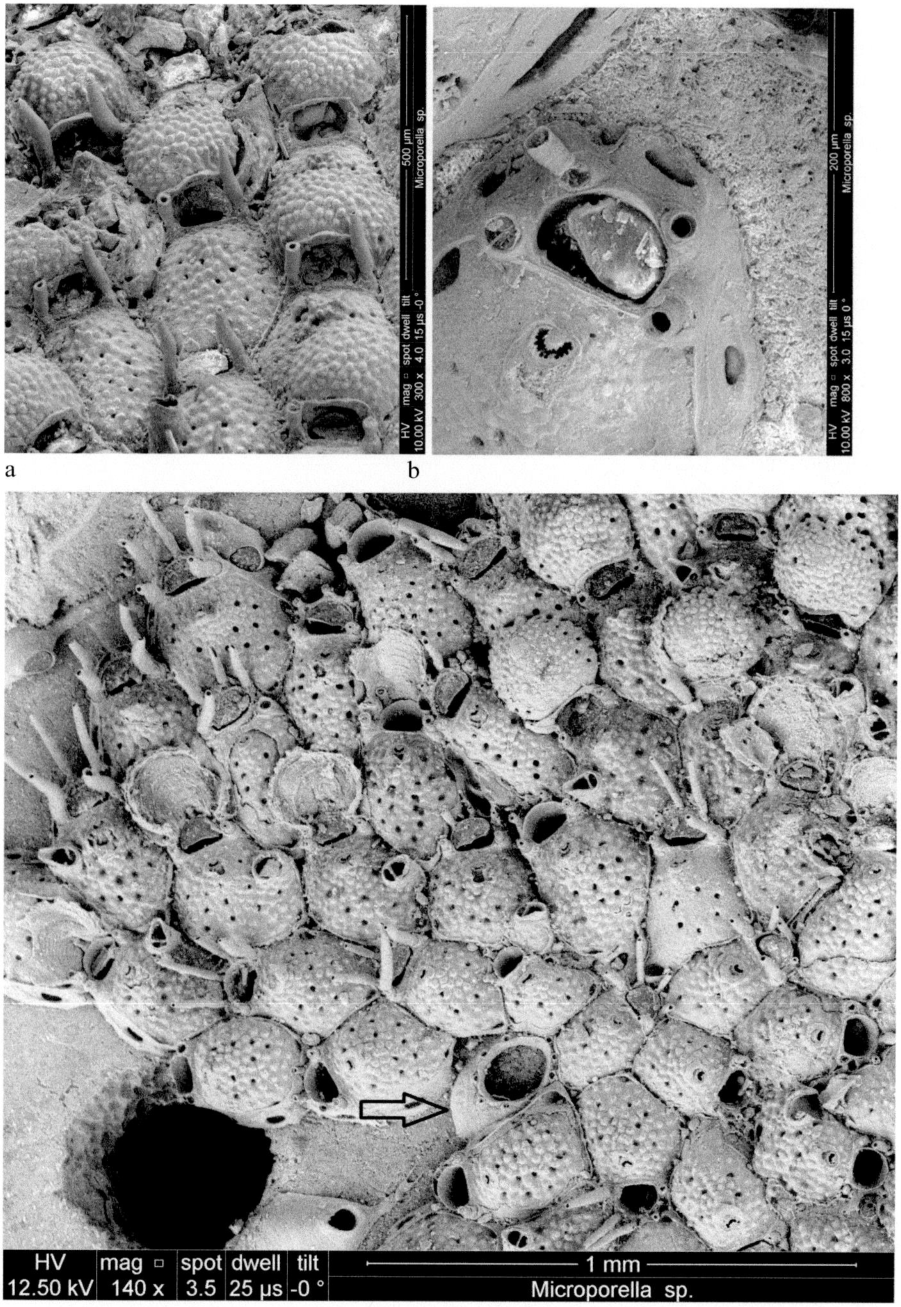

Fig. 4.168 *Microporella annae* nov. sp. (**a–c**) Colony growing in a dead shell, Pléneuf-le-Val-André (France). (**a**) Ovicellate zooids with a pair of thick oral spines. (**b**) Zooidal orifice showing displaced operculum, small triangular condyle, and bases of the oral spines. (**c**) Part of the colony including ancestrula (*arrow*), periancestrular region, and ovicellate and non-ovicellate autozooids with avicularia and spines

eight; the proximal pair of spines is not thicker than the others, standing distolateral to the condyles; and persists in ovicellate zooids. Ascopore field acute to the frontal plane, distal to a spiky umbo located proximal to the orifice; opening transversely C-shaped, with a small, subcircular tongue projecting from the distal edge and tiny radial denticles. Avicularia are sparse, single, elevated above the frontal surface, located laterally, on either side, at about half zooidal length; rostrum long lanceolate, channeled and open-ended, directed distolaterally.

Ovicellated zooids non-personate. Ovicell prominent, semispherical, ooecium produced by and continuous with a frontal shield of distal zooid; ectooecium membranous, calcification of endooecium granulated and with pits. Ancestrula not observed (Fig. 4.169).

Occurrence

Germany Two colonies were collected on *Laminaria* holdfasts at the lower beach of Heligoland in 2024.

Superfamily Celleporoidea

Family Celleporidae

Genus *Cellepora*

Cellepora pumicosa (Pallas, 1766)

Description Colony white/orange, developing as a lenticular patch, becoming domed or nodular, or forming a pumice stone-like cylinder around erect substrate as hydroids or fishing nets. Zooids at the colony margin oval and convex. Zooids in superposed laminae erect with little of the frontal shield visible. Despite the appearance of condyles and a sinus, the opening is more or less round. No oral spines. Peristome as a thin, raided rim around the proximal and lateral borders of the orifice. A columnar spike develops proximal to the orifice, in which lies laterally an apex-facing avicularium, its rostrum perpendicular to the orifice. Frontal shield smooth with few areolae. Ovicells prominent, spherical, ectooecium smooth with three to six irregular frontal pseudopores in the proximal half. Embryos red. Ancestrula oval with a rounded frontal membrane surrounded by spines (Fig. 4.170).

Occurrence

Belgium Loppens (1906) mentions them as common on stones and shells and especially on the hydroid *Abietinaria abietina*. In the period 1899–1911, several colonies were collected on the Flemish Banks. Deceased colonies, as pumice stones with a length of 1–7 cm, are regularly encountered during sampling at the Kwinte Bank and the Hinder Banks. Fresh colonies often wash up on plastic, nets, and *Himanthalia* (De Blauwe 2005a).

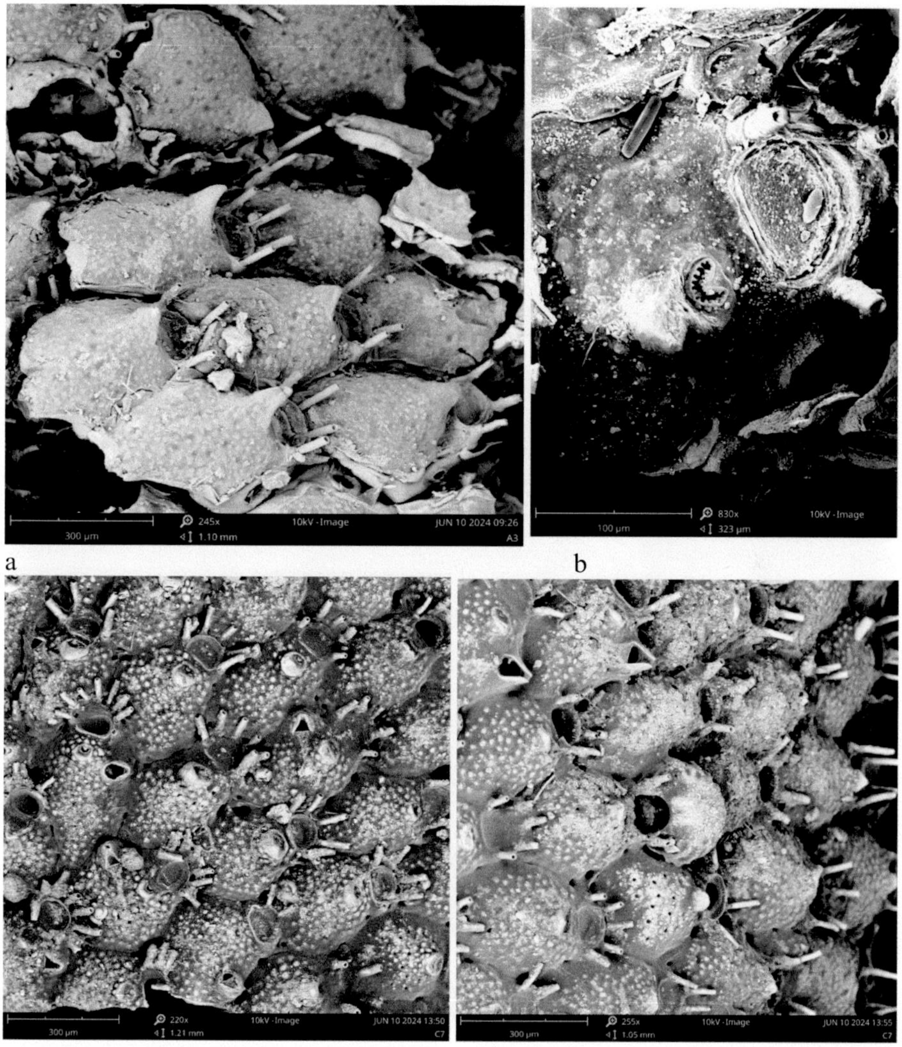

Fig. 4.169 *Microporella helgolandensis* nov. sp. (**a**) Oblique lateral view of unbleached colony area, showing zooids with umbo, on *Laminaria* holdfast, from Heligoland (Germany). (**b**) Same colony, showing proximal pair of oral spines. (**c**) Another unbleached colony showing non-ovicellate zooids with up to eight oral spines. (**d**) Same colony showing the first pair of oral spines persisting in ovicellate zooids

The Netherlands Reported as washed up on hydroids, cork, stones, worm tubes, and more recently on plastic and rope (Lacourt 1949; de Ruijter 2006b, 2007, 2012a, 2013a, 2018d). Collected in 1989 on the Dogger Bank, growing on *Abietinaria abietina* at object "Z," just west of the Dutch continental flat (Van Moorsel and

Fig. 4.170 *Cellepora pumicosa.* (**a**) Living colony growing on a stone in Brittany (France). (**b**) Loose colony, presumably once encompassing a hydroid, Westhinder Bank (Belgium). (**c**) Zooids with avicularia and developing ovicells

Waardenburg 1990). Found again in 2011 at probably the same object viz. the remains of the platform "Ocean Prince" (unpublished record Wouter Lengkeek). The first record within the dutch continental flat (DCF) originates from the Cleaver Bank (Van Moorsel 2003). In 2010, the species has been recorded by Wouter Lengkeek from the wreck of the "Elbe" closer inshore, west of Texel (Faasse et al. 2013). Collected during DDNZS in June 2018 from the wreck of the Somali.

Genus *Buffonellaria*

Buffonellaria muriella Berning and Kuklinski, 2008

Description Colony encrusting, multiserial, unilaminar to multilaminar. Zooids hexagonal, quadrangular, or oval, separated by shallow grooves that become obscured by secondary calcification. Frontal wall slightly convex, smooth, initially pseudoporous, later closing pseudopores or with only tiny ones; four to seven slit-like, marginal areolae. Primary orifice usually longer than wide; anter more or less semicircular, proximal margins short, straight, and slightly sloping, passing into rounded shoulders of poster, deep and rounded V-shaped to relatively shallow and broadly U-shaped, occupying more than half of the total width; condyles variable in shape and size, blunt. Ovicell initially prominent, globular; during later ontogeny, it becomes almost completely immersed by secondary calcification; ectooecium membranous, exposed calcified endooecium semicircular, relatively flat, marked with 11 to 13 radiating tubercular ribs slightly thickening toward margin, surrounded by a broad smooth secondary calcification. Small oral avicularia single or paired, usually proximolateral or lateral to orifice, occasionally distal to mid-distance of orifice, oval, situated on a slightly raised and swollen cystid, with a tiny, proximal, areolar pore, frontal plane at an acute angle to colony surface; rostrum semielliptical, directing proximally or proximolaterally; proximal uncalcified area semicircular; crossbar complete, occasionally with a small columella. Additional larger avicularia present in older areas of the colony, arising from a marginal pore, on a large swollen cystid with areolae; rostrum elongate triangular, pointing in various directions; proximal uncalcified area semicircular; crossbar complete with a median columella (Fig. 4.171).

Occurrence

The Netherlands Washed up in 1947, reported under the old name *Stephanosella biaperta* in Bloklander and Leenhouts (1948). Then again in 2013 (de Ruijter 2014b) on plastic originating from the English Channel.

Until 2008 this species was considered *Buffonellaria divergens*. *Buffonellaria muriella* has a population in the English Channel and in the Adriatic Sea. The Atlantic population occurs west to the Iroian Sea (west of Brittany) and not to the Irish Sea as Berning and Kuklinski (2008) mistakenly stated.

Genus *Celleporaria*

Celleporaria brunnea (Hincks, 1884)

Description Colony encrusting, unilaminar or multilaminar. Autozooids rectangular, in radiating linear series in the first layer, more irregular in position and morphology in areas of frontal budding. Opercula and mandibles dark brown in color,

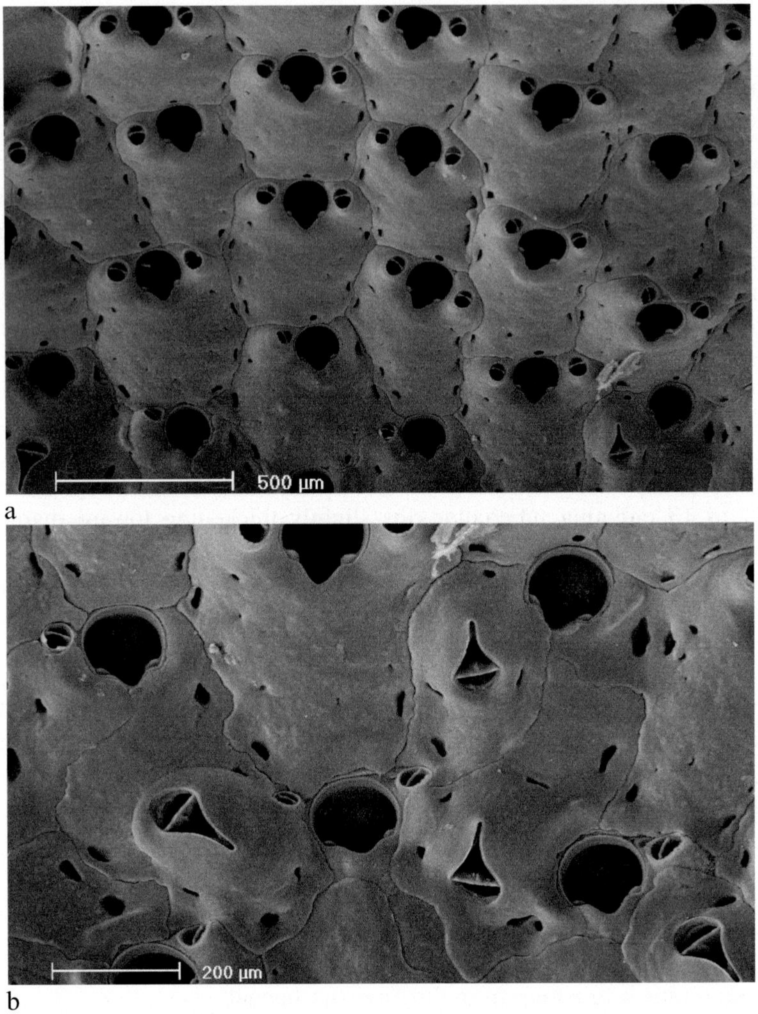

Fig. 4.171 *Buffonellaria muriella*. (**a**) Autozooids with typical avicularia. (**b**) Non-ovicellate autozooids with small and enlarged avicularia. Colony collected on a dead shell in a fishing net in the harbor of Le Conquet (France)

zooids white to gray. Lophophore tentacles brown. Frontal wall granular, with areolae. Primary orifice slightly wider than long. Distal margin rounded, and proximal margin lightly bowed with a rounded sinus, often pinched together at the top so that the sinus is almost closed. Two widely spaced, distal, oral spines present in early ontogeny, later lost, and their bases obscured by calcification. Orifice rounded with a slight peristome, raised proximally, with an umbo that supports an oval suboral avicularium, with toothed distal rim, perpendicular to frontal plane. Vicarious avicularia large, spatulate, with dark brown spade shaped columella and parallel-side rostrum. Ovicells prominent, noncleithral, with widely open orifice; ectooecium covered by the granular secondary calcification (Fig. 4.172).

Fig. 4.172 *Celleporaria brunnea.* (**a**) Part of the colony growing on beached plastic showing autozooids, ovicells, and avicularium, Bergen aan Zee (the Netherlands). (**b**, **c**) Colony areas showing zooidal orifices with sinus, oral spines, suboral avicularia, and vicarious avicularium; operula and mandibles are dark brown, Banyuls-sur-mer (France, Mediterranean). (**d**) Spade-like mandible of vicarious avicularium. (**e**) Non-ovicellate zooids with suboral avicularia and their toothed rostrum, perpendicular to the frontal plane, and vicarious avicularium. (Photo by G. Droual and N. Gayet)

Occurrence Originally described from British Columbia and is a very common species on the Pacific coast of North America. The west coast distribution extends from British Columbia to the Galapagos Islands, Ecuador. Recently, *Celleporaria* invaded the Mediterranean (Aegean Sea) and Portugal (Cascais) (Canning-Clode et al. 2013; Lodola et al. 2015). The nearest known location where *Celleporaria brunnea* occurs is Arcachon on the Atlantic coast of France (Harmelin 2024). Figure 4.172b–e shows details of a large colony collected by the author in the marina of Banuyls-sur-Mer, France, in the Mediterranean in 2024.

The Netherlands Rien de Ruijter found a colony on beached plastic in Bergen aan Zee in November 2023, together with the barnacle *Amphibalanus amphitrite* and the bryozoan *Biflustra tenuis*.

Genus *Celleporina*

1 (a) Colonies extensive. Opening with clear U-shaped sinus. Peristome a concave rim on the proximal side of the opening. .... *Celleporina hassallii* (p. 392)
(b) Colonies very small. Orifice with deep V-shaped sinus or with wide shallow sinus. 2

2 (a) Opening with deep V-shaped sinus. Proximally, the rim of the aperture has a rounded raised mucro, often as high as the avicularia. Typically on hydroids and fine nylon nets .... *Celleporina decipiens* (p. 390)
(b) Opening with wide shallow U-shaped sinus. High peristome. Typically on shells. 3

3 (a) Zooid shorter than 0.5 mm. Normally a high peristome with one or two tiny avicularia on the edge. .... *Celleporina pygmaea* (p. 392)
(b) Zooid longer than 0.6 mm. High peristome without avicularia on the rim. .... *Celleporina* sp. (p. 393)

Celleporina decipiens Hayward, 1976

Description Inconspicuous small colonies. Zooids resembling a small *Celleporina hassallii*, the sinus is deep, rounded V-shaped. The peristome forms a rounded raised mucro at the proximal edge of the aperture, often as high as the avicularia. Avicularia on either side of the aperture, columnar with terminal rostrum. Usually no vicarious avicularia. Ovicells prominent, globose, terminal, with smooth ectooecium having narrow lunar membranous area and porous part of calcified endooecium beneath it, visible in cleaned colonies. Embryos red (Fig. 4.173).

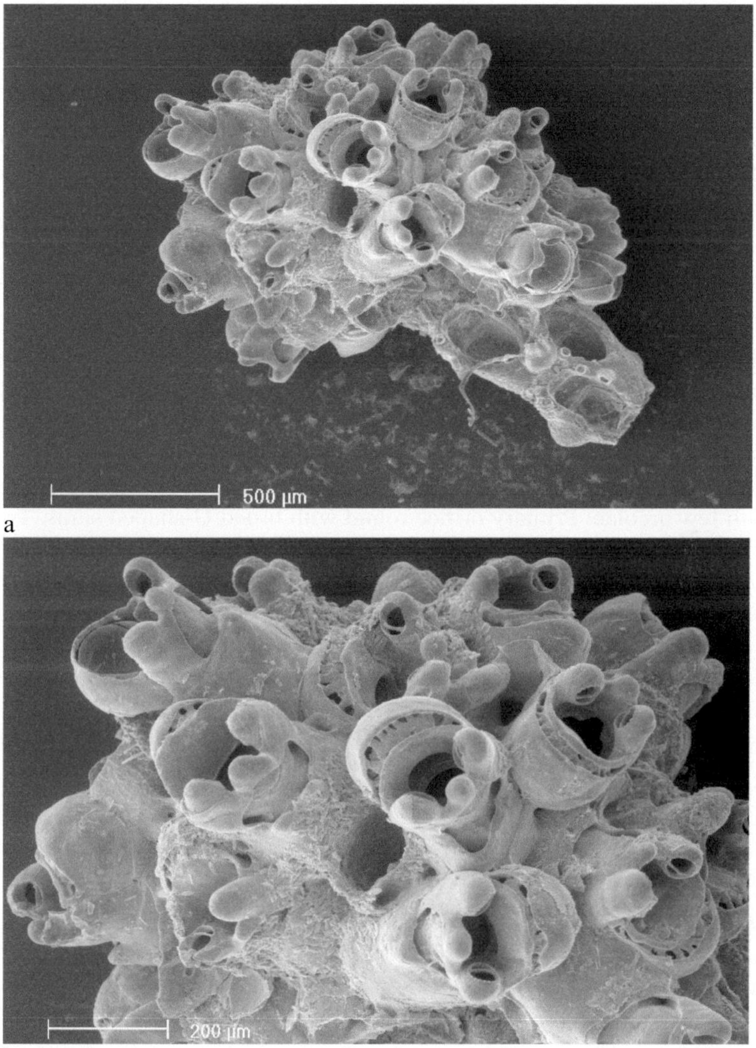

Fig. 4.173 *Celleporina decipiens*. (**a**) Entire colony detached from unknown substrate, beached at the Belgian coast. (**b**) Detail of the same colony showing ovicellate and non-ovicellate zooids

Occurrence A colony on *Flustra foliacea* was dredged in September 1900 (51°17′ N, 2°44′ E, collection RBINS). Beached on a thin net in Middelkerke in January 2000 and on a lobster pot between Oostduinkerke and De Panne in Oktober 2004 (De Blauwe 2005a).

The Netherlands On a beached lobster pot at Westkapelle in 2004 and in 2012 on plastic (de Ruijter 2013c). To be expected on beached hydroids as *Sertularia* or *Hydrallmania*.

UK In 2015, five colonies on *Cellaria fistulosa* at 7 km southeast of Folkestone (51°02.416 N, 01°16.269 E), identified by Kris Spurrier. On the English part of the Dogger Bank (pers. comm. Joël Cuperus) .

Celleporina hassallii (Johnston, 1847)

The opinion of Jean-Loup d'Hondt (2000), without scanning evidence, that this is *Celleporina caliciformis* (Lamouroux, 1816) is not accepted by Dennis Gordon and present author. *C. caliciformis* may be treated as a nomen oblitum.

Description Colony orange, developing as a lenticular patch, becoming domed or nodular, or forming a pumice stone-like cylinder around erect substrate as hydroids or fishing nets. Colony resembling *Cellepora pumicosa*, often more glistening. Zooids at the colony margin oval and convex. Zooids in the middle of the colony more erect with little of the frontal shield visible. Frontal shield smooth, often vitreous, with few areolae. Primary orifice round with broad U-shaped sinus. Two erect lateral oral avicularia as small cylinders proximolateral to the orifice often bent somewhat to each other. Their rostrum is oval and small, crossbar complete, without columella. Peristomial rime proximal between the avicularia deeply concave. Larger, vicarious aviculara with spatulate rostrum, are often present on the colony. Ovicells prominent, terminal, globose, with smooth ectooecium having narrow lunar membranous area. Embryos red. Ancestrula tatiform, surrounded by about seven spines (Fig. 4.174).

Occurrence Found in Belgium, the Netherlands, and Heligoland on washed-up plastic and *Himanthalia* buttons.

Celleporina cf. *pygmaea* (Norman, 1868)

Description Forms a small colony on a shell. Zooids 0.3–0.4 mm long, oval and convex with a smooth surface and few inconspicuous areolae. Opening round with small hinge points. Peristome large and cylindrical and somewhat flared, often the edge is provided with blunt projections. There are one or two tiny avicularia on the edge of the peristome. Ovicells terminal, ooecium wider than long, with narrow lunar membranous area, and porous part of calcified endooecium beneath it (Fig. 4.175).

Occurrence Rarely reported from Western Norway, the Faroes, and the Shetlands deeper than 50 m.

Belgium One damaged colony was collected on the Westhinder Bank in 2005, which corresponds well to the above description. The zooids are slightly longer here (about 0.5 mm) and ridges between the pores on the frontal field on the ovicell are not really visible.

a

b

Fig. 4.174 *Celleporina hassallii.* (**a**) Colony on beached *Himanthalia*, Oostduinkerke (Belgium). (**b**) Cylindrical avicularia and a membranous area hiding endooecium of the ovicell in unbleached material

Celleporina sp.

Description Forms a small colony on a shell. Zooids large, 0.8 mm long, oval and convex with conspicuous areolae and pores around the peristome. Primary orifice hidden in a high peristome so no sinus or condyles are seen. In zooids without ovicell, the peristome forms a large cylinder. In zooids with ovicells, the edge of the peristome is provided with blunt projections. Avicularia have not been observed. Ovicell terminal, globose, ooecium wider than long, ectooecium with lunar membranous area, and porous part of calcified endooecium beneath it (Fig. 4.176).

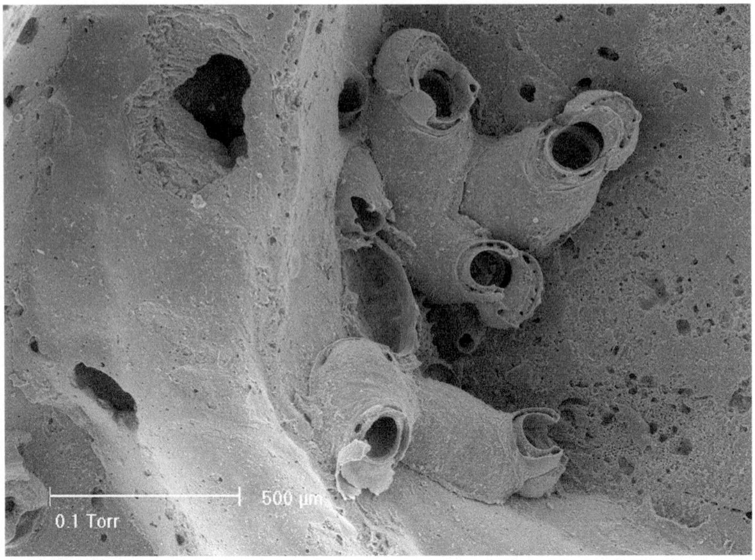

Fig. 4.175 *Celleporina* cf. *pygmaea*. A colony with ovicellate zooids growing in a dead shell, Westhinder Bank (Belgium)

Occurrence

Belgium One colony was collected on the Westhinder Bank in 2005.

Genus *Lagenipora*

Lagenipora lepralioides (Norman, 1868)

Description Colony encrusting, unilaminar, forming a small patch up to 3 mm in diameter. Zooids up to 0.4 mm long. Primary orifice round, with minute condyles, hidden by a tall peristome. Frontal shield pustulose, with three to seven areolae. No avicularia. Ovicell prominent, terminal, globose, ectooecium with large frontal membranous area, lost in dry colonies, calcified endooecium smooth. Zooids without clear orientation, empty spaces are filled in by kenozooids that are as large as or smaller than autozooids (Fig. 4.177).

Occurrence Offshore on shell fragments or pebbles, known from Shetland, the English Channel, the Channel Islands, the western Mediterranean, and Faroe (Hayward and Ryland 1999).

Belgium Three colonies were found on shells from the Westhinder Bank in 2005, of which two in the same whelk.

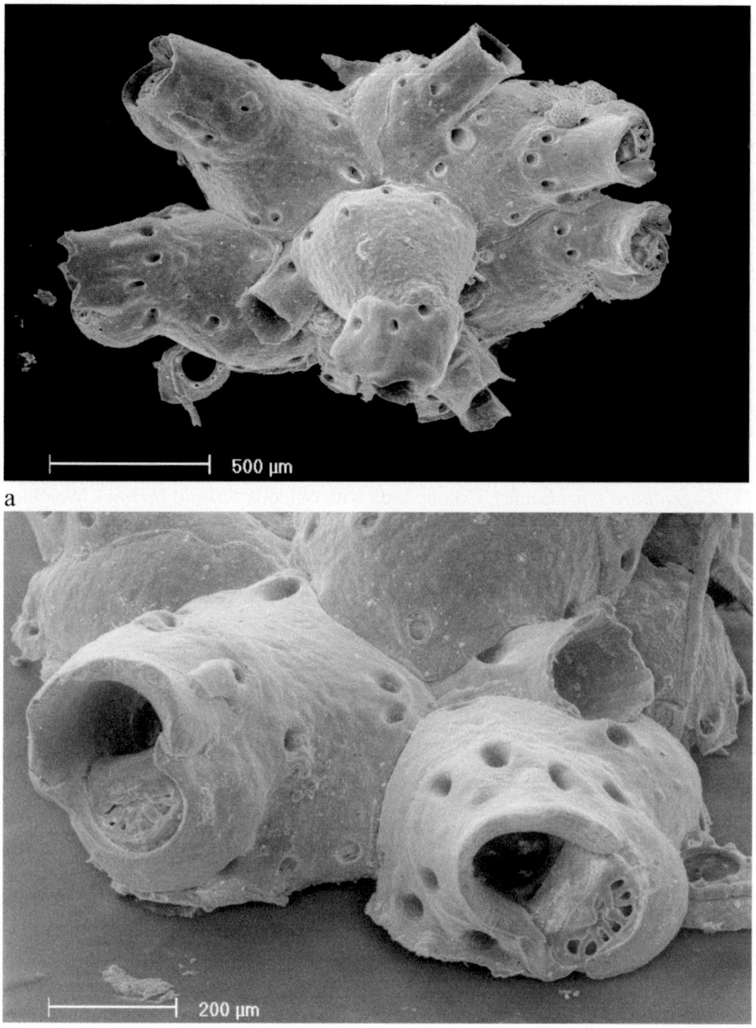

Fig. 4.176 *Celleporina* sp. (**a**) Small colony, Westhinder Bank (Belgium). (**b**) Distal view of two ovicellate zooids showing terminal ovicells having lunar area

Genus *Omalosecosa*

Omalosecosa ramulosa (Linnaeus, 1767)

Description Colony encrusting at first, later developing a branched staghorn growth, up to 4 cm high. Autozooids disposed around the branch axis, oval, convex, with few small areolae. No oral spines. Primary orifice wider than long, hidden by the peristome, a large hollow spike, and a lateral suboral pseudospiramen. Avicu-

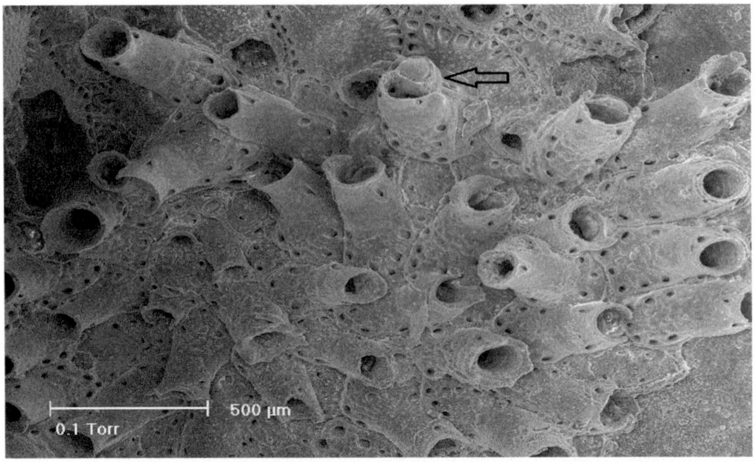

Fig. 4.177 *Lagenipora lepralioides*. Colony area with ovicellate zooid (*arrow*), Westhinder Bank (Belgium)

larium laterally facing in the suboral spike. Rostrum hooked distally, crossbar complete, small median columella present. Ovicell terminal, globose, noncleithral. Ectooecium fully calcified, smooth (Fig. 4.178).

Occurrence

Belgium A colony was dredged in 1899 (51°20′ N, 2°53′ E, RBINS collection). Washed ashore once on plastic in Heist on December 20, 2005.

The Netherlands Washed ashore on plastic and rope in 2011 at Egmond aan Zee (de Ruijter 2011b), 2012 (de Ruijter 2013a), at Noordwijk (de Ruijter 2014c), Ouddorp (de Ruijter 2014c), six colonies near Petten (de Ruijter 2016) and again six colonies on November 24, 2020, on Texel, (pers. comm. Rob Dekker, de Ruijter 2021a).

Genus *Turbicellepora*

1 (a) Orifice with narrow U- or V-shaped sinus. Ovicell with multiple pseudopores. Vicarious spatulate avicularia present. .... *Turbicellepora avicularis* (p. 397)
(b) Orifice with broad shallow sinus. Ovicell with a few large pseudopores. Vicarious avicularia with broadly triangular or elliptical mandible. .... *Turbicellepora magnicostata* (p. 399)

Fig. 4.178 *Omalosecosa ramulosa.* (**a–c**) Colonies detached from a lobster pot, Le Conquet (France). (**a**) Various colony forms. (**b**) Close-up of the branch. (**c**) Ovicellate zooid with avicularia, one laterally facing in a suboral spike; its complete crossbar and a small median columella are visible

Turbicellepora avicularis (Hincks, 1860)

Description Colony lenticular at first, later pumice-like as *Cellepora pumicosa*, often quite large (up to 5 cm). Decumbent zooids convex with areolae, which are especially visible at the tip of erect zooids in the center of the colony. Orifice circular with a V-shaped sinus, occupying one-third of the length of the orifice. Peristome mainly developed proximally with a single suboral avicularium to the left or right. The columnar cystid containing the avicularium is less pronounced than the columnar spike of *C. pumicosa*. Mandible obtuse triangular and obliquely directed upward. Larger vicarious avicularia either with elongate spatulate mandible or obtuse triangular mandible. Ovicell prominent, globose, ectooecium smooth, with about 6 to 12 frontal pseudopores, laterally overgrown by the secondary calcification. Embryos yellow.

Ancestrula tatiform, surrounded by about ten spines (Fig. 4.179).

Fig. 4.179 *Turbicellepora avicularis*. (**a**) Large colony, Westhinder Bank (Belgium). (**b**) Part of colony growing on a lobster pot, Le Conquet (France). (**c**) Ovicellate and non-ovicellate zooids and different types of avicularia and orifice with V-shaped sinus

Occurrence

Belgium The RBINS contains a lot of old material, remarkably much on *Tubularia indivisa* and some on *Abietinarina abietina* (De Blauwe et al. 2006). Vincent Zintzen recently found huge amounts of *Tubularia indivisa* on shipwrecks off the Belgian coast, there was no *Turbicellepora avicularis* to be found. It seems that this species is extinct here. Deceased colonies, resembling pumice stones, with a length of 1 to 4 cm are regularly found during sampling at the Kwinte Bank and the Hinder Banks (see also *Cellepora pumicosa*). Washed up on plastic and nets (De Blauwe 2005a; Vanhaelen et al. 2006) and once on the hydroid *Hydrallmania falcata* in 2023; on plastic in 2024 at De Panne.

The Netherlands Lacourt (1949) reports *T. avicularis* (as *C. armata*) from Terheyde, Scheveningen, Katwijk, and Zandvoort-IJmuiden on (washed up?) hydroids and cork and stones. Washed up on plastic and nets (de Ruijter 2012b, 2013a, 2014c, d, 2015a, d, 2016, 2021b, c, 2023).

Germany Collected in the Tiefe Rinne (Kuhlenkamp and Kind 2012).

Turbicellepora magnicostata (Barroso, 1919)

Description Colony encrusting, forming an orange domed patch or nodular crust. Recumbent zooids broad and convex. Frontal surface smooth with large areolae separated by ridges. These ridges sometimes extend over the frontal surface to a suboral umbo. In erect zooids in the center of the colony, the pores and ridges are also evident. Orifice round with broad shallow sinus. The low peristome includes an avicularium to the left or right of the sinus. Mandible wide, tapering bluntly and directed upward. Large, vicarious avicularia have a semielliptic mandible. Ovicell prominent, smooth, usually with four or five large membranous foramina, often converging. Embryos from deep yellow to orange (Fig. 4.180).

Occurrence Southern species just reaching the English Channel. Common at the Isles of Scilly on rocky beaches and on subtidal seaweeds. Washed up on a seaweed holdfast in Zeebrugge in September 2001 (De Blauwe 2003a).

Family Hippoporidridae

Genus *Hippoporidra*

Hippoporidra lusitania Taylor and Cook, 1981

Description Colony encrusting gastropod shell, almost always inhabited by a hermit crab; nodular and forming conical processes. Live colonies are bright red. The first layer of zooids is clearly oriented, later the orientation is lost. Zooids oval,

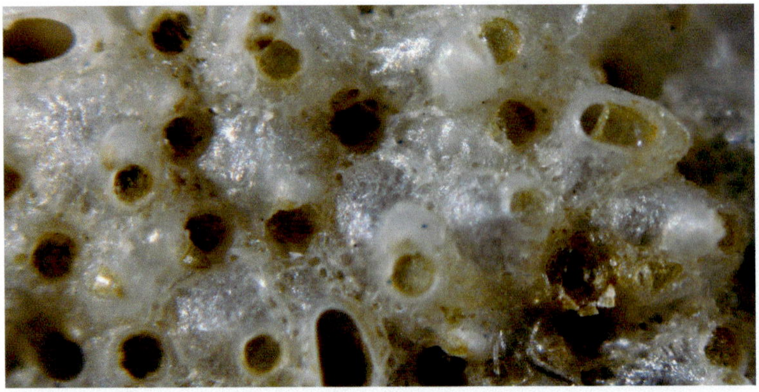
a

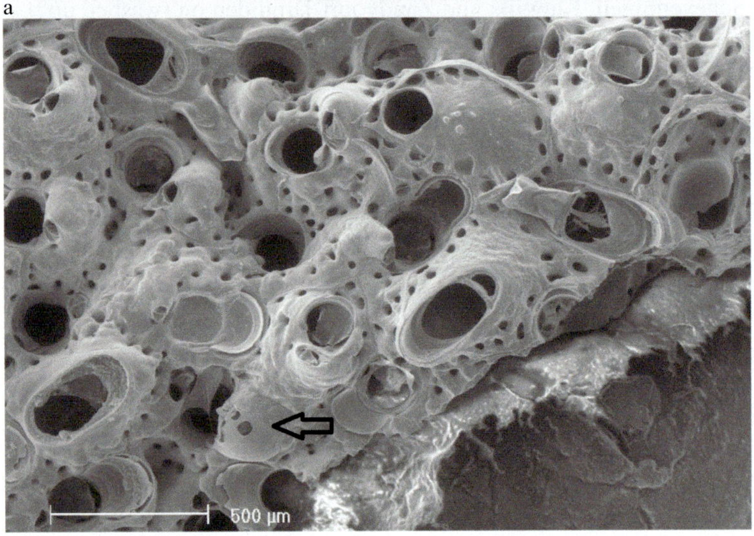

b

Fig. 4.180 *Turbicellepora magnicostata.* (**a**, **b**) Colony growing on beached *Himanthalia*, Zeebrugge. (**b**) Ovicellate and non-ovicellate zooids; avicularia and ovicell with large pseudopores (*arrow*)

angular, sometimes irregular, convex, boundaries sometimes unclear. Frontal surface rough-grained, often with a large umbo proximal to the aperture. Distinctive areolae that may develop tubularly. Orifice keyhole-shaped, distal part orbicular, separated by the condyles from an elliptical proximal part. Special zooids develop at the periphery of the colony or in clusters at the base of the conical humps in the colony. They are larger than the regular zooids and have a smaller opening. Avicularia small and elliptic, near the areolae, but often absent. Vicarous avicularium rare, with transversely elliptical rostrum and mandible. Ovicell prominent, later hidden by progressive calcification. A frontal area of endooecium remains uncovered by secondary calcification (Fig. 4.181).

a b

Fig. 4.181 *Hippoporidra lusitania.* (**a**) Colony encrusting a gastropod shell near lightship "Westhinder" (Belgium). (**b**) Non-bleached colony area showing two autozooids with keyhole-shaped orifices

Occurrence

Belgium In 1904 a colony was dredged near the lightship "Westhinder" on *Pleurotoma* sp. (RBINS collection).

The Netherlands De Ruijter (2019c) found a colony on *Gibbula tumida* from a whelk egg case.

Family Phidoloporidae

Genus *Rhynchozoon*

Rhynchozoon bispinosum (Johnston, 1847)

Description Colony encrusting, unilaminar, forming extensive pale pink to silvery white patch. Zooids oval to hexagonal, strongly convex. The thick frontal wall is smooth and bears a row of areolae. Orifice almost circular, slightly wider than long, the distal rim apparently finely serrated, a small U-shaped sinus proximal to the obtuse condyles. Young zooids may bear two widely spaced oral spines. The thickened peristome is mainly developed proximally and laterally and hides the primary orifice. Proximally, the peristome has a large conical umbo with an irregular notch next to it. Avicularia often replacing the conical umbo of the peristome. Other avicularia may originate at the level of the areolae. Ovicell initially prominent, later embedded to the secondary calcification, noncleithral, ectooecium with the membranous area, ectooecium beneath it smooth (Fig. 4.182).

Occurrence

Belgium On washed-up plastic in 1999 in Blankenberge and Zeebrugge and in 2000 in Zeebrugge, De Haan and Ostend.

The Netherlands On Schouwen (?) (Bloklander and Leenhouts 1948) and in Westkapelle in 2004. Washed up on plastic (de Ruijter 2012b, 2015a, d, 2017a, 2018a, 2019a, b, c, 2021c).

Genus *Schizotheca*

1 (a) Ooecium with a slender slit frontally. Avicularia rare, with a small semielliptic mandible. .... *Schizotheca divisa* (p. 402)
  (b) Ooecium with a triangular notch proximally. Avicularia common, with a large pointed mandible. .... *Schizotheca fissa* (p. 405)

Schizotheca divisa (Norman, 1864)

Description Colony encrusting, unilaminar, forming a small patch. Autozooids oval to hexagonal, convex. Frontal shield smooth and vitreous, with only few marginale pores. Primary orifice wider than long, distal border denticulate, proximal edge straight with blunt condyles. Peristome partially hides the primary orifice, with a proximal, deep, U-shaped notch left or right of the center. Six delicate oral spines, four spines in ovicellate zooids, are generally broken off. Avicularia occasionally between the zooids or absent, as large as a zooid, but with a small elliptical rostrum. Ovicells prominent, ooecia smooth and elongate, frontally somewhat flattened with

a

b

Fig. 4.182 *Rhynchozoon bispinosum.* (**a**) Living colony from Brittany showing spines in young autozooids (France, photo by F. Gully). (**b**) Non-ovicellate zooids, growing on a dead bivalve shell collected at the fishing harbor of Le Conquet (France) a narrow longitudinal slit (whether it is open or plugged with a cuticle is unknown). Ancestrula tatiform, surrounded by about eight spines (Fig. 4.183).

Occurrence

Belgium Found in 2004 on shells from the Kwinte Bank. In 2005, 13 deceased colonies were found at the Westhinder Bank.

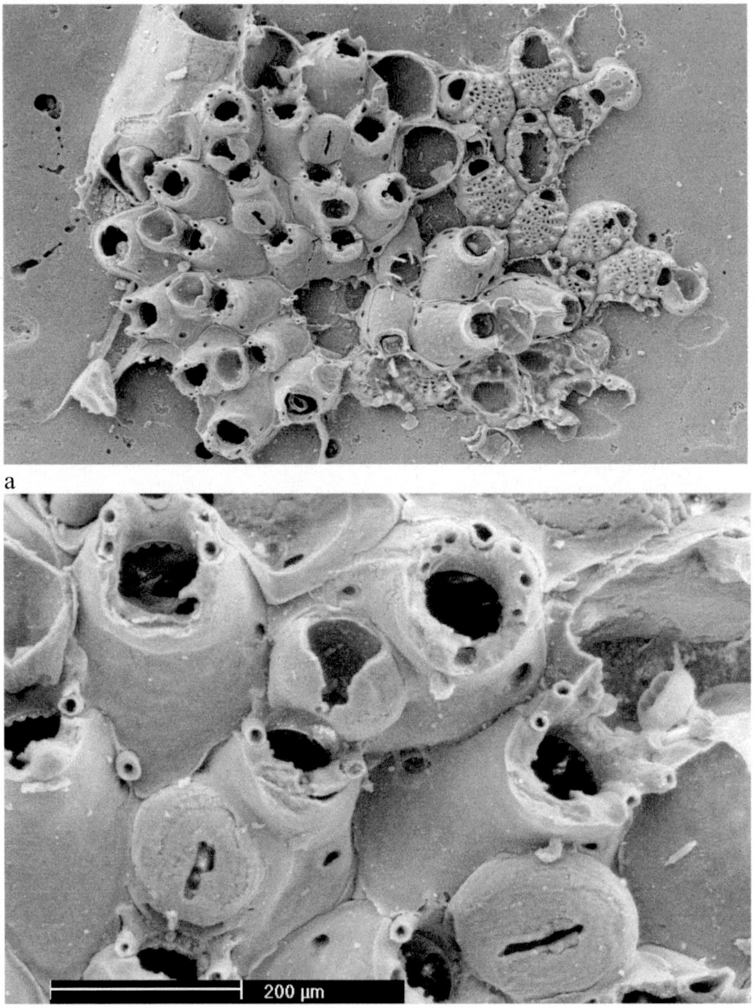

a

b

Fig. 4.183 (**a**) *Schizotheca divisa* (left), *Puellina praecox* (top right), and *Neolagenipora collaris* with ancestrula (bottom right) on a dead shell, Kwinte Bank (Belgium). (**b**) Detail of the colony showing ovicells with a slender slit frontally

Schizotheca fissa (Busk, 1856)

Description Colony encrusting, unilaminar, forming a small patch. Autozooids oval to hexagonal, convex. Frontal shield smooth and vitreous, later coarser, with only a few marginal pores. The nearly circular primary orifice has a serrated distal rim and a small U-shaped sinus proximally. The thickened peristome has a U-shaped notch proximally. Six delicate oral spines, four in ovicellate zooids, are frequently broken off. Avicularia vicarious, as large as an autozooid, with triangular rostrum. Ovicells prominent, noncleithral, ectooecium with a wide triangular proximal fissure. Embryos red (Fig. 4.184)

Occurrence

Belgium A colony was dredged in 1906 at 51°26′ N, 2°33′ E (RBINS collection). Found in 2004 on shells from the Kwinte Bank and 52 colonies, 14 of which were alive on the Westhinder Bank in 2005.

The Netherlands De Ruijter (2014b) reports the washing-up in 2013 of a colony on plastic near Egmond aan Zee.

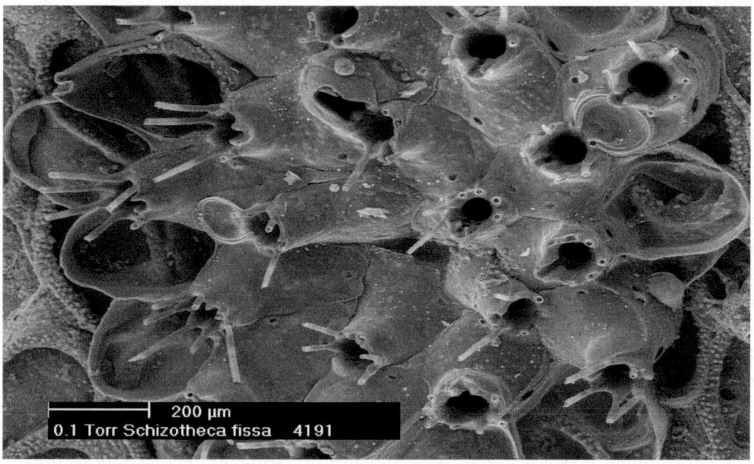

Fig. 4.184 *Schizotheca fissa,* colony growing over *Conopeum reticulum* on a dead shell, Kwinte Bank (Belgium)

References

Aleman A, Faasse M, De Blauwe H (2018) Het exotische bloedrode mosdiertje *Watersipora subatra* gevestigd in de Grevelingen. Het Zeepaard 78(5):197–201

Almeida ACS, Souza FBC, Vieira LM (2017) Malacostegine bryozoans (Bryozoa: Cheilostomata) from Bahia State, northeast Brazil: taxonomy and non-indigenous species. Mar Biodivers 48:1463. https://doi.org/10.1007/s12526-017-0639-x

Banta WC, Perez FM, Santagata S (1995) A setigerous collar in *Membranipora chesapeakensis* n. sp. (Bryozoa); implications for the evolution of cheilostomes from ctenostomes. Invertebr Biol 114(1):83–88

Berning B, Kuklinski P (2008) North-east Atlantic and Mediterranean species of the genus *Buffonellaria* (Bryozoa, Cheilostomata): implications for biodiversity and biogeography. Zool J Linnean Soc 152:537–566

Berning B, Tilbrook KJ, Rosso A (2008) Revision of the north-eastern Atlantic and Mediterranean species of the genera *Herentia* and *Therenia* (Bryozoa: Cheilostomata). J Nat Hist 42(21–22):1509–1547. Available online at https://doi.org/10.1080/00222930802109140

Berning B, Spencer Jones ME, Vieira LM (2021) Revision of the European species of the genus *Hincksina* Norman, 1903 (Bryozoa, Cheilostomatida, Flustridae). Zootaxa 5081(3):333–352. https://doi.org/10.11646/zootaxa.5081.3.2

Beukhof ED, Coolen JWP, van der Weide BE, Cuperus J, De Blauwe H, Lust J (2016) Records of five bryozoan species from offshore gas platforms rare for the Dutch North Sea. Mar Biodivers Rec 9:91

Bijma J, Boekschoten GJ (1985) Recent bryozoan reefs and stromatolites development in brackish inland lakes, SW-Netherlands. Senckenbergiana maritima 17(1/3):163–185

Bishop JDD, Househam BC (1987) Puellina (Bryozoa; Cheilostomatida; Cribrilinidae) from British and adjacent waters. Bull Br Mus (Nat Hist) Zool 53:1–63

Bloklander G, Leenhouts P (1948) Wat de najaarsstormen ons brachten. Het Zeepaard 8:9-14 and 24-25

Canning-Clode J, Souto J, McCann L (2013) First record of Celleporaria brunnea (Bryozoa: Lepraliellidae) in Portugal and inthe East Atlantic. Mar Biod Rec 6:1–5. https://doi.org/10.1017/S175526721300082

Canu F, Bassler RS (1929) Bryozoa of the Philippine region. U S Natl Mus Bull 100(9):1–685

Cook PL (1977) Early development in *Aetea* (Bryozoa). Am Zool 17:55–61

Coolen JWP, Bos OG, Glorius S, Lengkeek W, Cuperus J, Van der Weide BE, Agüera A (2015) Reefs, sand and reef-like sand: a comparison of the benthic biodiversity of habitats in the Dutch Borkum Reef Grounds. J Sea Res. 103:84–92

d'Hondt J-L (2000) Discovery of some types of cheilostomatous Bryozoa described from 1812 to 1824 by J.-V.-F. Lamouroux. Proceedings of the 11th International Bryozoology Association conference (Herrera Cubilla, Amalia and Jackson, Jeremy B.C., editors), Smithsonian Tropical Research Institute, Balboa. 211-218

d'Hondt J-L, Cadée GC (1994) *Bugula stolonifera* nieuw voor Nederland en enkele andere Bryozoën van Texel. Het Zeepaard 54(2):33–37

Dankers NMJA, van Moorsel GWNM (2001) Schelpenbanken als ecotoop; de fauna van schelpenbanken in de Waddenzee. Alterra, Wageningen. (Alterra-rapport 202)

Davoult D, Dewarumez J-M, Glaçon R (1993) Nouvelles signalisations d'espèces macrobenthiques sur les côtes françaises de la Manche orientale et de la Mer du Nord. 4. Groupes divers. Cah Biol Mar 34(1):55–64

Davoult D, Dewarumez J-M, Luczak C, Migné A (1999) Nouvelles signalisations d'espèces benthiques sur les côtes françaises de la Manche orientale et de la Mer du Nord. Cah Biol Mar 40(2):121–127

De Blauwe H (2000a) Impressies van twee strandwandelingen tussen Oostende en Zeebrugge. De Strandvlo 20(1):19–22

De Blauwe H (2000b) *Chlamys distorta* en andere allochtonen op aangespoelde gasfles te Wenduine. De Strandvlo 20(1):30–32

De Blauwe H (2000c) Riemwiervoetjes en hun begroeiing. De Strandvlo 20(3):130–137

De Blauwe H (2000d) Mosdiertjes gevonden tijdens de SWG-reis naar Bretagne 1999. De Strandvlo 20(2):58–72

De Blauwe H (2002) Determinatie en verspreiding van *Tricellaria inopinata* d'Hondt and Occhipinti Ambrogi (Bryozoa, Cheilostomatida), een recente immigrant uit het noorden van de Stille Oceaan. Het Zeepaard 62(3):73–88

De Blauwe H (2003a) Aanvoer van zuidelijke soorten aan de Belgische kust in september 2001. De Strandvlo 23(1):5–8

De Blauwe H (2003b) Bryozoa verzameld op weg SWG-reis van 2002 naar de Cotentin, Frankrijk. s 23(1):21–25

De Blauwe H (2005a) Bryozoa op wieren en plastic uit de herfstvloedlijn van het najaar 2004. De Strandvlo 25(1):14–16

De Blauwe H (2005b) Bryozoa verzameld tijdens de SWG-reis naar Bretagne in April 2005. De Strandvlo 25(3–4):76–82

De Blauwe H (2006a) On the taxonomy and distribution of the family Pacificincolidae Liu and Liu, 1999 (Bryozoa, Cheilostomata), with the description of a new genus. Bull Kon Belg Inst Natuurwet Biol 76:139–145

De Blauwe H (2006b) Bryozoa on shells from the Kwintebank, southern bight of the North Sea (Belgium). Bulletin van het Koninklijk Belgisch Instituut voor Natuurwetenschappen, Biologie 76:125–138

De Blauwe H (2008) *Fenestrulina delicia* Winston, Hayward and Craig, 2000, een nieuw mosdiertje (Cheilostomata, Bryozoa) in Europa. De Strandvlo 28(4):154–157

De Blauwe H (2009a) Mosdiertjes van de Zuidelijke bocht van de Noordzee: Determinatiewerk voor België en Nederland. Vlaams Instituut voor de Zee (VLIZ), Oostende, 445 pp. ISBN 978-90-812-9003-6

De Blauwe H (2009b) Diversity and long-term changes in the bryozoan fauna of the hinder banks, Belgium. In: Mees J et al (eds) Book of abstracts: an overview of marine research in Belgium anno 2009. 10th VLIZ young scientists' day. Special edition at the occasion of 10 years VLIZ. VLIZ special publication, vol 43. VLIZ, Oostende, p 39

De Blauwe H (2011) Riemwiervoetjes en hun begroeiing - eerste melding van *Watersipora subtorquata* (d'Orbigny, 1852) op het Belgische strand. De Strandvlo 31:93–95

De Blauwe H (2012) Zandsuppleties afkomstig van de Buiten Ratel in maart 2012 op het strand te Knokke. De Strandvlo 32(1):24–27

De Blauwe H (2017) *Biflustra grandicella* (Canu and Bassler, 1929) een warmteminnend mosdiertje (Bryozoa, Cheilostomata), nieuw voor Nederland en Europa. De Strandvlo 37(1):7–17

De Blauwe H (2019) Bryozoans on disarticulated shells from Pléneuf-Val-André, Brittany, France. Australas Palaeontol Mem 52:57–65

De Blauwe H, Faasse MA (2001) Extension of the range of the bryozoans *Tricellaria inopinata* and *Bugula simplex* in the north-East Atlantic Ocean (Bryozoa: Cheilostomatida). Ned Faunist Meded 14:103–112

De Blauwe H, Faasse MA (2004) *Smittoidea prolifica* Osburn, 1952 (Bryozoa, Cheilostomatida), a Pacific bryozoan introduced to The Netherlands (Northeast Atlantic). Bulletin Koninklijk Belgisch Instituut voor Natuurwetenschappen, Biologie 74:33–39

De Blauwe H, Kerckhof F, Houziaux JS (2006) Een opmerkelijke diversiteit aan Bryozoa (mosdiertjes) op de Hinderbanken: een tussentijds verslag. De Strandvlo 26(4):125–134

De Blauwe H, Kind B, Kuhlenkamp R, Cuperus J, van der Weide B, Kerckhof F (2014) Recent observations of the introduced *Fenestrulina delicia* Winston, Hayward and Craig, 2000 (Bryozoa) in Western Europe. Studi Trentini di Scienze Naturali 94:45–51

De Blauwe H, Witbaard R, Dijksen S, Perk F (2021) Massale stranding van het harig mosdiertje *Electra pilosa* aan de kust van Nederland in 2020/2021. Het Zeepaard 81(3):141–152

De Blauwe H, Gittenberger A, Kerckhof F (2024) Zijn *Pacificincola perforata* (Okada and Mawatari, 1937) en *Escharella hozawai* (Okada, 1929) (Bryozoa, Cheilostomatida) dezelfde soort? De Strandvlo 44(2):44–51

De Boer J, Bennema F, Kuijper W, van Leeuwen S (2020) Harlingen. Deel 1: Binnen de sluizen. Het Zeepaard 80(2):83–96

De Grave S, d'Udekem d'Acoz C (1988) Waarnemingen betreffende de mariene fauna in het noorden van Frankrijk gedurende 1987. De Strandvlo 8(2):86–99

de Kluijver MJ (1989) Sublittoral hard substrate communities of the southern Delta area, SW Netherlands. Bijdragen tot de Dierkunde 59(3):141–158

de Kluijver MJ (1997) Sublittoral communities of North Sea hard-substrata. Proefschrift, Amsterdam
de Kluijver MJ (2006) Non-indigenous species in the Oosterschelde. In: Wijsman JWM, Smaal AC (ed) Risk analysis of mussel transfer. IMARES, Wageningen, Report C044/06, 103pp
de Ruijter R (2003a) CS-verslag. Het Zeepaard 63(3):70–74
de Ruijter R (2003b) CS-verslag. Het Zeepaard 63(6):171–175
de Ruijter R (2004) CS-verslag. Het Zeepaard 64(6):165–169
de Ruijter R (2005a) CS-verslag. s 65(3):74–79
de Ruijter R (2005b) CS-verslag. Het Zeepaard 65(6):177–181
de Ruijter R (2006a) Naar het strand, of toch maar niet? Het Zeepaard 66(1):29–30
de Ruijter R (2006b) CS-verslag. Het Zeepaard 66(4):118–121
de Ruijter R (2007) CS-verslag. Het Zeepaard 67(2):34–38
de Ruijter R (2008) CS-verslag. Het Zeepaard 68(2):38–43
de Ruijter R (2010) CS-verslag. Het Zeepaard 70(2):35–41
de Ruijter R (2011a) CS-verslag. Het Zeepaard 71(1):8–14
de Ruijter R (2011b) CS-verslag. Het Zeepaard 71(4):107–114
de Ruijter R (2011c) CS-verslag 1. Het Zeepaard 71(5–6):150–155
de Ruijter R (2011d) CS-verslag 2. Het Zeepaard 71(5–6):169–175
de Ruijter R (2012a) CS-verslag. Het Zeepaard 72(1):3–10
de Ruijter R (2012b) CS-verslag. Het Zeepaard 72(3–4):70–77
de Ruijter R (2013a) CS-verslag. Het Zeepaard 73(1):3–11
de Ruijter R (2013b) CS-verslag. Het Zeepaard 73(3):78–84
de Ruijter R (2013c) CS-verslag. Het Zeepaard 73(5–6):156–165
de Ruijter R (2014a) CS-verslag. Het Zeepaard 74(1):3–12
de Ruijter R (2014b) CS-verslag. Het Zeepaard 74(2):38–45
de Ruijter R (2014c) CS-verslag. Het Zeepaard 74(3):74–85
de Ruijter R (2014d) CS-verslag. Het Zeepaard 74(5):146–156
de Ruijter R (2015a) CS-verslag. Het Zeepaard 75(1):3–12
de Ruijter R (2015b) CS-verslag. Het Zeepaard 75(2):42–51
de Ruijter R (2015c) CS-verslag. Het Zeepaard 75(3):80–90
de Ruijter R (2015d) CS-verslag. Het Zeepaard 75(4/5):111–119
de Ruijter R (2016) CS-verslag. Het Zeepaard 76(1):3–11
de Ruijter R (2017a) CS-verslag. Het Zeepaard 77(1):3–11
de Ruijter R (2017b) CS-verslag. Het Zeepaard 77(4):142–151
de Ruijter R (2017c) CS-verslag. Het Zeepaard 77(5):182–194
de Ruijter R (2018a) CS-verslag. Het Zeepaard 78(1):3–11
de Ruijter R (2018b) CS-verslag. Het Zeepaard 78(2):42–49
de Ruijter R (2018c) CS-verslag. Het Zeepaard 78(3):86–96
de Ruijter R (2018d) CS-verslag. Het Zeepaard 78(5):173–183
de Ruijter R (2019a) CS-verslag. Het Zeepaard 79(1):6–15
de Ruijter R (2019b) CS-verslag. Het Zeepaard 79(2):43–55
de Ruijter R (2019c) CS-verslag. Het Zeepaard 79(3):75–86
de Ruijter R (2019d) CS-verslag. Het Zeepaard 79(4):114–122
de Ruijter R (2020a) CS-verslag. Het Zeepaard 80(1):5–13
de Ruijter R (2020b) CS-verslag. Het Zeepaard 80(3):103–110
de Ruijter R (2020c) CS-verslag. Het Zeepaard 80(5):199–209
de Ruijter R (2021a) CS-verslag. Het Zeepaard 81(1):4–13
de Ruijter R (2021b) CS-verslag. Het Zeepaard 81(2):55–61
de Ruijter R (2021c) CS-verslag. Het Zeepaard 81(4):179–188
de Ruijter R (2023) CS-verslag. Het Zeepaard 83(2):42–49
de Ruijter R (2024a) CS-verslag. Het Zeepaard 84(1):5–14
de Ruijter R (2024b) CS-verslag. Het Zeepaard 84(2):41–51
Dekker R, Drent J (2013) The macrozoobenthos in the subtidal of the western Dutch Wadden Sea in 2008 and a comparison with 1981–1982. NIOZ-report 2013-5, 98pp

Di Martino E, Rosso A (2021) Seek and ye shall find: new species and new records of *Microporella* (Bryozoa, Cheilostomatida) in the Mediterranean. Zookeys 1053:1–42. https://doi.org/10.3897/zookeys.1053.65324

Dick MH, Grischenko AV, Gordon DP, Ostrovsky AN (2021) The "Cribrilina annulata" problem and new species of Juxtacribrilina (Bryozoa: Cheilostomata: Cribrilinidae) from the North Pacific. Zootaxa 5016(3):333–364

Faasse MA (1996) Meer over de fauna van het kanaal door Walcheren. Het Zeepaard 56(4):78–79

Faasse MA (1998) Vindplaatsen van het mosdiertje *Bugula stolonifera* Ryland 1960 in Nederland. Het Zeepaard 58(2):48–51

Faasse MA (2007) Het exotische mosdiertje *Bugula neritina* (Linnaeus, 1758) in Nederland. Het Zeepaard 67(6):190–192

Faasse MA, De Blauwe H (2004) Faunistisch overzicht van de mariene mosdiertjes van Nederland (Bryozoa: Stenolaemata, Gymnolaemata). Ned Faunist Meded 21:17–54

Faasse MA, de Ruijter R (2010) Het exotisch mosdiertje *Watersipora subtorquata* (d'Orbigny, 1852) aangespoeld in Nederland. Het Zeepaard 70:57–60

Faasse MA, Ligthart M (2016) Paarse invasie (*Bugula neritina*). Het Zeepaard 76(2):62–63

Faasse MA, Van Moorsel GWNM, Tempelman D (2013) Moss animals of the Dutch part of the North Sea and coastal waters of The Netherlands (Bryozoa). Ned Faunist Meded 41:1–14

Faasse MA, Coolen JWP, Gittenberger A, Schrieken N (2016) Nieuwe mosdiertjes van Noordzeewrakken (Bryozoa). Ned Faunist Meded 46:43–48

Fehlauer-Ale KH, Mackie JA, Lim-Fong GE, Ale E, Pie MR, Waeschenbach A (2014) Cryptic species in the cosmopolitan *Bugula neritina* complex (Bryozoa, Cheilostomata). Zool Scr 43:193–205

Fehlauer-Ale K, Winston JE, Tilbrook KJ, Nascimento KB, Vieira LM (2015) Identifying monophyletic groups within *Bugula* sensu lato (Bryozoa, Buguloidea). Zoologica Scripta 44(3):334–347. https://doi.org/10.1111/zsc.12103

Gittenberger A, Rensing M, Niemantsverdriet P, Schrieken N, Stegenga H (2013) Native and non-native species of hard substrata in the Schleswig-Holstein Wadden Sea 2012-2013. Commissioned by Erzeugerorganisation Schleswig-Holsteinischer Muschelzüchter eV. GiMaRIS rapport 2013_23, 105pp

Gittenberger A, Rensing M, Dekker R, Niemantsverdriet P, Schrieken N, Stegenga H (2015a) Native and non-native species of the DutchWadden Sea in 2014. Report nr.: GiMaRIS 2015_08, 93pp

Gittenberger A, Rensing M, Niemantsverdriet P, Schrieken N, D'Hont, A and Stegenga H (2015b) Soorteninventarisatie oesterputten en oesterpercelen, 2015. I.o.v. Bureau Risicobeoordeling and onderzoeksprogrammering, Nederlandse Voedsel- en Warenautoriteit, Ministerie van Economische Zaken. GiMaRIS rapport 2015_19, 23pp

Gittenberger A, Rensing M, Wesdorp KH, D'Hont A (2018) Monitoring non-native species in the port of Antwerp in 2017 conform the joint HELCOM/OSPAR port survey protocol. Commissioned by Antwerp Port Authority GiMaRIS rapport 2018_01, 46pp

Gittenberger A, Rensing M, van der Veer HW, Philippart CJM, van der Hoorn B, D'Hont A, Wesdorp KH, Schrieken N, Klunder L, Kleine-Schaars L, Holthuijsen S, Stegenga H (2019) Native and non-native species of the Dutch Wadden Sea in 2018. Commissioned by Office for Risk Assessment and Research, The Netherlands Food and Customer Product Safety Authority of the Ministry of Economical Affairs (BuRO). GiMaRIS rapport 2019_09: 124 pp

Gittenberger A, Rensing M, Faasse MA, Keeler Perez H, Gittenberger E (2021) Native and non-native species of the Veerse Meer 2020 - 2021. GiMaRIS rapport 2021_07: 153 pp

Gittenberger A, Rensing M, Keeler-Perez H (2022a) Marine alien species detection network of The Netherlands: 2021 results. Issued by the Office for Risk Assessment and Research, The Netherlands Food and Customer Product Safety Authority of the Ministry of Agriculture, Nature and Food Quality. GiMaRIS report 2022_15, 16pp

Gittenberger A, Rensing M, Keeler-Perez H (2022b) Non-native species focused assessment in the port of IJmuiden in 2022Issued by: the Office for Risk Assessment and Research, The

Netherlands Food and Customer Product Safety Authority of the Ministry. GiMaRIS rapport 2022_52, p 32

Gittenberger A, Rensing M, Keeler-Perez H (2023a) Marine alien species detection network of The Netherlands: 2022 results. Issued by the Office for Risk Assessment and Research, The Netherlands Food and Consumer Product Safety Authority of the Ministry of Agriculture, Nature and Food Quality. GiMaRIS report 2023_14, 46pp

Gittenberger A, Rensing M, Keeler-Perez H (2023b) Monitoring non-native species in the port of Amsterdam in 2023. Issued by The Office for Risk Assessment and Research, The Netherlands Food and Consumer Product Safety Authority of the Ministry of Agriculture, Nature and Food Quality GimaRIS rapport 2023_36, 29pp

Gittenberger A, Rensing M, Keeler Perez H, Coppis S, Gittenberger E 2023c) Native and non-native species of the Dutch Wadden Sea in 2022. Commissioned by Office for Risk Assessment and Research, The Netherlands Food and Consumer Product Safety Authority of the Ministry of Agriculture, Nature and Food Quality. GiMaRIS rapport 2023_08, 179pp

Gittenberger A, Rensing M and Keeler-Perez H (2024) Marine alien species detection network of The Netherlands: 2023 results. Issued by the Office for Risk Assessment and Research, The Netherlands Food and Consumer Product Safety Authority of the Ministry of Agriculture, Nature and Food Quality GiMaRIS report 2024_04, 49pp

Gomez A, Hughes RN, Wright PJ, Carvalho G, Lunt DH (2007) Mitochondrial DNA phylogeography and mating compatibility reveal marked genetic structuring and speciation in the NE Atlantic bryozoans *Celleporella hyalina*. Mol Ecol 16:2173–2188. https://doi.org/10.1111/j.1365-294x.2007.03308.x

Grange KR, Gordon DP (2005) The spread and potential impact of the introduced bryozoan *Biflustra grandicella*. Final research report for Ministry of Fisheries Research Project ZBS2003-07. National Institute of Water and Atmospheric Research (NIWA), Wellington, 30p

Grischenko AV, Mawatari SF (2005) *Oshurkovia*: a new genus of Umbonulidae (Bryozoa: Cheilostomata) from the northwest Pacific. In: Bryozoan Studies 2004: Proceedings of the 13th International Bryozoology Association conference, Dublin pp 99–106

Harmelin J-G (2024) Alien bryozoans in the eastern Mediterranean Sea – new records from the coast of Lebanon. Zootaxa 3893(3):301–338

Harmelin J-G (2024) Biodiversity of bathyal coral gardens—portrait of a uniserial bryozoan endemic to the South Azorean Seamount Chain: an unexpected evolutionary testbed? Zoosystema 46(30):749–774

Harmelin J-G, Boronat J, Moisette P, Rosso A (1989) *Distansescharella seguenzai* Cipolla, 1921 (Bryozoa, Cheilostomata), nouvelles données morphologiques et écologiques tirées de spécimens fossiles (Miocène, Pliocène) et actuels de Méditerranée. Geobios 22(4):485–501

Harmelin J-G, Bishop JDD, Madurell T, Souto J, Spencer Jones ME, Zabala M (2019) Unexpected diversity of the genus *Collarina* Jullien, 1886 (Bryozoa, Cheilostomatida) in the NE Atlantic-Mediterranean region: new species and reappraisal of *C. balzaci* (Audouin, 1826) and *C. fayalensis* Harmelin, 1978. Zoosystema 41(21):385–418

Harmer SF (1923) On cellularine and other polyzoa. J Linn Soc Lond Zool 35(235):293–361. https://doi.org/10.1111/j.1096-3642.1923.tb01731.x

Hayward PJ (1976) The marine fauna and flora of the Isles of Scilly: Bryozoa II. J Nat Hist 10:319–330

Hayward PJ, McKinney FK (2002) Northern Adriatic Bryozoa from the vicinity of Rovinj, Croatia. Bull Am Mus Nat Hist 270:139pp

Hayward PJ, Ryland JS (1998) Cheilostomatous Bryozoa: 1. Aeteoidea—Cribrilinoidea: notes for the identification of British species, Synop Br Fauna (new series), 10, 2nd edn. Field Studies Council, Shrewsbury, 366pp

Hayward PJ, Ryland JS (1999) Cheilostomatous Bryozoa: 2. Hippothooidea—Celleporoidea: notes for the identification of British species, Synop Br Fauna (new series), 14, 2nd edn. Field Studies Council, Shrewsbury, 416pp

Heerebout GR (1970) Verspreidingsecologie van de Bryozoa in het Deltagebied, speciaal in relatie tot het brakke water. Delta Instituut voor Hydrobiologisch Onderzoek, Yerseke. Doctoraal verslag D1-1970

Holsteijn H (1998) Een dagje op het strand—Strandwandelverslag, Schiermonnikoog 21-3-1998. Het Zeepaard 58(3):65–68

Horst R (1885) Eerste Vervolg op den Catalogus van de Verzameling van de Nederlandsche Dierkundige Vereeniging. Tijdschrift Nederlandsche Dierkundige Vereeniging 6:CXCIV–CXCIX

Jebram D (1968) Zur Bryozoen-Fauna der Niederlande. Neth J Sea Res 4(1):86–94

Kelso A, Jackson PNW (2012) Invasive bryozoans in Ireland: first record of *Watersipora subtorquata* (d'Orbigny, 1852) and an extension of the range of *Tricellaria inopinata* d'Hondt and Occhipinti Ambrogi, 1985. BioInvasions Records 1:209–214

Kerckhof F (2000) Waarnemingen van de mosdiertjes *Cryptosula pallasiana* (Moll, 1803), *Bugula stolonifera* Ryland, 1960 en *Bugula neritina* (Linnaeus, 1758), nieuw voor de Belgische fauna. De Strandvlo 20(3):114–126

Kerckhof F, De Blauwe H (2006) Over het voorkomen van *Membranipora tenuis* Desor, 1848 voor de Belgische kust, foutieve identificaties en het nut van museumcollecties. De Strandvlo 26(4):141–143

Kind B, Kuhlenkamp R (2016) Discovery of the non-indigenous bryozoan *Smittoidea prolifica* Osburn, 1952 near Heligoland: first record in 2011 for the German North Sea. Mar Biodivers 48:1237–1240

Kind B, De Blauwe H, Faasse MA, Kuhlenkamp R (2015) *Schizobrachiella verrilli* (Bryozoa, Cheilostomata) new to Europe. Mar Biodivers Rec 8(e43):6. https://doi.org/10.1017/S1755267215000160

Kuhlenkamp R, Kind B (2012) Makrozoobenthos Heligoland 2011, WRRL Maßnahmen und Bewertungen. Report of the State Agency for Agriculture, Nature and Rural Areas (LLUR) of Schleswig-Holstein, Germany, 55pp

Kuhlenkamp R, Kind B (2013) Arrival of the invasive *Watersipora subtorquata* (Bryozoa) at Heligoland (Germany, North Sea) on floating macroalgae (*Himanthalia*). Mar Biodivers Rec. https://doi.org/10.1017/S1755267213000481 74:33-39

Kuijper W, Cadée M (2018) Een invasie van het paars vogelkopmosdiertje *Bugula neritina* (Linnaeus, 1758) op het Hollandse strand. Het Zeepaard 78(1):27–37

Lacourt AW (1949) Bryozoa of The Netherlands. Arch Néerl Zool 8(3):1–33

Lacourt AW (1978) De Nederlandse mariene mosdiertjes: Bryozoa (Ectoprocta, Gymnolaemata). Wetenschappelijke Mededelingen van de Koninklijke Nederlandse Natuurhistorische Vereniging, 129. KNNV/Strandwerkgemeenschap, Hoogwoud, 21 p

Lengkeek W, Didderen K, Dorenbosch M, Bouma S, Waardenburg H (2013) Bureau Waardenburg bv and Stichting De Noordzee, rapport nr. 13–226, 76 ppx. Bureau Waardenburg bv and Stichting De Noordzee, Culemborg, rapport nr 2013, p. 13

Liu X (1992) On the genus Membranipora (Anasca: Cheilostomata: Bryozoa) from south Chinese seas. Raffles Bull Zool 40(1):103–144

Liu X, Liu H (1999) Systematic position of *Mucronella perforata* Okada et Mawatari 1937. Chin J Oceanol Limnol 17(4):338–343

Lodola A, Ferrario J, Occhipinti-Ambrogi A (2015) Further Mediterranean expansion of the non-indigenous bryozoan *Celleporaria brunnea*: multiple records along the Italian coasts. Sci. Mar. [Internet]. 2015 June 30 [cited 15 Aug 2024];79(2):263–274. https://scientiamarina.revistas.csic.es/index.php/scientiamarina/article/view/1582

Lombardi C, Taylor PD, Cocito S (2010) Systematics of the Miocene-recent bryozoan genus Pentapora (Cheilostomata). Zool J Linn Soc 160:17–39

Loppens K (1906) Bryozoaires marins et fluviatiles de la Belgique. Ann Soc R Zool Malacol Belg 41:286–321

Loxton J, Wood CA, Bishop JDD, Porter JS, Jones MS, Nall CR (2017) Distribution of the invasive bryozoan *Schizoporella japonica* in Great Britain and Ireland and a review of its European distribution. Biol Invasions 19(8):2225–2235. https://link.springer.com/article/10.1007/s10530-017-1440-2

Maitland RT (1851) Fauna Belgii septentrionalis: systematische beschrijving der dieren welke in Noord-Nederland of aan deszelfs kusten voorkomen, met bijvoeging derzelver synonymen, benevens der plaatsen waar dezelve gevonden zijn, gerangschikt naar Prof. J. van der Hoeven's Handboek der Dierkunde (2de Uitgave). C.G. van der Hoek, Leyden, Nederland. XII, XXXVIII, 234p

Maitland RT (1897) Prodrome de la faune des Pays-Bas et de la Belgique flamande ou énumération systématique de tous les animaux y observés depuis 1679-1897 excepté les araignées et les insectes. EJ Brill : Leiden, X, 62p

Markert A, Matsuyama K, Rohde S, Schupp P, Wehrmann A (2015) First record of the non-native Pacific bryozoan Smittoidea prolifica Osburn, 1952 at the German North Sea coast. Mar Biodivers 46:717. https://doi.org/10.1007/s12526-015-0415-8

McCuller MI, Carlton JT (2018) Transoceanic rafting of Bryozoa (Cyclostomata, Cheilostomata, and Ctenostomata) across the North Pacific Ocean on Japanese tsunami marine debris. Aquat Invasions 13(1):137–162. https://doi.org/10.3391/ai.2018.13.1.11

Min BS, Chae HS, Yang HJ, Noh GW, Lee DH, Seo JE (2021) A new species and new record of Pacificincolidae (Bryozoa: Cheilostomata) from Korea. J Species Res 10(3):276–286. https://doi.org/10.12651/JSR.2021.10.3.276

Mulder T (1983) De Nederlandse mariene Bryozoa. Universiteit van Amsterdam, Amsterdam. (Verslag doctoraalstage ITZ)

Nauta RW, Gittenberger A, Claassen J, Philippart CMJ (2023) Mosdiertjes op het strand van Nederland: bronnen, soorten en handelingsperspectieven: quickscan 2023-01. Waddenacademie, Leeuwarden. 22p

Nikulina EA (2007) Einhornia, a new genus for electrids formerly classified as the *Electra crustulenta* species group (Bryozoa, Cheilostomata). Schriften des Naturwissenschaftlichen Vereins für Schleswig-Holstein 69:29–40

Okada Y (1929) Report of the biological survey of Mutsu Bay. 12. Cheilostomatous Bryozoa of Mutsu Bay. Science Reports of the Tohoku Imperial University, Series 4 Biology 4, 11–35

Okada Y, Mawatari S (1937) On the collection of Bryozoa along the coast of Onagawa Bay and its vicinity, the northern part of Honshu, Japan. Sci Rep Tôhoku Univ 11:433–445

Ostrovsky AN (2013) Evolution of the sexual reproduction in marine invertebrates: example of gymnolaemate Bryozoa. Springer, Dordrecht/Heidelberg/New York, 356pp. http://link.springer.com/book/10.1007%2F978-94-007-7146-8

Ostrovsky AN, Dick MH, Mawatari SF (2007) The internal-brooding apparatus in the bryozoan genus *Cauloramphus* (Cheilostomata: Calloporidae) and its inferred homology to ovicells. Zool Sci 25(1):36–52

Otten BG (1992) Epifyten op de Japanse zakpijp, *Styela clava*. Het Zeepaard 52(3):62–63

Perk F (2000) CS-verslag 1941–1950. Het Zeepaard 60(1):37–51

Pica D, Berning B, Calicchio R (2022) Cheilostomatida (Bryozoa) from the Ionian Apulian coast (Italy) with the description of new species. Eur Zool J 89(1):371–422. https://doi.org/10.1080/24750263.2022.2032849

Platvoet D, de Bruijne RH, Gmelig Meyling AW (1995) Description of a new Caprella-species from The Netherlands: *Caprella macho* nov.spec. (Crustacea, Amphipoda, Caprellidea). Bull Zool Mus Univ Amsterdam 15(1):1–4

Polk P (1976) Inventarisatie plankton: fauna en flora. In: Nihoul JCJ and De Coninck L (ed.) Project Sea final report: 7. Inventory of fauna and flora. Project Sea final report, vol 7, pp 233–311

Prenant M, Bobin G (1966) Bryozoaires: 2. Chilostomes Anasca. Faune de France, 68. Fédération Française des Sociétes des Sciences Naturelles, Paris, 647p

Prigge H (1967) Über eine Massenanspülung kugelförmiger Electra-pilosa-Kolonien an den Küsten der südlichen Nordsee in den Jahren 1965 und 1966. Abh Verh naturw Ver Hamburg 11(1966):63–78

Reverter O, Fernandez E (1996) Cribrilinidae (Bryozoa : Cheilostomatida) from the Ria de Ferrol (NW Spain). J Nat Hist 30:1247–1260

Reverter O, Fernandez E (2007) A new name for *Puellina parva* Reverter and Fernandez 1996 (Bryozoa, Cheilostomatida). J Nat Hist 41(9–12):729–730

Reverter-Gil O, Souto J (2019) Watersiporidae (Bryozoa) in Iberian waters: an update on alien and native species. Mar Biodivers 49:2735. https://doi.org/10.1007/s12526-019-01003-4

Reverter-Gil O, Berning B, Souto J (2015) Diversity and systematics of *Schizomavella* species (Bryozoa: Bitectiporidae) from the bathyal NE Atlantic. PLoS One 10(10):e0139084. https://doi.org/10.1371/journal.pone.0139084

Robertson A (1905) Non-incrusting Chilostomatous Bryozoa of the west coast of North America. University of California Publications. Fortschr Zool 2(5), 235–322, pls 4–16

Rosso A, Beuck L, Vertino A, Sanfilippo R, Freiwald A (2018) Cribrilinids (Bryozoa, Cheilostomata) associated with deep-water coral habitats at the great Bahama Bank slope (NW Atlantic), with description of new taxa. Zootaxa 4524(4):401–439. https://doi.org/10.11646/zootaxa.4524.4.1

Rosso A, Di Martino E, Ostrovsky AN (2021) Cribrilinid bryozoans from Pleistocene Mediterranean deep-waters, with the description of new species. J Paleontol 95(2):268–290

Ryland JS, Hayward PJ (1977) British Anascan bryozoans. Academic Press for the Linnaean Society, London, pp 1–188

Ryland JS, De Blauwe H, Lord R, Mackie JA (2009) Recent discoveries of alien *Watersipora* (Bryozoa) in Western Europe, with redescriptions of species. Zootaxa 2093:43–59

Ryland JS, Holt R, Loxton J, Spencer Jones ME, Porter JS (2014) First occurrence of the non-native bryozoan *Schizoporella japonica* Ortmann (1890) in Western Europe. Zootaxa 3780(3):481–502

Schack CR, Gordon DP, Ryan KG (2018) Modularity is the mother of invention: a review of polymorphism in bryozoans. Biol Rev 94:773–809

Schüler L, Leitinger J, Faasse M, Kind B, De Blauwe H (2022) First record of the recently described Korean bryozoan *Primavelans glabricollaris* Min, Chae, Yang, Noh, Lee and Seo, 2021 (Cheilostomatida: Pacificincolidae) from the north-East Atlantic. Bioinvasions Rec 11(3):659–666. https://doi.org/10.3391/bir.2022.11.3.08

Slager GS (1983) C.S.-verslag. Het Zeepaard 42:148–157

Staal A (2020) Fossiele sporen van *Electra monostachys* (Busk, 1854). Het Zeepaard 80(2):72–75

Stebbing ARD (1971a) Growth of *Flustra foliacea* (Bryozoa). Mar Biol 9:267–273

Stebbing ARD (1971b) The epizoic fauna of *Flustra foliacea* (Bryozoa). J Mar Biol Assoc UK 51:283–300

Taylor PD, Monks N (1997) A new cheilostome bryozoan genus pseudoplanktonic on molluscs and algae. Invertebr Biol 116(1):39–51

Tilbrook KJ (2012) Bryozoa, Cheilostomata: first records of two invasive species in Australia and the northerly range extension for a third. Check List 8(1):181–183

Tilbrook KJ, Gordon DP (2016) Checklist of Singaporean Bryozoa and Entoprocta. Raffles Bull Zool Suppl 34:593–603

van der Sleen WGN (1920) Lijst der aan de Nederlandsche kust aangetroffen Nederlandsche evertebraten. Tijdschrift der Nederlandsche Dierkundige Vereeniging serie 2(18): XXXIII–XXXIX

Van Leeuwen S (2023) De biodiversiteit op wrakken in de Noordzee, waargenomen tijdens drie duikexpedities in 2021. Het Zeepaard 83(3):99–113

van Leeuwen S, Faasse MA, Driessen F, Leeuw D, van Heugten B, Ligthart M, Heerebout G (2017) Vondsten in de visnetten van Stichting Duik de Noordzee Schoon, september 2016. Het Zeepaard 77(5):201–213

van Moorsel GWNM (1996) Biomonitoring van levensgemeenschappen op sublitorale harde substraten in Grevelingenmeer, Oosterschelde, Veerse Meer en Westerschelde, resultaten t/m 1995. Bureau Waardenburg, Culemborg. [Rapp. 96.14]

van Moorsel GWNM (2003) Ecologie van de Klaverbank. Biotasurvey 2002. ecosub, Doorn van Moorsel GWNM, Waardenburg HW (1990) De fauna op en rond wrakken in de Noordzee in 1989. Bureau Waardenburg, Culemborg

van Moorsel GWNM, Waardenburg HW (1999) Biomonitoring van levensgemeenschappen op sublitorale harde substraten in Grevelingenmeer, Oosterschelde, Veerse Meer en Westerschelde. Resultaten t/m 1998. Bureau Waardenburg, Culemborg. (Raport nr. 99.011)

Vanagt T, Van de Moortel L, Faasse M (2013) Development of hard substrate fauna in the Princess Amalia wind farm. ecoast, Oostende. [ecoast report 2011036]

Vanhaelen M-T, Jonckheere I, De Blauwe H (2006) Grote wierstranding aan de Belgische Westkust tijdens de zomer van 2005. De Strandvlo 26(1):5–11

Verhaeghe F (2010) Het mosdiertjes *Fenestrulina delicia* (Winston, Hayward and Craig, 2000) voor het eerst aangespoeld aan de Belgische kust, 29 december 2009. Lombardsijde De Strandvlo 30(2):36–38

Verkuil J (1998a) CS-verslag. Het Zeepaard 58(1):3–9

Verkuil J (1998b) CS-verslag. Het Zeepaard 58(6):150–157

Vieira LM, Spencer Jones ME (2012) The identity of *Sertularia reptans* Linnaeus, 1758 (Bryozoa, Candidae). Zootaxa 3563:26–42

Vieira LM, Spencer Jones M, Taylor PD (2014a) The identity of the invasive fouling bryozoan *Watersipora subtorquata* (d'Orbigny) and some other congeneric species. Zootaxa 3857(2):151–182. https://doi.org/10.11646/zootaxa.3857.2.1

Vieira LM, Spencer Jones ME, Winston JE, Migotto AE, Marques AC (2014b) Evidence for polyphyly of the genus *Scrupocellaria* (Bryozoa: Candidae) based on a phylogenetic analysis of morphological characters. PLoS ONE 9(4):e95296. https://doi.org/10.1371/journal.pone.0095296

Wasson B, De Blauwe H (2014) Two new records of cheilostome Bryozoa from British waters. Mar Biodivers Rec 7(e123):4. https://doi.org/10.1017/S1755267214001213

Westendorp GD (1843) Recheches sur les polypiers flexibles de la Belgique, et particulièrement des environs d'Ostende. Annales de la Société Médico-Chirurgicale de Bruges IV:5–48

Winston JE (1982) Marine bryozoans (Ectoprocta) of the Indian River area (Florida). Bull Am Mus Nat Hist 173:99–176

Winston JE (2012) Dispersal in marine organisms without a pelagic larval phase. Integr Comp Biol 52(4):447–457. https://doi.org/10.1093/icb/ics040

Winston JE, Hayward PJ (2012) The marine bryozoans of the northeast coast of the United States: Maine to Virginia. Virginia Mus Nat Hist Mem 11:1–180

Yang HJ, Seo JE, Min BS, Grischenko AV, Gordon DP (2018) Cribrilinidae (Bryozoa: Cheilostomata) of Korea. Zootaxa 4377(2):216–234

Zettler ML, Beermann J, Dannheim J, Ebbe B, Grotjahn M, Günther CP, Gusky M, Kind B, Kuhlenkamp R, Orendt C, Rachor E, Schanz A, Schröder A, Schüler L, Witt J (2018) An annotated checklist of macrozoobenthic species in German waters of the north and Baltic seas. Heligol Mar Res 72:5. https://doi.org/10.1186/s10152-018-0507-5

Zintzen V (2007) Biodiversity of shipwrecks from the southern bight of the North Sea. PhD Thesis, University of Louvain, Department of Biology, 343p

Glossary

Adventitious avicularia Polymorphic zooids budded on the frontal, lateral, and/or basal wall of an autozooid. Smaller than autozooids and occurring upon them (i.e., their basal walls do not touch the substrate).

Ancestrula First zooid of a colony formed by metamorphosis of a free-swimming larva.

Anter Part of the orifice distal to the condyles.

Apertural bar Fused pair of costae immediately proximal to the orifice in cribrimorph Cheilostomatida.

Areolae Openings (often in rows) on the periphery of the umbonuloid and lepralioid frontal shield connecting the hypostegal and zooidal coeloms.

Ascopore Median frontal pore that serves as an inlet of an ascus, occurs in *Microporella*, *Haplopoma,* and *Fenestrulina.*

Ascus Sac-like hydrostatic organ serving for polypide protrusion.

Autozooid Feeding zooid having a filtering lophophore (funnel of ciliated tentacles).

Avicularium Transformed zooid in Cheilostomatida in which the operculum is modified into a mandible and the polypide is reduced to a vestige with a ciliary tuft. The mandible is snapped shut against the palatal surface of the rostrum via adductor muscles.

Basal The side opposite to the frontal wall (of the autozooid).

Brown body Colored spheroid in a zooid resulting from the aggregation of the residue of a degenerated polypide.

Condyle One of a pair of protuberances on which the operculum pivots in the orifice of ascophorine Cheilostomatida.

Costa One of the modified spines overarching the frontal membrane in cribrimorph Cheilostomatida; usually united with neighboring costae to form a costate frontal shield.

Costate shield Several costae fuse with their neighbors and with the tips of the spines across them to form a costate shield over the frontal membrane.

H. De Blauwe, *Marine and Brackish Bryozoans from the Southern Bight of the North Sea*, https://doi.org/10.1007/978-3-031-92513-9

Cryptocyst Horizontal calcareous lamina below the frontal membrane in some cheilostomate bryozoans.

Cryptocystidean Ascophorines in which the calcified frontal shield is overlain by a hypostegal coelom (a **hypostegal coelom** is an extension of the main zooidal coelom overlying the frontal calcification).

Cystid Zooidal wall, receptacle of the polypide and gonads.

Distal Toward the growth end of the colony, away from the ancestrula.

Ectooecium Outer, calcified, or membraneous wall of the ooecial fold.

Endooecium Inner calcified wall of ooecial fold.

Endotoichal ovicell Brood chamber with incubation cavity immersed in the distal zooid, characteristic of *Cellaria*.

Fenestra Uncalcified area in the ectooecium, in which the endooecium is visible after cleaning.

Frontal The top zooidal surface, which contains the orifice.

Gonozooid Cyclostome zooid modified as an incubation chamber.

Gymnocyst An exterior calcified wall (thus not overlain by a membrane), constituting part or most of the zooidal wall in cheilostomate zooids.

Gymnocystidean Ascophorines in which most of the zooidal wall is constituted by gymnocyst.

Heterozooid Polymorphic non-feeding zooid (stolonial kenozooid in Ctenostomatida, avicularium, and vibraculum in Cheilostomatida).

Immersed ovicell Ovicell with small cap-like ooecium and brood cavity totally embedded to the distal part of the maternal zooid.

Internode Segment of a jointed colony consisting of one or more autozooids.

Intertentacular organ Tubular organ between the medio-dorsal tentacles that serves for entering alien sperm and release of zygotae, present mostly in broadcasting gymnolaemates.

Interzooidal avicularium Distally budded, occurring in-between autozooids, their basal wall touch the substratum in encrusting forms.

Joint Uncalcified connection between calcified colony segments (internodes).

Jointed spine Spine with cuticular articulation near spine base.

Kenozooid Heterozooid without a polypide, usually without either orifice or muscles.

Lacuna Perforation (true pore) between costae in cribrimorph Cheilostomatida.

Lateral The side walls.

Lyrula Median tooth, often anvil-shaped, on the proximal side of the orifice in some ascophorine Cheilostomatida.

Mandible Motile part of an avicularium, moved by muscles, and homologous with the operculum of an autozooid.

Marginal Surrounding the opesia (e.g., marginal spines).

Mucro A blunt or spinous elevation of the proximal lip of the orifice of ascophorine autozooid, usually median suboral.

Nanozooid Dwarf non-feeding zooid with a single tentacle and reduced gut (in cyclostomate *Diplosolen obelium* and *Plagioecia sarniensis*).

Ooeciostome Opening of the cyclostomate gonozooid.

Ooecium Protective fold of the ovicell (brood chamber in cheilostomates).
Operculum Flap that closes the opening in Cheilostomatida.
Opesia The opening below a frontal membrane that remains after the development of a cryptocyst in zooids of Neocheilotomatina (former anasca).
Oral Surrounding the orifice (oral spines).
Orifice Opening in the autozooid wall through which the tentacles are exserted.
Ovicell Globular brood chamber in Cheilostomatida.
Ovicellate zooid Autozooid with ooecium distal to it.
Peristome In Cheilostomatida, an elevated rim in relation to the orifice; in Cyclostomatida, peristome refers to the erect cylindrical part of the autozooid.
Polypide Retractile ciliated tentacle crown (lophophore) with associated gut, muscles, and nerve ganglion.
Pore chamber Small chamber containing communication pore(s) near the base of vertical walls of certain Cheilostomatida.
Proximal Toward the origin of the colony.
Rhizoid Root-like kenozooid that attaches established colonies to the substrate.
Rostrum Prominent distal part of an avicularium, covered by the closed mandible.
Scutum A broad and sometimes lobed marginal spine bent over the opesia in some species of Cheilostomatida.
Sinus Slit, notch, or bend in the proximal edge of the orifice in some Cheilostomatida.
Spiramen Single or paired frontal pore that serves as a passage for water entering and leaving an ascus, occurs in adeonellids.
Stolon Tube-like kenozooid-bearing autozooids in some Ctenostomatida, attached to the substrate or not.
Subimmersed ovicell Ovicell in which more than half of the brood cavity is below the colony surface.
Umbo Central elevation on the frontal zooidal wall or ovicell.
Vibraculum Heterozooid (in *Cradoscrupocellaria*, *Scrupocellaria* and *Caberea*) with operculum in the form of a long seta.
Vicarious avicularium Heterozooid that replaces an autozooid in a series (as large as an autozooid or even larger).
Zoecium Protective wall of a zooid, membranous or calcified.
Zoöeciule Filiform heterozooid with small opening met in some Hippothoidae.
Zooid Single bryozoan module, various types are distinguished by prefixes.

Index

A

Aetea anguina, 3, 120, 152–154
Aetea sica, 120, 152–154
Aetea truncata, 120, 152, 154–156
Alcyonidium albidum, 2, 55, 63–65
Alcyonidium ambulans, 3, 55, 58, 78–79
Alcyonidium condylocinereum, 3, 55, 57, 62–67, 69, 71
Alcyonidium diaphanum, 55, 63–65, 67–69, 71, 83
Alcyonidium gelatinosum, 55, 58, 62–64, 67, 69–70
Alcyonidium hirsutum, 55, 63, 70–71
Alcyonidium hydrocoalitum, 3, 55, 63, 72–73
Alcyonidium mamillatum, 55, 63, 73–74
Alcyonidium mytili, 55, 58, 64, 69, 73–76
Alcyonidium parasiticum, 55, 63, 76–77, 83
Alcyonidium polyoum, 55, 58, 64, 71, 77–78
Amathia cf. *gracilis* sp.1, 56, 107–108
Amathia cf. *gracilis* sp.2, 56, 107–109
Amathia citrina, 2, 56, 57, 104–106
Amathia imbricata, 56, 104, 107–110, 112
Amathia lendigera, 3, 56, 104, 110–112
Amathia spinosa, 56, 60, 61, 63, 72, 73, 76, 102–104
Amathia tertia, 56, 104, 105, 112–113
Ammatophora nodulosa, 120, 137, 138, 208–209
Amphiblestrum auritum, 120, 196, 205, 206
Amphiblestrum flemingii, 120, 205–208
Anguinella palmata, 56, 58, 59, 76, 86–87
Annectocyma major, 16, 20, 21, 46, 47
Arachnidium fibrosum, 3, 55, 58, 76, 81, 83–84
Arachnidium lacourti, 3, 55, 81–82
Aspidelectra melolontha, 120, 150, 151, 176, 184–186

B

Beania mirabilis, 3, 121, 131, 132, 224–226
Bicellariella ciliata, 121, 134, 135, 224, 225
Biflustra grandicella, 119, 128, 138, 164–167, 214
Biflustra tenuis, 119, 126, 138, 163–164, 167, 390
Buffonellaria divergens, 387
Buffonellaria muriella, 125, 126, 387–388
Bugula neritina, 3, 121, 133, 134, 167, 208, 210, 213–214
Bugulina avicularia, 121, 208, 210–212, 214–215
Bugulina flabellata, 121, 158, 187, 210–212, 215–217
Bugulina fulva, 121, 208, 210–212, 217–218
Bugulina simplex, 121, 208, 210–212, 217–219
Bugulina stolonifera, 121, 208–212, 215, 218–221
Bugulina turbinata, 121, 208, 210–212, 221–222
Buskia nitens, 56, 60, 113–115

C

Caberea boryi, 121, 225–227
Callopora craticula, 3, 120, 193–194

H. De Blauwe, *Marine and Brackish Bryozoans from the Southern Bight of the North Sea*, https://doi.org/10.1007/978-3-031-92513-9

Callopora discreta, 9, 120, 137, 138, 193–195, 304
Callopora dumerilii, 120, 193, 195–197, 205
Callopora lineata, 120, 193, 197–199
Callopora rylandi, 120, 150, 151, 193, 197–200, 273
Calvetomavella, 326
Cauloramphus spinifer, 120, 139, 140, 200–202
Cellaria fistulosa, 121, 241–242, 392
Cellaria salicornioides, 121, 241–243
Cellaria sinuosa, 121, 241, 243–244
Cellepora pumicosa, 125, 384–386, 392, 397–399
Celleporaria brunnea, 4, 125, 148, 149, 387–390
Celleporella hyalina, 3, 122, 146, 147, 277–278
Celleporina cf. *pygmaea*, 125, 392–394
Celleporina decipiens, 125, 390–392
Celleporina hassallii, 125, 390, 392–393
Celleporina sp., 125, 390, 393–395
Chartella papyracea, 121, 135, 136, 188–191
Chorizopora brongniartii, 122, 146, 147, 281–282
Collarina gautieri, 122, 126, 150, 151, 244–247
Conopeum chesapeakensis, 6, 120, 126, 136, 171–172
Conopeum reticulum, 3, 8, 76, 120, 136, 139, 170–174, 176, 183, 187, 405
Conopeum seurati, 120, 171, 173–175, 181
Cradoscrupocellaria, 121, 135, 187, 228–232, 235
Cradoscrupocellaria ellisi, 121, 228–231
Cradoscrupocellaria reptans, 121, 229, 231–232
Crassimarginatella solidula, 120, 137, 138, 141, 200, 201
Cribrilaria arrecta, 122, 254–255
Cribrilaria bifida, 122, 254, 256, 257, 267
Cribrilaria directa, 122, 253, 256–258
Cribrilaria innominata, 122, 254, 258–260
Cribrilaria venusta, 122, 253, 259–261
Cribrilina belgica nov. sp., 122, 246, 249–250
Cribrilina cryptooecium, 122, 246, 247, 249
Cribrilina punctata, 122, 246–250
Crisia aculeata, 15, 24–26
Crisia calyptostoma, 29
Crisia denticulata, 15, 17, 24, 26, 27
Crisia eburnea, 15, 17, 24, 26–29, 187
Crisia klugei, 29
Crisia ramosa, 29
Crisia sp., 9, 15, 18, 24, 29, 30
Crisidia cornuta, 15, 19, 22–24
Crisularia, 134, 135, 208, 210
Crisularia plumosa, 83, 121, 158, 208, 210, 211, 221–224
Cryptosula pallasiana, 112, 124, 143, 144, 324, 370–372

D

Desmeplagioecia amphorae, 16, 20, 21, 44–45
Diplosolen obelium, 16, 20, 42–44
Disporella hispida, 16, 19, 47–50
Distansescharella seguenzai, 122, 126, 151, 267–270
Distansescharella spirorbicola nov. sp., 122, 268, 270–271

E

Einhornia crustulenta, 120, 173–175, 181–183
Electra cf. *asiatica*, 3, 120, 175–177
Electra monostachys, 120, 132, 140, 173, 175–179, 184
Electra pilosa, 3, 8, 120, 135, 174, 179–182
Entalophoroecia deflexa, 16, 20, 21, 46–48
Escharella gilsoni, 123, 293–294
Escharella immersa, 123, 293–296, 299
Escharella labiosa, 123, 293, 296, 297
Escharella variolosa, 123, 293, 296–299
Escharella ventricosa, 123, 293, 299–300
Escharina haywardi, 124, 355–357
Escharina johnstoni, 124, 146, 147, 355–358
Escharina vulgaris, 124, 355, 358–360
Escharoides bishopi, 123, 146, 147, 299–302
Escharoides coccinea, 123, 299, 302–304
Escharoides mamillata, 3, 123, 299, 303–305
Eucratea loricata, 120, 134, 135, 159–161
Eurystrotos compacta, 38

F

Farrella repens, 56, 60, 61, 93–95
Fenestrulina delicia, 124, 352–355
Fenestrulina malusii, 351, 352
Figularia figularis, 122, 150, 151, 272
Filicrisia geniculata, 15, 19, 22, 23
Flustra foliacea, 3, 22, 26, 28, 29, 121, 135, 136, 139, 141, 187–188, 215–217, 221, 231, 234, 277, 377, 391
Flustrellidra hispida, 2, 55, 57, 58, 80–81

G

Glabrilaria corbula, 122, 254, 261–262

H

Hagiosynodos latus, 124, 143, 144, 372–374
Haplopoma bimucronatum, 283
Haplopoma graniferum, 122, 283–284
Haplopoma impressum, 122, 283–285
Herentia hyndmanni, 124, 130, 355, 359–361
Hincksina flustroides, 121, 191–192
Hippoporidra lusitania, 125, 148, 149, 399–401
Hippoporina pertusa, 123, 143, 144, 320–322
Hippothoa divaricata, 122, 273–276
Hippothoa flagellum, 122, 257, 273, 276
Hypophorella expansa, 56, 60, 96–97

I

Immergentia suecica, 56, 61, 62, 98–100

J

Jellyella eburnea, 3, 119, 136, 169–170
Jellyella tuberculata, 119, 138, 167–168
Juxtacribrilina mutabilis, 122, 150, 251–252

L

Lagenipora lepralioides, 125, 146, 147, 394–396

M

Membranipora membranacea, 119, 135, 136, 138, 161–162, 187
Membraniporella nitida, 122, 150, 151, 199, 273–274
Metroperiella cf. *biformis*, 3, 123, 144, 322–323
Micropora complanata, 341
Microporella annae nov. sp., 124, 375, 376, 380–383
Microporella ciliata, 374, 376–378
Microporella dirupoi nov. sp., 124, 374, 375, 378–379
Microporella helgolandensis nov. sp., 124, 376, 382–385
Microporella kindi nov. sp., 124, 374, 375, 379–381
Microporella vanoutryvei nov. sp., 124, 375–378
Monobryozoon ambulans, 78–79

N

Neolagenipora collaris, 123, 146, 147, 304–307, 404
Nolella pusilla, 85

O

Omalosecosa ramulosa, 125, 133, 395–397
Oncousoecia dilatans, 15, 20, 21, 38–40
Oshurkovia littoralis, 123, 306–308

P

Pacificincola hozawai, 4, 363, 366–369
Pacificincola perforata, 124, 363–366, 368
Palmiskenea skenei, 123, 133, 287–289
Panolicella nutans, 56, 60, 84–85
Parasmittina trispinosa, 123, 145, 146, 318–319
Parellisina curvirostris, 120, 140, 141, 201–203
Patinella verrucaria, 16, 48–51
Penetrantia concharum, 56, 61, 62, 96, 98, 99
Pentapora foliacea, 123, 138, 143, 144, 164, 324–325
Phaeostachys spinifera, 124, 143–145, 361–363
Phylactella labrosa, 123, 142, 143, 304, 318–320
Plagioecia patina, 16, 19, 40–42
Plagioecia sarniensis, 16, 40, 42, 43
Plesiothoa gigerium, 122, 132, 133, 278–279
Porella chadleyi nov. sp., 123, 290–293
Porella concinna, 123, 290–291
Porella minuta, 292
Prenantia cheilostoma, 123, 309–311
Primavelans glabricollaris, 4, 124, 363, 364, 369–371
Puellina modica, 122, 253, 262–263
Puellina monostachys nov. sp., 122, 131, 254, 264, 265
Puellina nana, 122, 254, 264–267
Puellina praecox, 122, 254, 267–268, 404
Pyripora catenularia, 120, 132, 139, 140, 155, 184–185

R

Reptadeonella insidiosa, 122, 147, 285–286
Reptadeonella violacea, 122, 141, 142, 147, 285, 287–288
Rhynchozoon bispinosum, 4, 125, 145, 146, 402–403
Rosseliana rosselii, 239

S

Schizobrachiella verrilli, 4, 124, 142–144, 350–351
Schizomavella (Calvetomavella) biancae, 123, 326–328
Schizomavella cristata, 123, 128, 326, 330–331
Schizomavella discoidea, 327
Schizomavella (Schizomavella) auriculata, 123, 327–330
Schizomavella (Schizomavella) cornuta, 123, 327, 330–332
Schizomavella (Schizomavella) hastata, 123, 326, 333–334
Schizomavella (Schizomavella) linearis, 123, 326, 333–336
Schizomavella (Schizomavella) teresae, 123, 327, 330, 336–337
Schizoporella cf. *variabilis*, 124, 341, 345–348
Schizoporella errata, 345
Schizoporella hesperia, 124, 341–343
Schizoporella japonica, 124, 126, 341–344
Schizoporella magnifica, 124, 341–343
Schizoporella sp., 124, 347–350
Schizoporella unicornis, 124, 341, 342, 345–346
Schizotheca divisa, 125, 402–404
Schizotheca fissa, 125, 402, 405
Scruparia ambigua, 120, 156–158
Scruparia chelata, 120, 156, 158–159
Scrupocellaria scrupea, 121, 229, 232–234, 238
Scrupocellaria scruposa, 121, 229, 234–235
Securiflustra securifrons, 3, 121, 135, 136, 189–191
Setosella vulnerata, 121, 139, 140, 238, 239
Smittina affinis, 123, 309–312
Smittina crystallina, 123, 309, 311–313
Smittina landsborovii, 123, 310, 313–314
Smittoidea prolifica, 123, 314–316, 318
Smittoidea reticulata, 123, 130, 314–318
Spathipora sertum, 56, 101
Spathipora sp., 56, 62, 102
Stephanosella biaperta, 387
Steraechmella buski, 121, 138, 238–240
Stomatoporina incurvata, 15, 19, 20, 29–31

T

Tanganella mülleri, 3, 56, 86–88
Tegella unicornis, 3, 120, 140, 203–205
Terwasipora complanata, 124, 143, 144, 338–341
Tricellaria inopinata, 29, 112, 121, 234–237
Tricellaria ternata, 121, 234, 237–238
Triticella flava, 3, 56, 91–93
Triticella pedicellata, 3, 56, 92–94
Trypostega venusta, 122, 143, 280–281
Tubulipora aperta, 2, 15, 17, 32–34
Tubulipora liliacea, 15, 32–35
Tubulipora lobifera, 15, 32, 33, 36–37
Tubulipora phalangea, 15, 33, 34, 37, 38
Tubulipora plumosa, 15, 33, 37, 39
Turbicellepora avicularis, 125, 396–399
Turbicellepora magnicostata, 125, 396, 399–400

U

Umbonula ovicellata, 3, 123, 306–309

V

Vesicularia spinosa, 102–104
Victorella pavida, 56, 86, 89, 90

W

Walkeria uva, 3, 56, 61, 89–91
Watersipora subatra, 3, 124, 143, 338, 339

MIX
Papier aus verantwortungsvollen Quellen
Paper from responsible sources
FSC® C105338

If you have any concerns about our products,
you can contact us on
ProductSafety@springernature.com

In case Publisher is established outside the EU,
the EU authorized representative is:
Springer Nature Customer Service Center GmbH
Europaplatz 3, 69115 Heidelberg, Germany

Printed by Libri Plureos GmbH
in Hamburg, Germany